Lecture Notes in Computer Science　　　10348

Commenced Publication in 1973
Founding and Former Series Editors:
Gerhard Goos, Juris Hartmanis, and Jan van Leeuwen

More information about this series at http://www.springer.com/series/7410

Sylvain Guilley (Ed.)

Constructive Side-Channel Analysis and Secure Design

8th International Workshop, COSADE 2017
Paris, France, April 13–14, 2017
Revised Selected Papers

 Springer

Editor
Sylvain Guilley (iD)
Secure-IC S.A.S. and TELECOM-ParisTech
Paris
France

ISSN 0302-9743   ISSN 1611-3349 (electronic)
Lecture Notes in Computer Science
ISBN 978-3-319-64646-6   ISBN 978-3-319-64647-3 (eBook)
DOI 10.1007/978-3-319-64647-3

Library of Congress Control Number: 2017947503

LNCS Sublibrary: SL4 – Security and Cryptology

Printed on acid-free paper

This Springer imprint is published by Springer Nature
The registered company is Springer International Publishing AG
The registered company address is: Gewerbestrasse 11, 6330 Cham, Switzerland

Preface

The 8th International Workshop on Constructive Side-Channel Analysis and Secure Design (COSADE 2017) was organized by and held at Télécom ParisTech, Paris, France, during April 13–14, 2017. The host was the Paris 13th district site of Télécom ParisTech, which is also known as the LTCI (*Laboratoire Traitement et Communication de l'Information*) of Université Paris-Saclay. The workshop was financially sponsored by five golden sponsors, namely, AlphaNov, ANSSI, NewAE Technology, Riscure, and Secure-IC S.A.S. The company INVIA was a silver sponsor of the event.

The excellent arrangements were led by the COSADE 2017 general chair, Prof. Jean-Luc Danger, and organizing chair, Prof. Guillaume Duc. They were helped by a highly motivated team of PhD students from our SEN (*Systèmes Électronique Numérique*) research group, namely, Nicolas Bruneau, Sébastien Carré, Éloi de Chérisey, Margaux Dugardin, Khaled Karray, Damien Marion, Martin Moreau, Alexander Schaub, and Michaël Timbert. This year COSADE provided an open forum for exchanging and sharing of ongoing hot issues and results of research, development, and applications in the analysis of attacks and design of protection against attacks on embedded devices.

The Program Committee prepared for an interesting program, including two invited talks, namely, from Dr. Victor Lomné (ANSSI), talking on "Overview of Fault-based Cryptanalysis on Block Ciphers," and Prof. Philippe Maurine (LIRMM), about the question "Impacts of Technology Trends on Physical Attacks?". The technical program also included an industrial exhibition show, which allowed for fruitful discussions about applications of basic research for transfer to industry.

The workshop had seven sessions built from the contributed papers: on Thursday, "Side-Channel Attacks and Technological Effects," "Side-Channel Countermeasures," "Algorithmic Aspects in Side-Channel Attacks," and on Friday, "Side-Channel Attacks," "Fault Attacks," "Embedded Security," and "Side-Channel Tools."

We would like to thank all authors who submitted papers. Each paper was reviewed by at least three reviewers. The 25 external reviewers as well as the 26 Program Committee members contributed to the reviewing process from their particular areas of expertise. The reviewing and active discussions were facilitated by the EasyChair Web-based system. Through the system, we could check the amount of similarity between the submitted papers and previously published papers to prevent plagiarism and self-plagiarism. Following the strict reviewing processes, 16 outstanding papers from eight countries (Austria, Belgium, France, Germany, Japan, Korea, The Netherlands, and Switzerland) were accepted for publication in this volume of *Lectures Notes in Computer Science* by Springer (LNCS Vol. 10348). I would also like to thank the session chairs (Naofumi Homma, François-Xavier Standaert, Jens-Peter Kaps, Benoît Feix, Benoît Gérard, Yannick Téglia, Pierre-Yvan Liardet, Jean-Luc Danger, and Guillaume Barbu) for their commitment to COSADE.

The workshop featured a welcome reception on the evening of Wednesday April 12, and a social event on board the *Bateaux Mouches* (cruising dinner on the Seine) on Thursday, April 13. During this enjoyable event, François-Xavier Standaert was awarded for the nearest distance with respect to COSADE, and Werner Schindler received the random lottery special prize.

Many people contributed to the success of COSADE 2017. We would like to express our deepest appreciation to each of the COSADE Organizing and Program Committee members as well as the paper contributors. Without their endless support and sincere dedication and professionalism, COSADE 2017 would have been impossible.

May 2017   Sylvain Guilley

Organization

Program Committee

Shivam Bhasin	Temasek Labs@NTU, Singapore
Christophe Clavier	Université de Limoges, France
Hermann Drexler	Giesecke & Devrient, Germany
Cécile Dumas	CEA, France
Thomas Eisenbarth	WPI, USA
Wieland Fischer	Infineon, Germany
Christophe Giraud	Oberthur Technologies, France
Johann Groszschaedl	University of Luxembourg, Luxembourg
Sylvain Guilley	GET/ENST, CNRS/LTCI, France
Benoît Gérard	DGA-MI, France
Tim Güneysu	University of Bremen and DFKI, Germany
Johann Heyszl	Fraunhofer AISEC, Germany
Naofumi Homma	Tohoku University, Japan
Michael Hutter	Cryptography Research, USA
Thanh Ha Le	SAFRAN, France
Kerstin Lemke-Rust	Bonn-Rhein-Sieg University of Applied Sciences, Germany
Houssem Maghrebi	SAFRAN Idendity and Security, France
Marcel Medwed	NXP Semiconductors, Austria
Amir Moradi	Ruhr University Bochum, Germany
Debdeep Mukhopadhyay	IIT Kharagpur, India
Stjepan Picek	KU Leuven, Belgium
Francesco Regazzoni	ALaRI – USI, Switzerland
Matthieu Rivain	CryptoExperts, France
Kazuo Sakiyama	The University of Electro-Communications, Tokyo, Japan
Ruggero Susella	STMicroelectronics, Italy
Carolyn Whitnall	University of Bristol, UK

Additional Reviewers

Alessio, Davide	Chabrier, Thomas
Bauer, Sven	Chattopadhyay, Anupam
Bhasin, Shivam	Chen, Cong
Bocktaels, Yves	Dabosville, Guillaume
Breier, Jakub	De Mulder, Elke
Burri, Samuel	Debande, Nicolas
Cagli, Eleonora	Dinu, Daniel

Grosso, Vincent
Hoffmann, Lars
Jungk, Bernhard
Kavun, Elif Bilge
Le Corre, Yann
Murray, Bruce
Nikov, Ventzi
Patranabis, Sikhar
Pilato, Christian
Prouff, Emmanuel
Reynaud, Léo
Richter, Bastian
Rondepierre, Franck

Saha, Sayandeep
Schneider, Tobias
Seker, Okan
Stöttinger, Marc
Szerwinski, Robert
Tunstall, Michael
Ueno, Rei
Vadnala, Praveen Kumar
Valencia, Felipe
Wenger, Erich
Yamamoto, Dai
Yli-Mäyry, Ville

Contents

Does Coupling Affect the Security of Masked Implementations?

Thomas De Cnudde[1]([⊠]), Begül Bilgin[1], Benedikt Gierlichs[1], Ventzislav Nikov[2], Svetla Nikova[1], and Vincent Rijmen[1]

[1] ESAT-COSIC and imec, KU Leuven, Leuven, Belgium
{thomas.decnudde,begul.bilgin,benedikt.gierlichs,svetla.nikova,
vincent.rijmen}@esat.kuleuven.be
[2] NXP Semiconductors, Leuven, Belgium
venci.nikov@gmail.com

Abstract. Masking schemes achieve provable security against side-channel analysis by using secret sharing to decorrelate key-dependent intermediate values of the cryptographic algorithm and side-channel information. Masking schemes make assumptions on how the underlying leakage mechanisms of hardware or software behave to account for various physical effects. In this paper, we investigate the effect of the physical placement on the security using leakage assessment on power measurements collected from an FPGA. In order to differentiate other masking failures, we use threshold implementations as masking scheme in conjunction with a high-entropy pseudorandom number generator. We show that we can observe differences in—possibly—exploitable leakage by placing functions corresponding to different shares of a cryptographic implementation in close proximity.

Keywords: Masking · Threshold Implementations · Crosstalk · Non-independent leakage · Leakage detection · TVLA

1 Introduction

Side-Channel Analysis (SCA) is a powerful menace against embedded cryptosystems. Whether timing information [24], instantaneous power consumption [25] or electromagnetic radiation [19,36] is exploited, extracting sensitive information (e.g. secret keys) from cryptographic devices is feasible in contrast to cryptanalytic or brute force techniques. Counteracting SCA has consequently been an active research topic and many countermeasures have been proposed. In this paper we focus on masking methods which provide provable security given certain assumptions on the implementation, physical behavior of the device and capability of an attacker.

Masking. Masking [9,21], which is based on secret sharing and multi-party computation, relies on randomizing the intermediate values and computations

© Springer International Publishing AG 2017
S. Guilley (Ed.): COSADE 2017, LNCS 10348, pp. 1–18, 2017.
DOI: 10.1007/978-3-319-64647-3_1

based on them. For this purpose each sensitive value $x \in GF(2^m)$ is uniformly and randomly split into s shares using a certain operation $\perp$ such that the following condition holds:

$$x = x_1 \perp x_2 \perp ... \perp x_{s-1} \perp x_s$$

Typically, it is assumed that the physical leakage of calculation and storage of each share is independent of the others. With this assumption, the security proofs of masking schemes consider an attacker capable of observing leakages coming from calculation or storage depending jointly on at most d shares and hence performing d^{th}-order attacks. Therefore, $s \geq d + 1$ is a natural bound as this implies incomplete information for the attacker. Besides these uniformly distributed input and independent leakage assumptions, different flavors of masking schemes may have additional computational or behavioral requirements on the implementation. Keeping the calculation order as is defined in Trichina AND gate [42] and having ideal nodes that do not toggle in private circuits [23] are well known examples. In this paper, we focus on the failure of the independent leakage assumption and satisfy further restrictions to the utmost.

Failure of Independent Leakage. The theoretical security of masking schemes degrades when the leakage of different shares get influenced by each other. The amount of this security reduction has been investigated theoretically in [16] with respect to the strength of joint leakage in comparison to independent leakages (called the flaw constant) and to noise level. It has been shown that mutual information increases together with the flaw constant. On the other hand, second-order leakage can become easier to detect than first-order leakage as the noise increases given enough dependent leakage.

In practice, Hamming Distance (HD) leakage from one share to another and glitchy gates are natural and visible examples of non-independent leakage. It is shown in [3] that a theoretically d^{th}-order secure implementation can be attacked using $d/2^{\text{th}}$-order attack in practice due to HD leakage if the security proofs assume Hamming Weight (HW) leakage. Moreover, classical Boolean masking is shown to be futile in circuits using CMOS-like technology [27]. The temporally separated masking scheme of Prouff and Roche [35], where shares are required to interleave their computations, have also been argued to be vulnerable when static leakage is measurable [29]. In order to distinguish undesired security degradation caused by HD leakage and redundant toggling of gates from other failures of independent leakage, we ensure not to have HD leakage between different shares of the same unmasked value and use threshold implementation (TI) masking scheme which provides security in glitchy circuits [5, 31–33].

Another example of non-independent leakage is crosstalk, which originates from coupling capacitors between circuit wires, and between circuit wires and ground. Coupling capacitance between two wires is influenced by the switching activity on that wire. Only a few publications have investigated the effect of crosstalk within the field of SCA attacks so far. In [11], Chen et al. showed that the leakage intensity of glitches and the leakage caused by inter-wire capacitance are comparable using SPICE simulations. Moreover, they retrieved the key successfully using first-order attacks on a masked implementation with dual-rail pre-charge logic. This logic style was thought to avoid non-independent leakages caused by glitching implying crosstalk to be the main leakage leading to the attack. However, the latter results based on real-world devices are considered to have measured the effect of early propagation issues in implementations using these logic styles [43] rather than crosstalk itself. Later, Dyrkolbotn considered the layout dependent phenomena of capacitive crosstalk in [17,18] in order to derive a more precise leakage model. They showed that the detection performance of values on an 8-bit data bus increases from 2.5-bits of information per sample with a Hamming Distance detector to a theoretical 5.7-bit and simulated 5-bit of information per sample with a crosstalk based detector by simulation. Power supply noise or IR drop, another coupling effect in circuits, was also shown to have a negative impact on the security of a countermeasure [46] by relating independent logic gates through the power supply line. Finally, Schmidt et al. performed successful key-retrieval attacks by measuring the power consumption on input or output peripherals instead of using the regular power supply lines [38]. The success of their method originates from the coupling between pins of an Integrated Circuit (IC).

To conclude, there is no definitive report on the observability of non-independent leakage originating from coupling on a real-world device when masking is considered. In order to distinguish between non-independent leakage originating from e.g. HD or glitches, and leakage originating from coupling, we will refer to the latter as out-of-model leakage.

Leakage Assessment. The security of a masked implementation is commonly assessed using side-channel evaluation platforms and techniques like Differential Power Analysis (DPA) [25], Correlation Power Analysis (CPA) [7] or Test Vector Leakage Assessment (TVLA) [14,20,39,40]. Unlike other evaluation methods, TVLA has the advantage of being very sensitive even if the detected leakage does not necessarily lead to key recovery. Therefore, it is a preferred tool to confirm the provable security of a masking schemes with high confidence [5,12,41]. In order to observe the possibly small differences in observable leakage caused by having or lacking coupling-like, out of model behavior, we opt for TVLA in this paper.

Contribution. In this work we further build on the observations of out-of-model leakage. In contrast to the WDDL enhanced masked AES S-box of Chen et al. [11], our focus is specifically directed towards masking schemes alone and the Threshold Implementations scheme is selected as test case. We choose the lightweight KATAN-32 [8] as our target block cipher as we expect coupling effects to be more prominent in a low noise setting. After showing a secure TI of the lightweight KATAN-32 block cipher, we investigate the out-of-model leakagewhen we induce coupling between shares on an FPGA.

Organization. In Sect. 2, we give an overview of the internal mechanism of two out-of-model leakagesources and revisit the KATAN-32 Threshold Implementation and briefly introduce FPGA concepts used throughout the paper. In Sect. 3, we theoretically evaluate the effect of out-of-model leakageon the conditions of a three share masking scheme. We describe and evaluate two leakage scenarios on the KATAN-32 Threshold Implementation in Sect. 4 and follow with a brief discussion and a conclusion in Sect. 5.

2 Preliminaries

2.1 Sources of Out-of-Model Leakage

We now revisit the conditions for masking from a power consumption point of view and give simplified models of physical phenomena that are known to lead to out-of-model leakage [26].

Power Consumption in Masking Schemes. From a power consumption perspective, a first-order masked implementation requires the following condition to hold: the mean power consumption for each unmasked sensitive value should be equal. One way to achieve this requirement is by using Boolean masking with masks drawn randomly from a uniform distribution.

If we mask a one-bit secret value x with a one-bit mask m as $\mathbf{x} = (s_1, s_2) = (x \oplus m, m)$ and denote the probability of $m = i$ by K_i, we can formalize the condition for the uniformity of the masks as:

$$K_0 = K_1 = \frac{1}{2}.$$

The expected power consumption P w.r.t. the unmasked value x can then be expressed as:

$$P(x = 0) = K_0 P(s_1 = 0, s_2 = 0) + K_1 P(s_1 = 1, s_2 = 1)$$
$$P(x = 1) = K_0 P(s_1 = 1, s_2 = 0) + K_1 P(s_1 = 0, s_2 = 1).$$

The condition for first-order Boolean masking is then formalized by the following equation:

$$P(s_1 = 0, s_2 = 0) + P(s_1 = 1, s_2 = 1) = P(s_1 = 0, s_2 = 1) + P(s_1 = 1, s_2 = 0).$$

In this example, first-order vulnerabilities occur in the masking scheme when this condition is violated. The effect of out-of-model leakagefrom coupling on the security of the masking scheme can be understood by analyzing the power consumption P [37]. The instantaneous power consumption P_{inst} represents a sample of a SCA measurement trace:

$$P_{inst} = I_{inst} V_{inst}.$$

Where I_{inst} and V_{inst} denote the instantaneous current and instantaneous voltage respectively.

Crosstalk. Crosstalk is the result of capacitive coupling between adjacent wires. Figure 1 shows two adjacent wires, each with a parasitic capacitance to the IC substrate and an inter-wire capacitance between them. When a wire (the aggressor) switches the value it carries, another wire in its vicinity (the victim) will be influenced through the inter-wire capacitance $C_{1,2}$ between the aggressor and the victim. This influence can range from increased delay of a signal to traverse the wire, through a wrong value being temporarily induced on the victim. The reduction in SCA security introduced by crosstalk can be explained as follows. A typical first-order masked implementation represents a sensitive variable by two randomized shares, such that the mean power consumption of either share is independent of the other share. If two wires belonging to different shares are coupled, the mean power consumption of one share depends jointly on both a neighboring aggressor share and itself. The masked implementation is hence rendered insecure.

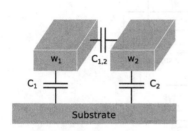

Fig. 1. Crosstalk originates from the inter-wire capacitance $C_{1,2}$.

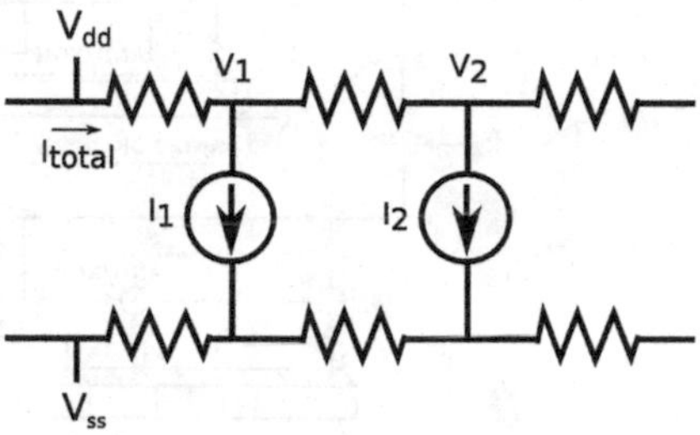

Fig. 2. Static and dynamic IR drop occurs from the non-zero resistance of conductive supply voltage and ground wires.

IR Drop. IR drop or power supply noise originates from the finite conductance of wires in i.a. the power distribution grid of ICs. Every wire segment has a small resistance associated with it leading to a drop in the power supply voltage when a current flows through that wire [37]. Both static and dynamic IR drop can lead to coupling between shares and hence to out-of-model leakage. A simplified model is given in Fig. 2.

As with crosstalk, the problem gets worse with shrinking technology nodes [37]. In the context of SCA, the effect of IR drop has not yet been investigated.

2.2   KATAN-32 and Its Threshold Implementation

KATAN is a set of block ciphers designed specifically for lightweight applications [8]. Its efficiency in hardware translates to a small area and a low power consumption. Three options are available for the state size: 32-, 48- or 64-bit. All options use an 80-bit key, making the security independent of the state size.

The diagram of the KATAN-32 round function is shown in Fig. 3. The 32-bit plaintext is stored in a state that consists of two shift registers: a 13-bit right shifting register $L1$ and a 19-bit left shifting register $L2$. The cipher processes the state by applying a round operation 254 times. The round operation relies on a small number of AND and XOR gates and is performed on several bits in order to update the first bits of $L1$ and $L2$. The function is of the form $A = f(X, Y, Z) = X \oplus YZ$. The IR (irregular update) bit represents the last bit of the round counter which enables or disables the fourth bit of $L1$ in the round operation. The bits k_{2i} and k_{2i+1} are the $2i^{th}$ and $(2i + 1)^{th}$ bits of the 80-bit key for rounds $i \le 40$. In rounds $i > 40$, they are derived from the original key by an LFSR. The full description can be found in [8].

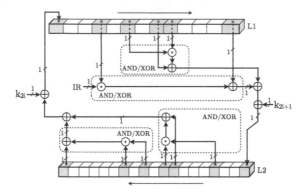

Fig. 3. KATAN-32 consists of two sets of shift registers and four nonlinear operation groups (Source: [5]).

The round operation is susceptible to glitching, making TI a natural choice for a masked implementation. This was shown by Bilgin et al. in [5] where a first-, a second- and a third-order Threshold Implementation of KATAN-32 were presented. We now revisit their first-order TI of KATAN-32.

The focus lies on the sharing of the state and its nonlinear round function since sharing nonlinear operations is more involved than sharing linear ones. An unshared key and key schedule are used, such that the key addition only needs to be performed on the first share of the state.

Since a first-order three share TI of a single AND gate with uniform outputs does not exist [31] without remasking, the AND gates of the round operations are always grouped with an XOR gate and masked using the uniform TI with $s_{in} = s_{out} = 3$ shares of the function $A = f(X, Y, Z) = X \oplus YZ$. This approach results in the following non-complete sharing:

$$a_1 = x_2 \oplus (y_2 z_2 \oplus y_2 z_3 \oplus y_3 z_2)$$
$$a_2 = x_3 \oplus (y_3 z_3 \oplus y_3 z_1 \oplus y_1 z_3)$$
$$a_3 = x_1 \oplus (y_1 z_1 \oplus y_1 z_2 \oplus y_2 z_1).$$

Since the round counter (and resultantly IR) is not key dependent and hence not shared, IR is added to the AND/XOR blocks in the following way:

$$a_i = x_i + IR \times y_i, \quad i \le s_{in}.$$

The number of state shares is chosen to be three following the number of shares of the nonlinear function.

2.3　Xilinx Virtex-II Pro FPGA Overview

In order to help the reader understand the FPGA related details, we first briefly review the hardware architecture and development flow, and highlight only the concepts we will use throughout this paper. We focus our discussion specifically on the Virtex-II Pro FPGA, since it forms the target device of our implementations.

Hardware Architecture. Field Programmable Gate Arrays (FPGAs) are a type of programmable ICs containing a regular grid of Configurable Logic Blocks (CLBs) and programmable routing resources. In the Virtex-II Pro FPGA, each CLB contains four slices, which are the primitive building blocks of the FPGA. Each slice contains amongst others two 4-input Look-Up Tables (LUTs) and two registers [45].

Design Flow. The classic design flow for FPGAs starts with the Hardware Description Language (HDL). During synthesis, the HDL files are compiled and transformed into an FPGA architecture-specific design netlist. Once synthesis is completed, the next step in the design flow is the implementation which consists of three phases: translate, map and place and route (PAR). During translation, the netlist is reduced to only contain Xilinx primitives. The mapping phase then maps the Xilinx primitives in the netlist to actual FPGA resources such as slice registers or LUTs. After mapping an NGC netlist file is output that corresponds to the physical components in the Xilinx FPGA and the constraints of the design. The final stage of the implementation is the PAR. The actual allocation of resources from the NCF file and their interconnections are decided upon here. Once the implementation is finished, the bitstream file for the FPGA configuration is generated using the Generate Programming File process.

3　Coupling in Threshold Implementations

3.1　Crosstalk

Since we are interested in the effect of out-of-model leakageon the first-order security of the KATAN-32 TI with three shares, we first provide a discussion of crosstalk in masking schemes with three shares.

Masking the secret value x yields $\mathbf{x} = (s_1, s_2, s_3) = (x \oplus m_1 \oplus m_2, m_1, m_2)$, where the masks m_i are drawn from a uniform random source to satisfy the condition of masking, i.e.

$$K_{0,0} = K_{0,1} = K_{1,0} = K_{1,1} = \frac{1}{4}$$

where $K_{i,j}$ denotes the probability of $m_1 = i$ and $m_2 = j$.

The masking condition $P(x = 0) = P(x = 1)$ on the expected power consumption P w.r.t. the unmasked value x is then expressed as:

$$P(s_1 = 0, s_2 = 0, s_3 = 0) + P(s_1 = 0, s_2 = 1, s_3 = 1)$$
$$+ P(s_1 = 1, s_2 = 0, s_3 = 1) + P(s_1 = 1, s_2 = 1, s_3 = 0)$$
$$= P(s_1 = 0, s_2 = 0, s_3 = 1) + P(s_1 = 0, s_2 = 1, s_3 = 0)$$
$$+ P(s_1 = 1, s_2 = 0, s_3 = 0) + P(s_1 = 1, s_2 = 1, s_3 = 1).$$

In order to examine the influence of out-of-model leakageon the security of the masking scheme, we need to find whether or not a dependence exists between the instantaneous power consumption $P_{inst} = I_{inst}V_{inst}$ and the unmasked value x. In order to perform this exemplary analysis, we rely on a data bus model [15]. The relation of the instantaneous power with the consumed energy can be seen from the expression for the energy required to charge a wire i from a bus from $V_i(t^-) = 0$ to $V_i(t^+) = V_{dd}$:

$$E_{rise,i} = \int_{t^-}^{t^+} V_{dd} \cdot I_j(t)dt.$$

The total energy consumption to change a three wire bus can be written as:

$$E_{total} = \sum_{i=0}^{3}(1 + 2\lambda - \lambda\delta_{i,i-1} - \lambda\delta_{i,i+1}) \cdot C_L \cdot V_{dd} \cdot V_i$$

where $\lambda = C_I/C_L$, the inter-wire capacitances are chosen as $C_I = C_{1,2} = C_{1,3}$ and the wire-substrate capacitances are chosen as $C_L = C_1 = C_2 = C_3$. The equality the inter-wire capacitances and of the wire-substrate capacitances are justified by the data bus model, where the wires are assumed to be equidistant both from each other and from the substrate. Furthermore, $\delta_{i,j} \in \{-1, 0, 1\}$ is the normalized relative voltage change of the jth line w.r.t. the ith line, V_{dd} is the supply voltage and V_j is the final voltage on the jth line.

We can now group and calculate the total energy transitions per unmasked value using values from [28] for $C_L = 400\,\mathrm{fF}$, $C_I = 250\,\mathrm{fF}$ and $V_{dd} = 3\,\mathrm{V}$:

$$E_{total,0 \to 0} = 0.430\,\mathrm{nJ} \qquad E_{total,1 \to 0} = 0.529\,\mathrm{nJ}$$
$$E_{total,0 \to 1} = 0.475\,\mathrm{nJ} \qquad E_{total,1 \to 1} = 0.498\,\mathrm{nJ}.$$

The difference in total energy per unmasked value is analytically distinguishable and hence the masking scheme is not secure in the presence of crosstalk.

3.2 IR Drop

Power supply noise or IR drop is a result of the finite conductance of wires from the power delivery network in ICs. Figure 4 shows a simplified version of IR drop that focuses on shared subcircuits [4]. The influence of IR drop on the security of the masking scheme is best understood by looking at the instantaneous power consumption $P_{inst} = I_{inst}V_{inst}$ on the voltage nodes V_1, V_2 and V_3:

$$V_1 = V_{dd} - (I_1 + I_2 + I_3)R_1$$
$$V_2 = V_{dd} - (I_1 + I_2 + I_3)R_1 - (I_2 + I_3)R_2$$
$$V_3 = V_{dd} - (I_1 + I_2 + I_3)R_1 - (I_2 + I_3)R_2 - I_3R_3.$$

We can now write the instantaneous power consumption of all the shares $P_{inst,Share1}$, $P_{inst,Share2}$ and $P_{inst,Share3}$ as:

$$P_{inst,Share1} = I_1V_1 = V_{dd}I_1 - I_1^2R_1 - I_1I_2R_1 - I_1I_3R_1$$

$$P_{inst,Share2} = I_2V_2 = V_{dd}I_2 - I_1I_2R_1 - I_2^2R_1 - I_2I_3R_1 - I_2^2R_2 - I_2I_3R_2$$

$$P_{inst,Share3} = I_3V_3 = V_{dd}I_3 - I_1I_3R_1 - I_2I_3R_1 - I_3^2R_1 - I_2I_3R_2 - I_3^2R_2 - I_3^2R_3.$$

The power consumption of any one share thus theoretically depends on adjacent shares and hence the masking scheme is not secure in the presence of IR drop.

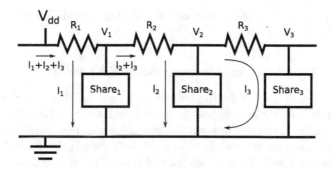

Fig. 4. Power supply noise or IR drop couples shares.

4 Coupling in Threshold Implementations of KATAN-32

We now investigate what effect coupling has on the first-order side-channel leakage of the KATAN-32 Threshold Implementation with three shares. In a first experiment, we measure the side-channel resistance of a regular Threshold Implementation of KATAN-32, for which we followed the design rules mentioned in the literature. In a second experiment, we show that placement has an influence on the leakage of the same (netlist-wise) KATAN-32 TI. Before we describe the actual experiments, we explain three constraints we use to guide the synthesis, map and place and route tools. Their full description and application are documented in [44].

Xilinx Constraints

Keep Hierarchy. "Keep Hierarchy" is a synthesis and implementation constraint and is commonly used in papers about Threshold Implementations [5,6,30,34]. HDL designs are generally a collection of hierarchical modules and submodules. In masked implementations the constraint is used to avoid optimizations over share boundaries, as its effect preserves the hierarchy throughout the implementation process and avoids the flattening of the design. In a masking context, the option is set globally as a synthesis option. Three values can be set for this option: true, soft and false. True preserves the design hierarchy throughout both synthesis and implementation, soft keeps the hierarchy during synthesis but not during the implementation phase while false allows all the submodules of the design to be merged within the top level module.

Keep. "Keep" is a constraint that influences the mapping phase of the implementation. It avoids nets from being merged into a single logic block. Taking the AND/XOR function $X \oplus YZ$ of KATAN-32 as example, the HDL code would explicitly declare an AND and an XOR operation while the mapper would merge both gates into a single LUT. This constraint is applied to signals in the HDL code.

Prohibit. "Prohibit" is a placement constraint that forbids the use of selected CLBs or Slices during PAR.

4.1 Secure Threshold Implementation of KATAN-32

To achieve a secure Threshold Implementation, we set the "Keep Hierarchy" synthesis option to true globally, as is done in related practical TIs [6,30,34]. Resulting from the hierarchy in both the synthesis and place and route phases, the individual shares are separately placed on the Virtex-II Pro floorplan and can be clearly distinguished. Figure 5 shows the separation of the three individual shares on the floorplan of the FPGA, the three different shares are shown in magenta, light green and dark green.

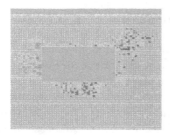

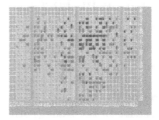

Fig. 5. The individual shares are placed far apart.

Fig. 6. All shares are placed in close proximity. (Color figure online)

Evaluation. We proceed with leakage assessment to evaluate whether or not the out-of-model leakagefrom coupling can be detected. To detect leakage in higher-order moments, we run the t-test on preprocessed traces. In all our evaluations we provide favorable measurement conditions for leakage detection: we use a very low noise platform (Virtex-II Pro FPGA on the SASEBO-G board [1]) clocked at a fixed frequency of 3.072 MHz while the instantaneous power consumption is measured with a Tektronix DPO 7254C oscilloscope at 1 GS/s.

Methodology. The evaluation methodology to check the masking scheme for leakage is as follows.

We first disable the masking scheme by turning off the masks. In that case, the first share equals the plaintext while the second and third shares are chosen to be zero. Leaks, i.e. t-values exceeding ±4.5, are expected in the leakage detection test as the masking scheme is effectively not applied. In their presence, this experiment gives us confidence that the measurement setup is sound. We proceed by assigning the masks from a uniform random distribution, i.e. we activate the masking scheme, and repeat the leakage detection test. Any decrease of leakages is accredited exclusively due to a proper masking scheme. If leaks are detected, the implementation of the masking scheme is concluded to be erroneous.

Masks Off. The result of the leakage detection test with the masks turned off is shown in Fig. 7. As expected, the t-value threshold of ±4.5 is exceeded meaning the design with disabled masks leaks with 20k traces.

Masks On. Turning the masks on results in the first- and second-order leakage detection tests in respectively the middle and bottom graphs shown in Fig. 7. The expected second-order leaks are present and suggest that we have enough

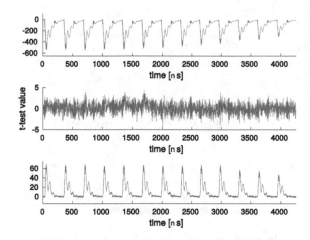

Fig. 7. Leakage detection test of a secure KATAN-32 TI, 20k traces masks off (top), 100M traces masks on 1st-order (middle), 100M traces masks on 2nd-order (bottom).

measurements to be able to detect leakage in lower-order moments, if any would be present.

The dashed line in Fig. 9 shows the evolution of the point of maximum first-order leakage in function of the number of traces in increments of 1M. The maximum of the absolute t-value fluctuates around the threshold but no steady increase in the maximum value is recognizable. We therefore conclude that no out-of-model leakageis observable with 100M traces.

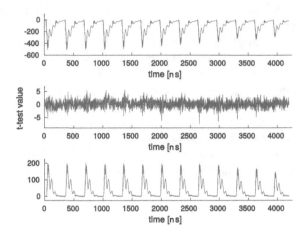

Fig. 8. Leakage detection test of an insecure KATAN-32 TI, 20k traces masks off (top), 100M traces masks on 1st-order (middle), 100M traces masks on 2nd-order (bottom).

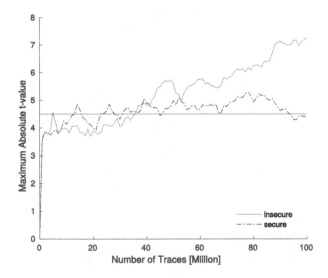

Fig. 9. Evolution of the points of maximum leakage with increasing number of traces for the secure and insecure KATAN-32 TI and plaintext value 0.

4.2   Leaking Threshold Implementation of KATAN-32

To investigate the effect of the placement and its inducing coupling, we first convert the NGC netlist back to an HDL file. Since the netlist is only produced after the synthesis step, and therefore is influenced by the "Keep Hierarchy" constraint, the resulting HDL file consists of Xilinx specific primitives grouped into separate modules that reflect the hierarchical structure of the secure KATAN-32 TI. By merging the resulting HDL modules and assigning the "Keep" constraint to all signals, we preserve the integrity of the secure implementation while dropping the placement constraints originating from the "Keep Hierarchy" constraint. We proceed by synthesizing the HDL file with "Keep Hierarchy" set to false and force the placement of the components to the lower right corner of the FPGA floorplan using the "Prohibit" constraints. Figure 6 shows the floorplan of the FPGA. The three individual shares are now placed in close proximity. The three different shares are again shown in magenta, light green and dark green.

Evaluation. We follow the same pattern for leakage detection tests.

Masks Off. The result of the leakage detection test with the masks turned off is shown in Fig. 8. Since the masks of the Threshold Implementation are set to zero, the t-value threshold of ±4.5 is exceeded with 20k traces.

Masks On. The middle and bottom graphs in Fig. 8 show the result of the first- and second-order leakage detection tests with 100M traces respectively. Small, periodic first-order leaks are visible and indicate the presence of out-of-model leakage.

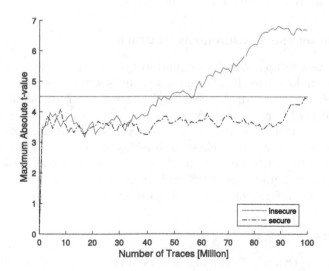

Fig. 10. Evolution of the points of maximum leakage with increasing number of traces for the secure and insecure KATAN-32 TI and plaintext value $087D2EC1_{hex}$.

The solid line in Fig. 9 shows the evolution of the point of maximum first-order leakage for the leaking KATAN-32 TI. Unlike the uncertain fluctuation around the ±4.5 threshold for the secure KATAN-32 TI, we now see a steady increase in the maximum of the absolute t-value. To increase the confidence in our observations, we repeated the experiments with a different fixed plaintext value (chosen randomly as $087D2EC1_{hex}$). Figure 10 shows the maximum value of the absolute t-value of the insecure design to be increasing, whereas this value does not exceed the ±4.5 threshold in the secure design. We conclude that out-of-model leakage, albeit small, is observable.

5   Discussion

5.1   A Note on "Keep Hierarchy"

In the majority of the Threshold Implementations literature, the "Keep Hierarchy" constraint is attributed with the function to keep the synthesis phase from optimizing over share boundaries. While this explanation is correct and different shares are indeed prevented from being merged in the same LUT, "Keep Hierarchy" also serves the purpose of not packing different shares in the same FPGA slice, which is shown to cause observable leakage in this work. When translating the "Keep Hierarchy" option to ASIC toolchains in order to avoid optimizations between shares, a common approach is to use "Compile Ultra" on different submodules and compiling the resulting netlists using the regular "Compile" process. While this effectively avoids optimizations across the boundaries of the shares, care might still be required to avoid standard cells belonging to different shares to be placed in the vicinity of each other, and routed wires of a share to be routed next to wires of other shares.

5.2   A Note on the Measurement Platform

Our measurement setup is a low-noise platform based on a Virtex-II Pro FPGA. The 90 nm technology node it uses delivers clean power traces with rather large amplitudes. As the out-of-model leakage effects we observed might not be as prominent with a 90 nm technology as with smaller nodes, other side-channel evaluation boards will need evaluation. As crosstalk and IR drop are known to become more prominent with smaller technology nodes, the 65 nm technology of the Virtex-5 on the Sasebo-GII platform [2], the 45 nm technology of the Spartan-6 on the Sakura-G board [22] and the 28 nm technology of the Kintex-7 on the Sakura-X platform form interesting targets for further investigation.

5.3   Conclusion

In this paper, we checked if coupling may be an issue in masking schemes. By using Threshold Implementations, we made sure the leakage we induced in our experiments originates from coupling, as the effects of glitches are ruled out.

We achieve a secure KATAN-32 TI using the state-of-the-art "Keep Hierarchy" implementation technique and show its security using state-of-the-art leakage detection methods. Afterwards, we induced out-of-model leakageby placing the gates and registers of the secure design in close proximity, as would be done in a real-world design. The leakage detection shows this new design to leak and leads us to the following conclusion. Leakage from coupling can be induced deliberately or "by accident" in masking schemes by placing shares in the vicinity of each other. As is shown from the related TI FPGA implementations using the "Keep Hierarchy" option that pass the leakage detection test [5,12,41], this does not necessarily happen as it has not been observed before. Since this problem can be caused on an FPGA however, we believe that careful examination of other environments is required when shares of a masking scheme might be densely packed, e.g. in cryptographic ASIC implementations, on which we can draw no conclusions. While the number of traces required for the leakage to be noticeable is high for our 90 nm platform, smaller process technologies are known to be more susceptible to crosstalk and IR drop coupling [37] and can lead to more leakage and hence possibly insecure designs. The actual source of the observed leakage (e.g. crosstalk, IR drop, ...) is nontrivial to isolate and more targeted experiments are required. Moreover, recent masked implementations have passed leakage detection tests using "Keep Hierarchy" even when the number of shares is decreased to the theoretical minimum number of shares $(d+1)$ [10,13]. Since these implementations use a lower number of shares, less noise is present and coupling can lead to leakages with a lower number of traces. These implementations are expected to be more favorable targets for actual key recovery attacks and are the subject of our future work.

Acknowledgments. This work was supported by NIST with the research grant 60NANB15D346, by the Research Council KU Leuven, OT/13/071 and by the Flemish Government through FWO project Cryptography secured against side-channel attacks by tailored implementations enabled by future technologies (G0842.13). Begül Bilgin and Benedikt Gierlichs are Postdoctoral Fellows of the Fund for Scientific Research - Flanders (FWO). Thomas De Cnudde is funded by a research grant of the Institute for the Promotion of Innovation through Science and Technology in Flanders (IWT-Vlaanderen).

References

1. Research Center for Information Security: National Institute of Advanced Industrial Science and Technology, Side-channel Attack Standard Evaluation Board SASEBO-G Specification. http://satoh.cs.uec.ac.jp/SASEBO/en/board/sasebo-g. html
2. Research Center for Information Security: National Institute of Advanced Industrial Science and Technology, Side-channel Attack Standard Evaluation Board SASEBO-GII Specification. http://www.rcis.aist.go.jp/special/SASEBO/ SASEBO-GII-en.html

3. Balasch, J., Gierlichs, B., Grosso, V., Reparaz, O., Standaert, F.-X.: On the cost of lazy engineering for masked software implementations. In: Joye, M., Moradi, A. (eds.) CARDIS 2014. LNCS, vol. 8968, pp. 64–81. Springer, Cham (2015). doi:10.1007/978-3-319-16763-3_5

4. Bhooshan, R., Rao, B.P.: Optimum IR drop models for estimation of metal resource requirements for power distribution network. In: VLSI-SoC, pp. 292–295. IEEE (2007)

5. Bilgin, B., Gierlichs, B., Nikova, S., Nikov, V., Rijmen, V.: Higher-order threshold implementations. In: Sarkar, P., Iwata, T. (eds.) ASIACRYPT 2014. LNCS, vol. 8874, pp. 326–343. Springer, Heidelberg (2014). doi:10.1007/978-3-662-45608-8_18

6. Bilgin, B., Gierlichs, B., Nikova, S., Nikov, V., Rijmen, V.: A more efficient aes threshold implementation. In: Pointcheval, D., Vergnaud, D. (eds.) AFRICACRYPT 2014. LNCS, vol. 8469, pp. 267–284. Springer, Cham (2014). doi:10.1007/978-3-319-06734-6_17

7. Brier, E., Clavier, C., Olivier, F.: Correlation power analysis with a leakage model. In: Joye, M., Quisquater, J.-J. (eds.) CHES 2004. LNCS, vol. 3156, pp. 16–29. Springer, Heidelberg (2004). doi:10.1007/978-3-540-28632-5_2

8. Cannière, C., Dunkelman, O., Knežević, M.: KATAN and KTANTAN — a family of small and efficient hardware-oriented block ciphers. In: Clavier, C., Gaj, K. (eds.) CHES 2009. LNCS, vol. 5747, pp. 272–288. Springer, Heidelberg (2009). doi:10.1007/978-3-642-04138-9_20

9. Chari, S., Jutla, C.S., Rao, J.R., Rohatgi, P.: Towards sound approaches to counteract power-analysis attacks. In: Wiener, M. (ed.) CRYPTO 1999. LNCS, vol. 1666, pp. 398–412. Springer, Heidelberg (1999). doi:10.1007/3-540-48405-1_26

10. Chen, C., Farmani, M., Eisenbarth, T.: A tale of two shares: why two-share threshold implementation seems worthwhile—and why it is not. In: Cheon, J.H., Takagi, T. (eds.) ASIACRYPT 2016. LNCS, vol. 10031, pp. 819–843. Springer, Heidelberg (2016). doi:10.1007/978-3-662-53887-6_30

11. Chen, Z., Haider, S., Schaumont, P.: Side-channel leakage in masked circuits caused by higher-order circuit effects. In: Park, J.H., Chen, H.-H., Atiquzzaman, M., Lee, C., Kim, T., Yeo, S.-S. (eds.) ISA 2009. LNCS, vol. 5576, pp. 327–336. Springer, Heidelberg (2009). doi:10.1007/978-3-642-02617-1_34

12. De Cnudde, T., Bilgin, B., Reparaz, O., Nikov, V., Nikova, S.: Higher-order threshold implementation of the aes s-box. In: Homma, N., Medwed, M. (eds.) CARDIS 2015. LNCS, vol. 9514, pp. 259–272. Springer, Cham (2016). doi:10.1007/978-3-319-31271-2_16

13. De Cnudde, T., Reparaz, O., Bilgin, B., Nikova, S., Nikov, V., Rijmen, V.: Masking AES with $d + 1$ shares in hardware. In: Gierlichs, B., Poschmann, A.Y. (eds.) CHES 2016. LNCS, vol. 9813, pp. 194–212. Springer, Heidelberg (2016). doi:10.1007/978-3-662-53140-2_10

14. Cooper, J., DeMulder, E., Goodwill, G., Jaffe, J., Kenworthy, G., Rohatgi, P.: Test vector leakage assessment (TVLA) methodology in practice. In: International Cryptographic Module Conference (2013). http://icmc-2013.org/wp/wp-content/uploads/2013/09/goodwillkenworthtestvector.pdf

15. Duan, C., LaMeres, B.J., Khatri, S.P.: On and Off-Chip Crosstalk Avoidance in VLSI Design. Springer, New York (2010). doi:10.1007/978-1-4419-0947-3

16. Barthe, G., Belaïd, S., Dupressoir, F., Fouque, P.-A., Grégoire, B., Strub, P.-Y.: Verified proofs of higher-order masking. In: Oswald, E., Fischlin, M. (eds.) EUROCRYPT 2015. LNCS, vol. 9056, pp. 457–485. Springer, Heidelberg (2015). doi:10.1007/978-3-662-46800-5_18

17. Dyrkolbotn, G.O., Wold, K., Snekkenes, E.: Security implications of crosstalk in switching cmos gates. In: Burmester, M., Tsudik, G., Magliveras, S., Ilić, I. (eds.) ISC 2010. LNCS, vol. 6531, pp. 269–275. Springer, Heidelberg (2011). doi:10.1007/978-3-642-18178-8_23

18. Dyrkolbotn, G.O., Wold, K., Snekkenes, E.: Layout dependent phenomena a new side-channel power model. JCP **7**(4), 827–837 (2012)

19. Gandolfi, K., Mourtel, C., Olivier, F.: Electromagnetic analysis: concrete results. In: Koç, Ç.K., Naccache, D., Paar, C. (eds.) CHES 2001. LNCS, vol. 2162, pp. 251–261. Springer, Heidelberg (2001). doi:10.1007/3-540-44709-1_21

20. Goodwill, G., Jun, B., Jaffe, J., Rohatgi, P.: A testing methodology for side-channel resistance validation. In: NIST Non-Invasive Attack Testing Workshop (2011). http://csrc.nist.gov/news_events/non-invasive-attack-testing-workshop/papers/08_Goodwill.pdf

21. Goubin, L., Patarin, J.: DES and differential power analysis the "duplication" method. In: Koç, Ç.K., Paar, C. (eds.) CHES 1999. LNCS, vol. 1717, pp. 158–172. Springer, Heidelberg (1999). doi:10.1007/3-540-48059-5_15

22. Guntur, H., Ishii, J., Satoh, A.: Side-channel attack user reference architecture board sakura-g. In: 2014 IEEE 3rd Global Conference on Consumer Electronics (GCCE), pp. 271–274, October 2014

23. Ishai, Y., Sahai, A., Wagner, D.: Private circuits: securing hardware against probing attacks. In: Boneh, D. (ed.) CRYPTO 2003. LNCS, vol. 2729, pp. 463–481. Springer, Heidelberg (2003). doi:10.1007/978-3-540-45146-4_27

24. Kocher, P.C.: Timing attacks on implementations of diffie-hellman, RSA, DSS, and other systems. In: Koblitz, N. (ed.) CRYPTO 1996. LNCS, vol. 1109, pp. 104–113. Springer, Heidelberg (1996). doi:10.1007/3-540-68697-5_9

25. Kocher, P., Jaffe, J., Jun, B.: Differential power analysis. In: Wiener, M. (ed.) CRYPTO 1999. LNCS, vol. 1666, pp. 388–397. Springer, Heidelberg (1999). doi:10.1007/3-540-48405-1_25

26. Mangard, S., Oswald, E., Popp, T.: Power Analysis Attacks - Revealing the Secrets of Smart Cards. Springer, Heidelberg (2007)

27. Mangard, S., Schramm, K.: Pinpointing the side-channel leakage of masked aes hardware implementations. In: Goubin, L., Matsui, M. (eds.) CHES 2006. LNCS, vol. 4249, pp. 76–90. Springer, Heidelberg (2006). doi:10.1007/11894063_7

28. Moll, F., Roca, M., Isern, E.: Analysis of dissipation energy of switching digital CMOS gates with coupled outputs. Microelectron. J. **34**(9), 833–842 (2003)

29. Moradi, A.: Side-channel leakage through static power. In: Batina, L., Robshaw, M. (eds.) CHES 2014. LNCS, vol. 8731, pp. 562–579. Springer, Heidelberg (2014). doi:10.1007/978-3-662-44709-3_31

30. Moradi, A., Poschmann, A., Ling, S., Paar, C., Wang, H.: Pushing the limits: a very compact and a threshold implementation of AES. In: Paterson, K.G. (ed.) EUROCRYPT 2011. LNCS, vol. 6632, pp. 69–88. Springer, Heidelberg (2011). doi:10.1007/978-3-642-20465-4_6

31. Nikova, S., Rechberger, C., Rijmen, V.: Threshold implementations against side-channel attacks and glitches. In: Ning, P., Qing, S., Li, N. (eds.) ICICS 2006. LNCS, vol. 4307, pp. 529–545. Springer, Heidelberg (2006). doi:10.1007/11935308_38

32. Nikova, S., Rijmen, V., Schläffer, M.: Secure hardware implementation of non-linear functions in the presence of glitches. In: Lee, P.J., Cheon, J.H. (eds.) ICISC 2008. LNCS, vol. 5461, pp. 218–234. Springer, Heidelberg (2009). doi:10.1007/978-3-642-00730-9_14

33. Nikova, S., Rijmen, V., Schläffer, M.: Secure hardware implementation of nonlinear functions in the presence of glitches. J. Cryptol. **24**(2), 292–321 (2011)

34. Poschmann, A., Moradi, A., Khoo, K., Lim, C., Wang, H., Ling, S.: Side-channel resistant crypto for less than 2, 300 GE. J. Cryptol. **24**(2), 322–345 (2011)
35. Prouff, E., Roche, T.: Higher-order glitches free implementation of the AES using secure multi-party computation protocols. In: Preneel, B., Takagi, T. (eds.) CHES 2011. LNCS, vol. 6917, pp. 63–78. Springer, Heidelberg (2011). doi:10.1007/978-3-642-23951-9_5
36. Quisquater, J.-J., Samyde, D.: ElectroMagnetic analysis (EMA): measures and counter-measures for smart cards. In: Attali, I., Jensen, T. (eds.) E-smart 2001. LNCS, vol. 2140, pp. 200–210. Springer, Heidelberg (2001). doi:10.1007/3-540-45418-7_17
37. Rabaey, J.M.: Digital Integrated Circuits: A Design Perspective. Prentice-Hall Inc., Upper Saddle River (1996)
38. Schmidt, J.-M., Plos, T., Kirschbaum, M., Hutter, M., Medwed, M., Herbst, C.: Side-channel leakage across borders. In: Gollmann, D., Lanet, J.-L., Iguchi-Cartigny, J. (eds.) CARDIS 2010. LNCS, vol. 6035, pp. 36–48. Springer, Heidelberg (2010). doi:10.1007/978-3-642-12510-2_4
39. Schneider, T., Moradi, A.: Leakage assessment methodology. In: Güneysu, T., Handschuh, H. (eds.) CHES 2015. LNCS, vol. 9293, pp. 495–513. Springer, Heidelberg (2015). doi:10.1007/978-3-662-48324-4_25
40. Schneider, T., Moradi, A.: Leakage assessment methodology - extended version. J. Cryptogr. Eng. **6**(2), 85–99 (2016)
41. Schneider, T., Moradi, A., Güneysu, T.: Arithmetic addition over boolean masking. In: Malkin, T., Kolesnikov, V., Lewko, A.B., Polychronakis, M. (eds.) ACNS 2015. LNCS, vol. 9092, pp. 559–578. Springer, Cham (2015). doi:10.1007/978-3-319-28166-7_27
42. Trichina, E., Korkishko, T., Lee, K.H.: Small size, low power, side channel-immune AES coprocessor: design and synthesis results. In: Dobbertin, H., Rijmen, V., Sowa, A. (eds.) AES 2004. LNCS, vol. 3373, pp. 113–127. Springer, Heidelberg (2005). doi:10.1007/11506447_10
43. Wild, A., Moradi, A., Güneysu, T.: Evaluating the duplication of dual-rail precharge logics on FPGAs. In: Mangard, S., Poschmann, A.Y. (eds.) COSADE 2014. LNCS, vol. 9064, pp. 81–94. Springer, Cham (2015). doi:10.1007/978-3-319-21476-4_6
44. Xilinx: Constraints guide 10.1. http://www.xilinx.com/itp/xilinx10/books/docs/cgd/cgd.pdf
45. Xilinx: Virtex-ii pro and virtex-ii pro x platform fpgas: Complete data sheet. http://www.xilinx.com/support/documentation/data_sheets/ds083.pdf
46. Zussa, L., Exurville, I., Dutertre, J., Rigaud, J., Robisson, B., Tria, A., Clédière, J.: Evidence of an information leakage between logically independent blocks. In: CS2@HiPEAC, pp. 25–30. ACM (2015)

Scaling Trends for Dual-Rail Logic Styles Against Side-Channel Attacks: A Case-Study

Kashif Nawaz[✉], Dinal Kamel, François-Xavier Standaert, and Denis Flandre

ICTEAM/ELEN/Crypto Group, Université catholique de Louvain,
Louvain-la-Neuve, Belgium
kashif.nawaz@uclouvain.be

Abstract. Dual-rail logic styles have been considered as possible alternatives to CMOS for the design of cryptographic circuits (more) secure against side-channel attacks. The state-of-the-art view on this approach is contrasted as they reduce the exploitable side-channel signal while not being sufficient to fully prevent the attacks. Since the limitations of dual-rail logic styles are essentially due to implementation challenges (e.g. the need of well-balanced capacitances), a natural question is to find out how they evolve with technology scaling. In this paper, we discuss this issue based on the relevant case study of an AES S-box implemented in CMOS and a dual-rail logic style, for two (65 nm and 28 nm) technologies. Our evaluations show that the security vs. performance tradeoff of our dual-rail logic style does not scale well compared to CMOS. It also shows that the scaling trends for CMOS are more positive (i.e. smaller technologies and supply voltages reduce the energy consumption and the side-channel signal). So these results suggest that dual-rail logic style may not be a sustainable approach for side-channel signal reduction as we move towards lower technology nodes.

1   Introduction

Following the first publications on power and electromagnetic analysis against cryptographic implementations, dual-rail (aka dynamic and differential) logic styles appeared as promising candidates to improve security against such attacks. Intuitively, these logic styles aim to solve the issue directly at the circuit level, by trying to reduce the side-channel Signal-to-Noise Ratio (SNR). For this purpose, they typically ensure that the switching activity of the circuits is independent of the manipulated data. However, despite constant switching activity, small data-dependent variations in the current traces can generally be observed, e.g. due to the unbalanced capacitances of the gates differential nodes and their interconnections. Therefore, a large body of work investigated the design of dual-rail logic styles in order to reach the best security vs. performance tradeoff, including but not limited to SABL [36], WDDL [35], DyCML [1], MCML [8] and MDPL [27]. Evaluations based on both simulations and actual measurements then confirmed that getting rid of these data dependencies is extremely challenging [19,26,29,34]. More recent works even showed that filtering effects in

© Springer International Publishing AG 2017
S. Guilley (Ed.): COSADE 2017, LNCS 10348, pp. 19–33, 2017.
DOI: 10.1007/978-3-319-64647-3_2

concrete measurement setups make these small data dependencies reasonably easy-to-exploit, e.g. thanks to linear regression [16]. To complete the picture, most of these dual-rail logic styles usually come with significant performance overheads, some of them additionally requiring full custom-design (which allows further control of the hardware, but is making development and deployment significantly more challenging/expensive).

In parallel, recent progresses have shown that mathematical countermeasures against side-channel analysis, and in particular the mainstream shuffling and masking techniques [6,13,21], can only lead to significant security improvements if the side-channel SNR has been sufficiently reduced beforehand [33,37]. This raises the problem of finding effective (and if possible efficient) hardware techniques allowing to fulfil this condition. Intuitively, it can be done by reducing the side-channel signal, which is what dual-rail logic styles achieve, or by increasing the noise.

Eventually, since the evaluation of secure hardware technologies goes together with technology scaling, another problem is to find out which of those approaches has more potential for the future.

In this paper, we therefore tackle this question of the comparative advantage of dual-rail logic styles as a mean of reducing the side-channel signal over standard CMOS in front of technology scaling. More precisely, we investigate how much the security vs. performance tradeoff of these design styles scales, based on the simple yet reflective case-study of CMOS and Dynamic and Differential Swing-Limited Logic (DDSLL) AES S-boxes, implemented in 65 and 28 nm technologies. DDSLL is yet another (full-custom) dual-rail logic style which has already been analyzed based on simulations and actual measurements [30]. Our choice of DDSLL arises from the fact that its design using 65 nm bulk technology shows 1.5× lower power consumption at the expense of 1.125× increase in area, while increasing the security 10× when compared to CMOS [15]. This compares positively with the typical power/energy and area costs obtained with the previously listed dual-rail logic styles. Therefore, DDSLL can be considered a good candidate to illustrate technology scaling trends (as discussed in conclusions, we expect other dual-rail logic styles to follow similar trends). In this respect, our main conclusions are twofold.

First, and looking at the tradeoff between the side-channel SNR and the implementation performances (here measured with the energy per operation, which is a quite reflective metric to compare cryptographic designs [17]), we see that the comparative advantage of DDSLL over CMOS is vanishing with technology scaling, and we explain this trend by the imbalances in DDSLL gates that gain impact with technology scaling.

Second, and more positively, we also see that technology and supply voltage scaling have a positive impact on the security vs. performance tradeoff of CMOS devices, essentially because such a scaling comes with energy gains and side-channel signal reductions.

Our case study therefore suggests that signal reduction using dual-rail logic styles may not be the best approach w.r.t technological scaling. It also suggests that the design of noisy CMOS implementations (which is a natural consequence of scaling [10]) appears as a promising strategy for ensuring a sufficiently small side-channel SNR allowing (e.g.) masking and shuffling to be effective in future technologies. Note that by noisy implementations, we do not mean measurement noise or additional external noise but intrinsic noise at the device level (i.e. transistor, interconnect, resistor, ...) as a result of technological scaling.

Cautionary note. The results in this paper are based on simulations. While we admit that in general, they can lead to shortcomings (e.g. regarding the shape/linearity of the leakage traces), the experiments in [30] showed that they can be used as a good predictor for the amount of information leakage in dual-rail logic styles. Since the goal in this paper is to discuss general scaling trends, we believe simulations can therefore be used as an interesting indication of how the comparative advantage of CMOS over DDSLL scales. Note that anyway, we do not expect the shape/linearity of the leakage traces to be significantly different in 65 and 28 nm technologies, nor for CMOS vs. DDSLL, since the main linearization factor is due to filtering effects in the measurement setup and not the internal transistor behavior. So as usual with simulations, they should be interpreted with care (which we try to do in the paper). But as usual with simulations as well, they are a useful tool to get some hints about the best solutions to investigate up to (more expensive) tape outs.

The rest of the paper is structured as follows. Preliminaries are in Sect. 2. Our target implementation and evaluation settings are in Sect. 3. The comparative study between CMOS and DDSLL is in Sect. 4. The positive impact of technology and supply voltage scaling for CMOS implementations is in Sect. 5. Finally, our conclusions and a discussion of the relevance of this case study are in Sect. 6.

2   Preliminaries

2.1   Logic Styles: CMOS and DDSLL

Traditional CMOS circuits have shown a data-dependency in the power consumption leading to exploitable side-channel information, e.g. thanks to Differential Power Analysis (DPA) [18], Correlation Power Analysis (CPA) [5] and Template attacks [7]. The power consumption of a CMOS circuit is modeled by the following equation:

$$P = P_{dyn} + P_{stat},$$
$$= \frac{1}{2} N_{nodes} \, \alpha_F \, C_L \, V_{DD}^2 \, f_{clk} + I_{leak} \, V_{DD}, \tag{1}$$

where P_{dyn} is the dynamic power consumption, P_{stat} the static power consumption, N_{nodes} is the number of nodes in the circuit, α_F represents the activity factor of the design, C_L is the load capacitance, V_{DD} is the supply voltage, f_{clk} represents the clock frequency and I_{leak} denotes the leakage current. We assume here that one operation is executed per clock cycle, and thus $f_{clk} = f_{op}$, which determines the target throughput of the application. The operation period $T_{op} = \frac{1}{f_{op}}$

should be more than the critical path delay T_{del} to guarantee correct functionality. The data-dependency of the CMOS logic comes from both its dynamic and static power consumptions. In the dynamic part, α_F directly depends on the data being processed. In the static part, the I_{leak} is the data-dependent parameter. Although the former is dominant, static power consumption can also be exploited if its value is sufficiently high and the operating frequency is low enough allowing the reduction of noise via simple averaging techniques [24, 28].[1] Yet, in our following experiments, dynamic power indeed dominates.

In comparison with CMOS, we investigate the Dynamic and Differential Swing-Limited Logic (DDSLL) which aims at low-power implementations and of which the dynamic power consumption is given by:

$$P_{dyn} = \frac{1}{2} N_{nodes} \, \alpha_F \, C_L \, V_{DD} \, V_{swing} \, f_{clk}, \tag{2}$$

where α_F equals 1 because all dynamic and differential logic styles ensure one output transition per clock cycle (independent of the data being processed), and V_{swing} is the output voltage swing. DDSLL gates are designed to have limited swing (i.e. $< V_{DD}$) in order to reduce the dynamic power consumption and hence the energy per operation.

Figure 1 shows the circuit of a generic DDSLL gate. A Differential Pull-Down Network (DPDN) is used to evaluate the required function. It mainly consists of NMOS transistors. The DDSLL gate employs a dynamic current source to significantly reduce the static power consumption similar to what is achieved in the DyCML logic style. The cut-off of this current source is performed via a feedback network which signals the end of an operation, so that the self-timing buffer creates a clock signal Clk_{i+1} declaring the termination of the current evaluation phase. This clock signal is used to feed the following DDSLL gate. The precharge transistors are used to precharge the differential outputs of a DDSLL gate to the supply voltage before an evaluation of the gate's function takes place, and the latch transistors preserve the evaluated voltage at the differential output nodes.

The operation of a DDSLL gate is quite simple. It works in two modes: precharge and evaluation. In the precharge mode, when the clock signal Clk_i is low, the outputs out and $\overline{out}$ are precharged to V_{DD}. There is no current path from V_{DD} to GND because transistor M_1 is switched off. However, transistors M_6 and M_2 are switched on. Next during the evaluation phase, Clk_i goes high turning on the transistor M_1 while M_2 is still on (both forming the dynamic current source) as node ENO was previously charged to VDD in the previous precharge phase, thus creating a path to discharge one of the output nodes to GND. The discharge path through the DPDN network is the one with the lowest impedance depending on the inputs being processed. As one of the outputs falls

---

[1] Note that advanced technologies usually provide multiple flavors such as low-power and high-performance along with different device choices such as high and low threshold voltages, providing circuit designers with various options to reduce the power consumption – and the leakage power as well – which may modify the respective importance of these source of leakages.

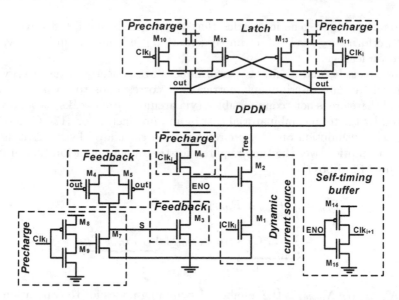

Fig. 1. Schematic of a generic DDSLL gate.

below the threshold voltage of the feed-back transistors (M_4, M_5), one of them will turn on which in turn will discharge the node ENO to GND thus starving the current source. The design of a DDSLL circuit comprising of several functions can benefit from resource sharing (the dynamic current source, parts of the precharge circuit, the feedback circuit and the self-timing buffer of functions that evaluate at the same time), therefore reducing the area cost and the overall power consumption.

In all differential design styles, the unbalanced capacitances are considered as a source of information leakage. They mainly come from either routing imbalances or from internal imbalances. In this paper, we only consider the latter ones. (As mentioned in Sect. 3.3, additionally considering post-layout simulations or variability should amplify the trend we put forward). In this respect, we note that the only way to eliminate internal imbalances is to design circuits with perfectly symmetrical differential gates, which is usually very expensive in terms of area and speed. Hence, our DPDN designs exploit the binary decision diagram technique from [11] in order to improve performances by exploiting more complex gates with minimum imbalance caused by internal capacitances (details can be found in [15]). So essentially, what we show next is that (here internal) imbalances have an increasing impact with technology scaling.

2.2   Evaluation Metrics

Evaluating logic styles across technology scaling and supply voltage scaling is a challenging task, since certain metrics may favour one logic style over another. In order to be as fair as possible in our performance and security evaluations, we therefore selected generic metrics that are generally more reflective of the "global performance level" of an implementation, and can capture any type of information leakage.

More precisely, and as far as performances are concerned, the energy per operation is a quite discriminant metric, as it corresponds to an integral over time, and therefore is not "compressible" (via architectural tweaks) beyond what is allowed by the total combinatorial cost of an implementation [17]. Concretely, the energy consumption of a logic circuit can be calculated by integrating the power consumption over the time required for the target operation (in our case-study, the AES S-box):

$$
\begin{aligned}
E_{op} &= \int_t (P_{dyn} + P_{stat})\, dt, \\
&= \underbrace{\frac{1}{2}\, N_{sw}\, C_L\, V_{DD}\, V_{swing}}_{\text{Dynamic}} + \underbrace{V_{DD}\, I_{leak}\, T_{del}}_{\text{static}},
\end{aligned}
\tag{3}
$$

where $N_{sw} = \alpha_F\, N_{nodes}$ is the number of switching nodes for the operation and T_{del} is the circuit delay (following the investigations in [4]). In the case of CMOS logic style, V_{swing} is equal to the supply voltage being used.

As for the security metrics, our choice was dictated by various constraints and features. First, the shape of the instantaneous dynamic power (i.e. the side-channel signal) changes significantly depending on the technology and the supply voltage used. Therefore, it is important to consider these changes while evaluating the CMOS and DDSLL logic styles, and to consider a multivariate analysis. Second, our target implementations are not masked, and therefore our simulations exclusively exploit first-order leakages. So while in general, a fair comparison of our logic styles would require to compute a mutual information metric [32], in this particular case we can simplify our evaluations by (i) applying a dimensionality reduction to our traces, namely a Principal Component Analysis (PCA) which will capture the shape of the noise-free simulated traces [2], and (ii) computing Mangard's SNR on the reduced traces [20], which is equivalent to the mutual information metric in this case [9, 22].

For this purpose, we denote a power trace as $\boldsymbol{l}$, its corresponding random variable as $\boldsymbol{L}$, and assume that this random variable is a function of the S-box input X, and a noise random variable $\boldsymbol{N}$. The multivariate traces are reduced to univariate ones thanks to PCA, which we denote as: $l = \mathrm{PCA}(\boldsymbol{l})$. Giving the trace l a subscript x corresponding to the input and a superscript i corresponding to an index (since the trace for a plaintext x can be measured multiple times), Mangard's SNR is defined as:

$$
\mathrm{SNR} = \frac{\hat{\mathrm{var}}_x(\hat{\mathsf{E}}_i(L_x^i))}{\hat{\mathsf{E}}_x(\hat{\mathrm{var}}_i(L_x^i))},
\tag{4}
$$

where $\hat{\mathsf{E}}$ (resp. $\hat{\mathrm{var}}$) denotes the sample mean (resp. variance) operator. In our following simulations, this SNR will be computed for noise-free traces. This amounts to maximizing the signal $\hat{\mathsf{var}}_x(\hat{\mathsf{E}}_i(L_x^i))$ (i.e. we ignore the denominator of Eq. 4 in this case).

Note that for readability, our following results only report the quantification of our experiments with this security metric. However, various other (heuristic) choices could be considered, e.g. computing the SNR for the most informative samples in the traces, or considering more dimensions after the application of the PCA. The same holds with performance metrics (e.g. the throughput over area ratio could be used as alternative efficiency metric). In our study, none of these variations (that we also browsed) lead to different conclusions regarding the two main trends outlined in introduction.

3   AES S-Box Implementations

3.1   AES S-Box

For the sake of simplicity, and in order to demonstrate the technology trends of CMOS and DDSLL, we chose an 8-bit AES S-box as the benchmark circuit. More specifically, we considered a combinatorial implementation of the S-box from [14], based on the architecture proposed in [23,31]. Thanks to mapping the elements of the original field $GF(2^8)$ to the composite field $GF(((2^2)^2)^2)$, the gate complexity and the power consumption can be greatly reduced. The adopted S-box consists of 3 stages: a transformation stage to map the elements to the $GF(((2^2)^2)^2)$ field, an inversion stage and an inverse transformation stage to map the elements back to $GF(2^8)$, grouped with the affine transformation.

3.2   Target Designs

The CMOS and DDSLL AES S-boxes were implemented in a full-custom fashion, using the CADENCE Virtuoso tools, in a low-power 65 nm bulk technology and 28 nm FDSOI technology. The CMOS implementation of the S-box is made only of 2-input AND/NAND and 2-input XOR/XNOR gates. The total number of transistors is 1,530, with a logic depth of 22 gates. On the other hand, the DDSLL S-box accounts for 1275 transistors, with a logic depth of 13 gates.

The gate design of CMOS and DDSLL S-boxes in both 65 nm bulk and 28 nm FDSOI technologies is kept identical, i.e. the gates used and the number of transistors remain unchanged. However, we respected the minimum feature size of each technology (to decrease the switching capacitance, hence the energy per operation) and resized the transistors' widths adequately to guarantee functionality. In 65 nm bulk technology, standard threshold voltage (SVT) transistors are used to reduce the static current while maintaining good performances with respect to the circuit delay. For benchmarking purposes, we also implemented the CMOS and DDSLL S-boxes using the low threshold voltage (LVT) transistors available from the same technology. In 28 nm FDSOI technology, both SVT and LVT transistors were again used to maintain a fair comparison with the 65 nm technology. Yet, for readability, our following results only report the

quantification of our experiments with SVT transistors (trends are again identical for LVT transistors). Eventually, the widths of the DDSLL S-box transistors
were chosen such that the voltage swing is sufficient for the circuit to operate
correctly at the lower limit of the supply voltages of each technology.

In our experiments, all the S-boxes are fed with buffered inputs to maintain
equal fan-ins and have realistic inputs, yet the DDSLL S-box additionally has a
buffered clock. Also, all the outputs of the S-boxes are loaded with equal fan-out
buffers. Each S-box is provided with a separate supply voltage than that of the
input/output buffers so that the buffers' energy consumptions are not taken into
account in our evaluations.

3.3 Simulation Settings

Simulations for the above designs are done at the schematic level (without any
extracted post-layout capacitances) using Eldo simulator based on SPICE models provided by the industrial foundries at room temperature of 25 °C. In this
respect, we note that any imbalance in the parasitic elements would affect the
difference between the delay of the differential routes of the DDSLL S-box,
which would impact its power consumption, leading to a higher (exploitable)
signal being observed by practical attacks [30]. And this is expected to get only
worse with technology scaling and variability, since balancing the capacitances
in an implementation naturally becomes more challenging with smaller circuits
and smaller routing capacitances. Therefore, and as previously mentioned, taking the parasitic routing capacitances into consideration could only amplify our
observation (which is that the comparative advantage of DDSLL over CMOS is
vanishing with technology scaling). A similar statement holds for other effects
that we did not take into consideration in this work such as crosstalk and the
influence of process, voltage and temperature (PVT) variations.

Consequently, we assume our results correspond to an ideal scenario and the
inclusion of more physical default(s) should only deteriorate the performance of
dual-rail logic styles compared to CMOS.

The frequency of operation is chosen to be 10 MHz, which is in accordance
with the usual operating frequencies for cryptographic applications (see, e.g. [3]).
The supply voltage is swept across a range of 500 mV, in steps of 100 mV starting
from the nominal voltage of each technology, namely, 1.2 V and 1 V for the 65 nm
bulk and the 28 nm FDSOI technologies, respectively. The lower limit of the supply voltages is imposed by the correct functionality of the circuit at the target
frequency for a given implementation.

Note that operating at high supply voltages and using the 28 nm technology
node allows the circuit implementation to run at higher frequencies. But the frequency choice has no impact on our results since we have chosen to use the energy
per operation as the evaluation metric and not the total power consumption.

To calculate the energy per operation, we considered 1000 random input signals. As for the security analysis, the S-box input signal has 256 possible values
whose transitions are chosen between 0 and an arbitrary input. Restricting the

inputs to a subset from the 256^2 possible inputs was mainly motivated by practical simulation constraints (memory and simulation time) and is not expected to strongly impact the comparison between the logic styles.   .

4   Comparative Scaling Trends: CMOS vs. DDSLL

In this section, we aim to compare the CMOS and DDSLL logic styles and study how their security versus performance tradeoff evolves with technology and supply voltage scaling. To be able to do that in a comprehensible manner, we plot the ratios between our (security and performance) metrics computed for both CMOS and DDSLL, one in function of the other. More precisely, Fig. 2 shows the PCA signal ratio between CMOS and DDSLL S-boxes versus the energy per operation ratio between these logic styles for the 65 nm bulk and the 28 nm FDSOI technologies (designed using SVT devices). The different points represent the supply voltages we used that span over a range of 0.5 V starting from the nominal supply of each technology (1.2 V and 1 V for the 65 and 28 nm technologies, respectively). This study allows us to make the following observations:

1 By reducing the supply voltage, the energy per operation ratio of CMOS with respect to DDSLL is decreasing for the 65 and 28 nm technologies. This reduction can be explained by the fact that the E_{op} value of the DDSLL style decreases almost linearly with the supply voltage, because it maintains nearly the same voltage swing while E_{op} of CMOS decreases quadratically (see Eq. 3).
2 As for the PCA signal ratio between CMOS and DDSLL (for both technologies), it also decreases with the supply voltage scaling. Again, this is due to the fact that the voltage swing of the DDSLL S-box is kept almost unchanged leading to a slow reduction rate of the transient power consumption. Hence the PCA signal with V_{DD} scaling is less compared to that of CMOS.
3 Most importantly, Fig. 2 illustrates that at 28 nm technology, the PCA signal ratio between CMOS and DDSLL is less than that of the 65 nm technology for similar E_{op} ratios. This figure neatly puts forward that the security vs. performance tradeoff between these logic styles does not scale positively for DDSLL, even though we do not consider any routing parasitics or PVT variations in our simulations.

It is worth emphasizing that similar observations were made by comparing CMOS to DDSLL S-boxes using LVT devices in both technologies. Therefore, changing the device type leads only to either better performance or more power savings, but the technology and supply voltage scaling trends remain the same. Also, and for the sake of completeness, we conducted the same experiments using the maximum SNR (before PCA) as a security metric and the technology and supply voltage scaling trends remained unchanged.

In addition, we note that changing the frequency of operation does not impact our conclusions as long as the dynamic power consumption dominates. We simulated the S-boxes down to 100 kHz and the technology and voltage scaling trends were again the same.

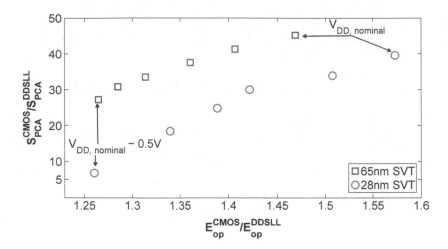

Fig. 2. Evolution of the tradeoff between the PCA signal ratio and the E_{op} ratio for CMOS vs. DDSLL using 65 and 28 nm technology nodes.

Eventually, we note that technology scaling is advancing at a fast pace and secure implementations will soon follow (given the fact that up until now applications such as smart cards tend to lag by one or two technologies). Also, circuit designers generally aim at scaling the supply voltage in order to further reduce the energy consumption of the digital circuits (sometimes operating below the transistor subthreshold voltage leading to minimum energy per operation). Both trends lead us to conclude that the observations in this section may rapidly have concrete relevance.

5   Technology and V_{DD} Scaling Trends for CMOS

Since the comparative advantage of DDSLL over CMOS vanishes in our case study, one natural complementary question is whether scaling trends for CMOS circuits lead to a more positive conclusion. In this section, we answer this question by focusing on the impact of technology and V_{DD} scaling on CMOS circuits only. For this purpose, Fig. 3(a) reports the energy per operation of the CMOS S-box for different supply voltages, using both 65 and 28 nm technologies. We recall that the minimum V_{DD} was chosen such that the CMOS S-box operates correctly at the 10 MHz target frequency for each technology. As expected, the energy per operation of CMOS decreases almost quadratically with the reduction of the supply voltage (see Eq. 3). The figure also shows clearly that technology scaling from 65 to 28 nm reduces the energy per operation (of the CMOS S-box) by a factor of 2.2×. This reduction is compliant with the expected technology scaling trend as explained in [12].

Similarly, Fig. 3(b) shows the signal after PCA of the CMOS S-box for different V_{DD} values, using both 65 and 28 nm technologies. The PCA signal of

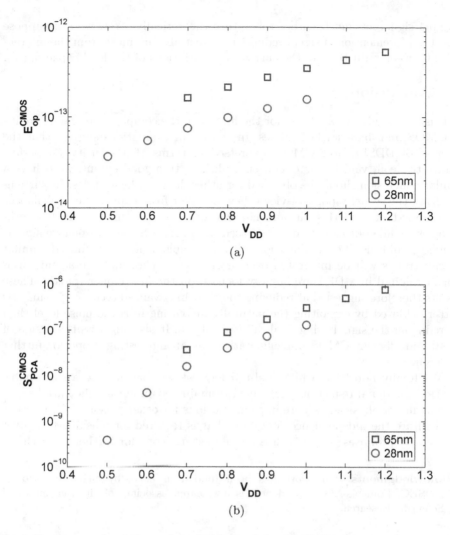

Fig. 3. Scaling trends of E_{op} and the PCA signal for the CMOS S-box across a supply voltage range of 500 mV for 65 and 28 nm tech.

CMOS decreases as the supply voltage scales down. In the 65 nm technology, the S_{PCA} reduction is more than one order of magnitude across the whole V_{DD} range (1.2 V to 0.7 V) and it reaches even more than two orders of magnitude in the 28 nm technology, by reducing the supply voltage from 1 V down to 0.5 V. As for the technology scaling from 65 to 28 nm, it is also clear that the signal after PCA of the CMOS S-box decreases by a factor of 2.3× at comparable supply voltages. So the scaling trends for PCA signal is also positive for CMOS. Here, we note that an analytical explanation of these observations is more challenging, since the small data-dependent current variations that lead to exploitable side-channel signal are much harder to capture theoretically. (Yet, we can suppose that the aforementioned signal reductions essentially originate from the interactions between reductions of the current and variations of the load capacitance).

6   Conclusion

In this case study, we analyzed for the first time the comparative scaling trends of CMOS and dual-rail logic styles. In short, our evaluations suggest that the interest of DDSLL over CMOS, expressed in terms of a security vs. performance tradeoff, vanishes as circuit sizes shrink. To a good extent, we believe a similar conclusion should be obtained for other dual-rail logic styles. In particular, from the security point-of-view, they all suffer from capacitance imbalances to some extent, and this phenomenon can only be magnified in smaller technologies. While our case study was based on an AES S-box, we believe similar trends should also be obtained for full AES implementations. Indeed, similar energy trends will be integrated over more clock cycles, and the signal variations exploited in a DPA are anyway focused on the first cipher rounds. These results therefore suggest that reducing the SNR in advanced technologies may be better achieved by exploiting the naturally increasing intrinsic noise level than by reducing the signal using dual-rail logic styles. It also suggests the design of noisy and efficient CMOS implementations as an interesting scope for further research.

We finally note that while dual-rail logic styles may not be a sustainable solution for signal reduction purposes, it remains possible that they are helpful ingredients of physically secure implementations for other reasons (e.g. in order to facilitate the independence condition that is required for masking to deliver its security promises [39]), which is an interesting scope for further research.

Acknowledgments. This work has been funded in parts by the ARC Project NANOSEC. François-Xavier Standaert is a research associate of the Belgian Fund for Scientific Research.

References

1. Allam, M., Elmasry, M.: Dynamic current mode logic (DyCML): a new low-power high-performance logic style. IEEE J. Solid-State Circ. **36**(3), 550–558 (2001)
2. Archambeau, C., Peeters, E., Standaert, F.-X., Quisquater, J.-J.: Template attacks in principal subspaces. In: Goubin, L., Matsui, M. (eds.) CHES 2006. LNCS, vol. 4249, pp. 1–14. Springer, Heidelberg (2006). doi:10.1007/11894063_1
3. Bellizia, D., Bongiovanni, S., Monsurro, P., Scotti, G., Trifiletti, A.: Univariate power analysis attacks exploiting static dissipation of nanometer CMOS VLSI circuits for cryptographic applications. IEEE Trans. Emerg. Top. Comput. **PP**(99), 1 (2016)
4. Bol, D., Kamel, D., Flandre, D., Legat, J.-D.: Nanometer MOSFET effects on the minimum-energy point of 45 nm subthreshold logic. In: Proceedings of the 2009 International Symposium on Low Power Electronics and Design, San Fancisco, CA, USA, 19–21 August 2009, pp. 3–8 (2009)

5. Brier, E., Clavier, C., Olivier, F.: Correlation power analysis with a leakage model. In: Joye, M., Quisquater, J.-J. (eds.) CHES 2004. LNCS, vol. 3156, pp. 16–29. Springer, Heidelberg (2004). doi:10.1007/978-3-540-28632-5_2

6. Chari, S., Jutla, C.S., Rao, J.R., Rohatgi, P.: Towards sound approaches to counteract power-analysis attacks. In: Wiener [38], pp. 398–412

7. Chari, S., Rao, J.R., Rohatgi, P.: Template attacks. In: Kaliski, B.S., Koç, K., Paar, C. (eds.) CHES 2002. LNCS, vol. 2523, pp. 13–28. Springer, Heidelberg (2003). doi:10.1007/3-540-36400-5_3

8. Deniz, Z.T., Leblebici, Y.: Low-power current mode logic for improved DPA-resistance in embedded systems. In: International Symposium on Circuits and Systems (ISCAS 2005), Kobe, Japan, 23–26 May 2005, pp. 1059–1062. IEEE (2005)

9. Duc, A., Faust, S., Standaert, F.-X.: Making masking security proofs concrete. In: Oswald, E., Fischlin, M. (eds.) EUROCRYPT 2015. LNCS, vol. 9056, pp. 401–429. Springer, Heidelberg (2015). doi:10.1007/978-3-662-46800-5_16

10. Ghosh, S., Roy, K.: Parameter variation tolerance and error resiliency: new design paradigm for the nanoscale era. Proc. IEEE 98(10), 1718–1751 (2010)

11. Giancane, L., Marietti, P., Olivieri, M., Scotti, G., Trifiletti, A.: A new dynamic differential logic style as a countermeasure to power analysis attacks. In: 15th IEEE International Conference on Electronics, Circuits and Systems, ICECS 2008, pp. 364–367, August 2008

12. Haensch, W., Nowak, E.J., Dennard, R.H., Solomon, P.M., Bryant, A., Dokumaci, O.H., Kumar, A., Wang, X., Johnson, J.B., Fischetti, M.V.: Silicon CMOS devices beyond scaling. IBM J. Res. Dev. 50(4–5), 339–362 (2006)

13. Herbst, C., Oswald, E., Mangard, S.: An AES smart card implementation resistant to power analysis attacks. In: Zhou, J., Yung, M., Bao, F. (eds.) ACNS 2006. LNCS, vol. 3989, pp. 239–252. Springer, Heidelberg (2006). doi:10.1007/11767480_16

14. Kamel, D., Standaert, F.X., Flandre, D.: Scaling trends of the AES S-box low power consumption in 130 and 65 nm CMOS technology nodes. In: 2009 IEEE International Symposium on Circuits and Systems, pp. 1385–1388, May 2009

15. Kamel, D., Renauld, M., Bol, D., F.-X., Standaert, D., Flandre, D.: Analysis of dynamic differential swing limited logic for low-power secure applications. J. Low Power Electron. Appl. 2(1), 98 (2012)

16. Kamel, D., Renauld, M., Flandre, D., Standaert, F.-X.: Understanding the limitations and improving the relevance of SPICE simulations in side-channel security evaluations. J. Cryptographic Eng. 4(3), 187–195 (2014)

17. Kerckhof, S., Durvaux, F., Hocquet, C., Bol, D., Standaert, F.-X.: Towards green cryptography: a comparison of lightweight ciphers from the energy viewpoint. In: Prouff, E., Schaumont, P. (eds.) CHES 2012. LNCS, vol. 7428, pp. 390–407. Springer, Heidelberg (2012). doi:10.1007/978-3-642-33027-8_23

18. Kocher, P.C., Jaffe, J., Jun, B.: Differential power analysis. In: Wiener [38], pp. 388–397

19. Macé, F., Standaert, F.-X., Quisquater, J.-J.: Information theoretic evaluation of side-channel resistant logic styles. In: Paillier and Verbauwhede [25], pp. 427–442

20. Mangard, S.: Hardware countermeasures against DPA – a statistical analysis of their effectiveness. In: Okamoto, T. (ed.) CT-RSA 2004. LNCS, vol. 2964, pp. 222–235. Springer, Heidelberg (2004). doi:10.1007/978-3-540-24660-2_18

21. Mangard, S., Oswald, E., Popp, T.: Power Analysis Attacks - Revealing the Secrets of Smart Cards. Springer, Heidelberg (2007)

22. Mangard, S., Oswald, E., Standaert, F.-X.: One for all - all for one: unifying standard differential power analysis attacks. IET Inf. Secur. 5(2), 100–110 (2011)

23. Mentens, N., Batina, L., Preneel, B., Verbauwhede, I.: A systematic evaluation of compact hardware implementations for the Rijndael S-Box. In: Menezes, A. (ed.) CT-RSA 2005. LNCS, vol. 3376, pp. 323–333. Springer, Heidelberg (2005). doi:10.1007/978-3-540-30574-3_22

24. Moradi, A.: Side-channel leakage through static power. In: Batina, L., Robshaw, M. (eds.) CHES 2014. LNCS, vol. 8731, pp. 562–579. Springer, Heidelberg (2014). doi:10.1007/978-3-662-44709-3_31

25. Paillier, P., Verbauwhede, I., (eds.) Proceedings of the 9th International Workshop Cryptographic Hardware and Embedded Systems - CHES 2007. LNCS, Vienna, Austria, 10–13 September 2007, vol. 4727. Springer, Heidelberg (2007)

26. Popp, T., Kirschbaum, M., Zefferer, T., Mangard, S.: Evaluation of the masked logic style MDPL on a prototype chip. In: Paillier and Verbauwhede [25], pp. 81–94

27. Popp, T., Mangard, S.: Masked dual-rail pre-charge logic: DPA-resistance without routing constraints. In: Rao, J.R., Sunar, B. (eds.) CHES 2005. LNCS, vol. 3659, pp. 172–186. Springer, Heidelberg (2005). doi:10.1007/11545262_13

28. Del Pozo, S.M., Standaert, F.-X., Kamel, D., Moradi, A.: Side-channel attacks from static power: when should we care? In: Proceedings of the 2015 Design, Automation & Test in Europe Conference & Exhibition, DATE 2015, Grenoble, France, 9–13 March 2015, pp. 145–150 (2015)

29. Regazzoni, F., Eisenbarth, T., Poschmann, A., Großschädl, J., Gürkaynak, F.K., Macchetti, M., Deniz, Z.T., Pozzi, L., Paar, C., Leblebici, Y., Ienne, P.: Evaluating resistance of MCML technology to power analysis attacks using a simulation-based methodology. Trans. Comput. Sci. 4, 230–243 (2009)

30. Renauld, M., Kamel, D., Standaert, F.-X., Flandre, D.: Information theoretic and security analysis of a 65-nanometer DDSLL AES S-Box. In: Preneel, B., Takagi, T. (eds.) CHES 2011. LNCS, vol. 6917, pp. 223–239. Springer, Heidelberg (2011). doi:10.1007/978-3-642-23951-9_15

31. Satoh, A., Morioka, S., Takano, K., Munetoh, S.: A compact rijndael hardware architecture with s-box optimization. In: Boyd, C. (ed.) ASIACRYPT 2001. LNCS, vol. 2248, pp. 239–254. Springer, Heidelberg (2001). doi:10.1007/3-540-45682-1_15

32. Standaert, F.-X., Malkin, T.G., Yung, M.: A unified framework for the analysis of side-channel key recovery attacks. In: Joux, A. (ed.) EUROCRYPT 2009. LNCS, vol. 5479, pp. 443–461. Springer, Heidelberg (2009). doi:10.1007/978-3-642-01001-9_26

33. Standaert, F.-X., Veyrat-Charvillon, N., Oswald, E., Gierlichs, B., Medwed, M., Kasper, M., Mangard, S.: The world is not enough: another look on second-order DPA. In: Abe, M. (ed.) ASIACRYPT 2010. LNCS, vol. 6477, pp. 112–129. Springer, Heidelberg (2010). doi:10.1007/978-3-642-17373-8_7

34. Tiri, K., Verbauwhede, I.: Securing encryption algorithms against DPA at the logic level: next generation smart card technology. In: Walter, C.D., Koç, Ç.K., Paar, C. (eds.) CHES 2003. LNCS, vol. 2779, pp. 125–136. Springer, Heidelberg (2003). doi:10.1007/978-3-540-45238-6_11

35. Tiri, K., Verbauwhede, I.: A logic level design methodology for a secure DPA resistant ASIC or FPGA implementation. In: 2004 Design, Automation and Test in Europe Conference and Exposition (DATE 2004), Paris, France, 16–20 2004, pp. 246–251. IEEE Computer Society, February 2004

36. Tiri, K., Verbauwhede, M.A.I.: A dynamic and differential CMOS logic with signal independent power consumption to withstand differential power analysis on smart cards. In: Proceedings of the 28th European Solid-State Circuits Conference, ESSCIRC 2002, pp. 403–406. IEEE (2002)

37. Veyrat-Charvillon, N., Medwed, M., Kerckhof, S., Standaert, F.-X.: Shuffling against side-channel attacks: a comprehensive study with cautionary note. In: Wang, X., Sako, K. (eds.) ASIACRYPT 2012. LNCS, vol. 7658, pp. 740–757. Springer, Heidelberg (2012). doi:10.1007/978-3-642-34961-4_44
38. Wiener, M.J. (ed.) 19th Annual International Cryptology Conference 1999 Proceedings Advances in Cryptology - CRYPTO 1999. LNCS, Santa Barbara, California, USA, 15–19 August 1999, vol. 1666. Springer, Heidelberg (1999)
39. Wild, A., Moradi, A., Güneysu, T.: GliFreD: Glitch-free duplication - towards power-equalized circuits on FPGAs. IACR Cryptology ePrint Archive 2015:124 (2015)

Dissecting Leakage Resilient PRFs
with Multivariate Localized EM Attacks
A Practical Security Evaluation on FPGA

Florian Unterstein[1]([⊠]), Johann Heyszl[1], Fabrizio De Santis[2],
and Robert Specht[1]

[1] Fraunhofer Research Institution AISEC, Munich, Germany
{florian.unterstein,johann.heyszl,robert.specht}@aisec.fraunhofer.de
[2] Technische Universität München, Munich, Germany
desantis@tum.de

Abstract. In leakage-resilient symmetric cryptography, two important
concepts have been proposed in order to decrease the success rate of differential side-channel attacks. The first one is to limit the attacker's data
complexity by restricting the number of observable inputs; the second one
is to create correlated algorithmic noise by using parallel S-boxes with
equal inputs. The latter hinders the typical divide and conquer approach
of differential side-channel attacks and makes key recovery much more
difficult in practice. The use of localized electromagnetic (EM) measurements has already been shown to limit the effectiveness of such measures in previous works based on PRESENT S-boxes and 90 nm FPGAs.
However, it has been left for future investigation in recent publications
based on AES S-boxes. We aim at providing helpful results and insights
from LDA-preprocessed, multivariate, localized EM attacks against a
45 nm FPGA implementation using AES S-boxes. We show, that even in
the case of densely placed S-boxes (with identical routing constraints),
and even when limiting the data complexity to the minimum of only
two inputs, the guessing entropy of the key is reduced to only 2^{48},
which remains well within the key enumeration capabilities of today's
adversaries. Relaxing the S-box placement constraints further reduces
the guessing entropy. Also, increasing the data complexity for efficiency,
decreases it down to a direct key recovery. While our results are empirical and reflective of one device and implementation, they emphasize the
threat of multivariate localized EM attacks to such AES-based leakage-
resilient constructions, more than currently believed.

1 Introduction

Differential Power Analysis (DPA) is one of the most powerful classes of side-
channel attacks against symmetric cryptographic implementations. It exploits
multiple measurements obtained under the same key and different inputs to recover
the secret key using statistical methods, and is particularly robust in the presence
of noise. Conventional side-channel countermeasures like protected logic styles [13]

© Springer International Publishing AG 2017
S. Guilley (Ed.): COSADE 2017, LNCS 10348, pp. 34–49, 2017.
DOI: 10.1007/978-3-319-64647-3_3

and masking schemes [3] typically come with significant overhead in terms of implementation complexity, area and time resources. Leakage-resilient and re-keying techniques aim at bounding the side-channel leakage to a level which is not computationally exploitable for the adversary, while having less area overhead than conventional countermeasures. For instance, most leakage-resilient and re-keying schemes [14] reduce the number of observable computations by changing the secret key according to a predefined mechanism, e.g. at every execution. In such cases, the implementation needs to be protected only against single observation attacks. To generate session keys, possible approaches for re-keying schemes e.g. require to update a secret internal state (stateful devices), or use an internal random number generator to realize some form of key-update agreement protocol among the parties (stateless devices), as e.g. in [16] or CIPURSE from Infineon AG [9]. However, many embedded devices require stateless and non-interactive solutions, e.g. for encrypted software updates, or do not have secure random number generators. Leakage-resilient Pseudo-Random Functions (PRFs) [23] provide a stateless method to derive session keys based on a public input. Arguably, the PRF tree construction by Goldreich et al. [11] is one of the most influential leakage-resilient PRFs currently investigated in literature. It can easily be instantiated from block ciphers as shown in [17]. This allows to thwart differential side-channel attacks in two ways: (1) by reducing the data complexity (number of different observable inputs) by construction; (2) by adding correlated algorithmic noise to the measurements (which cannot be averaged out) by exploiting parallel S-boxes which are provided with the same plaintext inputs. Note that measurement complexity (number of measurements allowed), on the contrary, is generally not restricted.

In 2012, Medwed et al. [17] showed that limiting the data complexity alone is not sufficient to achieve protection against differential side-channel attacks, even if it is as low as 2^1. Also, they concluded that the AES may not be a valid candidate for the construction of leakage-resilient PRFs (at least when the data complexity is > 2) due to the limited number of parallel S-boxes which can be instantiated, hence, leading to a remaining search complexity for enumeration of only $16! \approx 2^{44}$. Subsequently, Belaid et al. [2] investigated such a construction with 32 parallel PRESENT S-boxes and data complexity of 2^4, which led to a remaining search complexity of $32! \approx 2^{117}$ when faced with DPA. They also showed, however, that the security level can be reduced down to 2^{69} by employing univariate localized EM attacks [12] and breaking the equal leakages and correlated noise assumptions. Finally, in a recent contribution to ASIACRYPT 2016, Medwed et al. [18] used a Pseudo-Random Generator (PRG) (taken from Standaert et al. [22]) for the initialization of a novel unknown-inputs leakage-resilient PRF based on the AES block cipher. The security of the PRG part of the leakage-resilient construction is again based on minimal data complexity of two inputs and S-box parallelism to obtain correlated algorithmic noise. Their contribution explicitly mentions the lack of an empirical security evaluation using localized EM attacks, which is the main motivation for this work.

---

[1] Reducing the data complexity to 1 would mean that only a single observation (with possibly unlimited measurement complexity) would be available to adversaries. This corresponds to a simple power analysis attack scenario which is not generally considered in most contributions in the field of leakage-resilient cryptography.

Contributions. The main contribution of this paper is a laboratory evaluation of a leakage-resilient implementation based on AES using localized EM measurements. We provide answers to questions left open by Medwed et al. [18], who re-proposed the use of AES with data complexity 2, and Belaid et al. [2], who analyzed *unconstrained* S-boxes on a 90 nm FPGA with a data complexity of 2^4 and *univariate* localized EM attacks. In particular, we (1) employ state of the art profiled *multivariate* localized EM attacks using linear discriminant analysis (LDA) preprocessing for the identification of the Points of Interests (PoIs); and (2), investigate a design with carefully *constrained* S-boxes and a data complexity of 2 on a 45 nm Xilinx Spartan 6 device. Our results show that even when the lowest data complexity of 2, full parallelism of 16 S-boxes, and constrained placement are used, the practical achieved security level[2] is only 2^{48}. This suggests that leakage-resilient constructions [18,22,24] will not provide a sufficient level of implementation security in face of multivariate localized EM attacks, when implemented on FPGA devices similar to the one used for this evaluation.

2   Background

Leakage-Resilient PRFs. Leakage-resilient Pseudo-Random Functions (PRFs) have been introduced in [19,20,23] and essentially build on the tree construction of Goldreich, Goldwasser and Micali [11]. The input x to the PRF is split into parts of a small number of m bits, which are input to multiple subsequent block cipher operations using different keys, i.e. the result of every encryption iteration is used as the key for the next round. In each round, m bits are taken from x and are replicated for the plaintext input until all bits of x are processed, as depicted in Fig. 1. The replication of the input bits achieves what is referred to as carefully chosen plaintexts by Medwed et al. [17], i.e. the plaintext input to every S-box is the same. As a consequence, the data complexity in an attack on any intermediate key is restricted to 2^m possible plaintexts. The choice of m imposes a trade-off between data complexity and efficiency for designers, as the number of necessary block cipher iterations is $\frac{128}{m}$. Note that for $m = 1$, the data complexity equals 2 and the leakage of the PRF becomes equivalent to that of the PRG used in [18] and [22], where the data complexity is also limited to two. If the leakage of all S-boxes is assumed to be equal, then standard DPA attacks will recover all key bytes at once, but without any information about their correct order within the secret key. This leads to a search complexity of $16! \approx 2^{44}$ in case of AES, if measurements and attack have led to perfect results (all key bytes are ranked first). Results from Belaid et al. [2], however, have already shown that the equal and concurrent leakage assumption does not hold when localized EM measurements are performed.

---

[2] This corresponds to the ranking of the key after a practical laboratory evaluation using localized EM, where the order of the key bytes is discovered during the attacks, but not all correct subkeys are ranked first. In contrast, the previously mentioned 2^{44} corresponds to the remaining search complexity of global attacks, once all key bytes are assumed to be ranked first, despite the correlated algorithmic noise (theoretical best case).

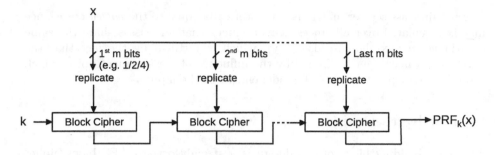

Fig. 1. Leakage resilient PRF.

Linear Discriminant Analysis. While multivariate template attacks are amongst the most powerful differential side-channel attacks, they are also computationally intensive and can face numerical issues when the number of time-samples per trace is large. A common way to deal with this is to reduce the number of time-samples included in the calculation of the templates, i.e. the dimensionality of the trace. Fisher's Linear Discriminant Analysis (LDA) [8] has been proposed for template attacks by Archambeau et al. [1] and then specialized for EM attacks by Standaert et al. [21]. It has later been shown by Bruneau et al. [4] that this is in fact the optimal strategy to reduce the dimensionality of leakage traces. LDA also has the advantage that the transformation makes templates more robust against measurement campaign-dependent variations caused by temperature or environmental noise [7]. LDA stems from statistical classification and is a linear transformation of a dataset onto a lower-dimensional subspace with good class-separability. It calculates a transformation matrix $\mathbf{W}$, which maximizes the ratio of between-class to within-class scatter. In our case, we calculate one transformation matrix for each S-box and use the S-box input values as classes. Let $t_{i,j}$ be all traces with S-box input value i with $j \in [0, N_i-1]$, $\mu_i = \frac{1}{N_i} \sum_{j=0}^{N_i-1} t_{i,j}$ the estimated class mean and $\mu = \frac{1}{256} \sum_{i=0}^{255} \mu_i$ the estimated overall mean. Then LDA calculates the within-class scatter matrix $\mathbf{S_w}$, between class scatter matrix $\mathbf{S_b}$ and $\mathbf{W}$, such that criterion J is maximized:[3]

$$\mathbf{S_w} = \sum_{i=0}^{255} \sum_{j=0}^{N_i-1} (t_{i,j} - \mu_i)(t_{i,j} - \mu_i)^T \tag{1}$$

$$\mathbf{S_b} = \sum_{i=0}^{255} N_i(\mu_i - \mu)(\mu_i - \mu)^T \tag{2}$$

$$J(\mathbf{W}) = \frac{\mathbf{W}^T \mathbf{S_b} \mathbf{W}}{\mathbf{W}^T \mathbf{S_w} \mathbf{W}} \tag{3}$$

---

[3] The equations show the calculation of the transformation matrix for one S-box. We omitted an additional identifier for the S-box number for the sake of clarity.

The within-class scatter matrix is asymptotically equal to the *pooled* covariance matrix calculated over all traces. This assumes that all classes share the same covariance matrix (homoscedasticity), which is justified by the fact that the covariance values are determined by the influence of measurement noise, which should in most practical cases be independent of the inputs.

3   Hardware Design

The main building block of our design is a straightforward AES block cipher implementation with *full parallelism*, hence, 16 S-boxes (Canright S-boxes [5]) in the data path and 4 S-boxes in the key schedule as shown in Fig. 2. There is one state register, each AES round is computed in one clock cycle, and key scheduling is computed in parallel. As in the case of many other leakage-resilient constructions, this block cipher is used in two different stages, or modes: first the block cipher is used to generate a secret IV or session key from a public input, i.e. PRF mode, then, it is used for encryption (block cipher mode of operation). We emphasize that the block cipher remains the same (it shares the same hardware), while the input and key are different in those two modes. This is reasonable to avoid unnecessary overhead from duplication of the AES hardware block, but helps adversaries during profiling, since they may build profiles using either of the two modes.

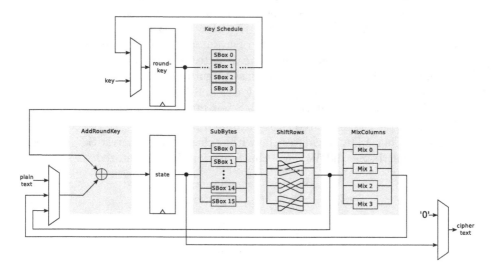

Fig. 2. AES hardware design.

The design is configured into a 45 nm Xilinx Spartan 6 XC6SLX9-3TQG144C FPGA. We synthesized the design in two ways, (1) without any routing constraints, and (2) with 16 hard-macro S-boxes placed as dense as possible. For the densely placed design, we first placed and routed one S-box (Fig. 6 in the appendix depicts the FPGA layout). Then we utilized Xilinx's relative location (RLOC) and area constraints to clone and place this 'hardmacro' as dense as possible (Fig. 7 in the appendix shows the placed S-boxes). This should help to fulfill the equal leakage assumption of the S-boxes as closely as possible because S-boxes are equal to a higher degree (apart from the routing to/from the S-box) and the area is generally smaller, which should make localized EM attacks more difficult. In addition, we constrained the placement of the rest of the AES to a confined area (black box in Fig. 7) in an attempt to make the routing, e.g. to the mix-columns logic, as short as possible. Based on the reports of the design tools, the estimated die area occupied by the AES is about $0.5\,\mathrm{mm}^2$. Under these circumstances and for both placement options, we synthesized designs with $m = 1$ (data complexity of 2), and $m = 4$ (data complexity of 16).

4   Side-Channel Analysis

Our contribution evaluates the implementation security of the previously described techniques for leakage resilience. Since the parallelism of S-boxes and their ideally equal leakage characteristics are crucial to the idea of the construction, a high-precision EM measurement setup is especially relevant. Our assumption is that the localization capability thereof allows a spatial separation of the leakage of the individual S-boxes and the exploitation of even subtle differences in their characteristics. We use a state of the art high-end setup with a Langer ICR HH 100-27 100 μm diameter EM probe which is positioned about 10 μm over the decapsulated die surface. In addition to the built-in 30 dB amplifier of the probe, another Langer PA303 30 dB pre-amplifier is employed. A LeCroy WavePro 725Zi oscilloscope with 2.5 GHz bandwidth and a sampling rate of 5 GS/s records the measurements. The FPGA-based design is clocked at 20 MHz and this clock is synchronized to the oscilloscope. An X-Y-table is used to collect measurements on multiple locations over the die surface. The measurement positions are located within an area of about 2.8 mm by 2.8 mm, which should cover most, but not all of the floorplan (shown in Fig. 7 in the appendix) because the probe movement is limited by the bonding wires.

It is common practice to allow profiling for a meaningful implementation security analysis which is representative of the fact that adversaries may use their own devices where they could choose keys for profiling. In this profiled setting, the adversary is able to compute *all* internal states of the implementation. Based on this, we performed profiling using the block cipher mode of operation[4] of our implementation instead of the PRF mode. Analogously, an adversary would use stage 2 in the construction of Medwed et al. [18]. Our analysis is split into three tasks: (1) the localization of the measurement positions with the

---

[4] We used OFB mode, but other modes would work as well.

maximum leakage for each S-box, (2) the profiling phase on these positions, and (3), the attack phase. The first task is the most time consuming since it requires a full scan of the die surface. Considering that the measurement time grows quadratically when reducing the step size, we partitioned the measurement area in a grid of 20×20 for the unconstrained design and 40×40 for the dense design, which corresponds to a step size of $140\,\mu m$ and $70\,\mu m$, respectively. We used a larger step size for the unconstrained design since we expect it to spread over a bigger area. On each position, we acquired $10,000$ traces. With our setup, the measurement takes roughly 1 day for the 20×20 grid and 4 days for the finer grid. For each S-box, we calculated the signal-to-noise ratio (SNR) by partitioning the traces according to the input values of the S-box [15]:

$$SNR^b = \frac{Var(Signal^b)}{Var(Noise^b)} = \frac{Var(\mu_0^b \ldots \mu_{255}^b)}{Mean(\sigma_0^{2^b} \ldots \sigma_{255}^{2^b})}, \tag{4}$$

with b being the index of the S-box and μ_i^b and $\sigma_i^{2^b}$ being the estimated mean and variance traces computed over all traces with input value i at this S-box. The result is a trace of many SNR values (SNR trace) which we evaluated within the timespan where the first AES round is computed. In our case one clock cycle corresponds to 250 samples, and the interesting part, i.e. the part where there is activity after the clock edge, is around 50 samples (10 ns) wide. There are several options how to chose positions from this part of the SNR trace. We found that in some cases, the positions with the highest peak SNR value gave the best results, and in others, the positions with the highest mean SNR (calculated over the 50 samples in the interesting region of the SNR trace) performed better. In cases where those metrics gave different positions or were ambiguous due to multiple peaks of similar amplitude, we conducted the rest of the analysis on all such positions for this S-box and kept the best result.

In two separate acquisition campaigns, we collected the profiling and attack traces. The attack traces were acquired with limited data complexity, i.e. 16 for $m = 4$, and 2 for $m = 1$. We cut the traces and limited our analysis to the time span where the first AES round is calculated. To reduce the number of samples included in the templates, we use LDA [8] as dimensionality reduction algorithm.

We compute full estimated Gaussian templates for each S-box and each of their S-box input values. As stated earlier, for LDA to be applicable, the traces belonging to one S-box are assumed to share a common covariance matrix, regardless of the input value. In this case, it suffices to calculate a single pooled covariance matrix for all templates belonging to one S-box. This gives a better estimate of the actual distribution and drastically reduces the computational effort for the template matching. Our experiments suggest that the assumption holds in our case and gave generally better results when using the pooled matrix when compared to separate covariance matrices. Thus, all our presented attacks were conducted using the pooled covariance matrix.

During the attack phase, the traces are matched against the templates in a template based DPA. Since we are using the pooled covariance matrix, we can make use of simplifications detailed by Choudary et al. [6] and calculate the logarithmic score. To combine the score of multiple attack traces, we sum the scores and calculate the average. This results in a list with scores for each subkey candidate. In order to calculate the overall key rank we used the key rank estimator proposed by Glowacz et al. [10]. The estimated key rank is, within its error boundaries, equivalent to the metric of guessing entropy used in other publications.

5   Results and Discussion

Using the SNR analysis, we were able to localize useful measurement positions for all S-boxes on both tested designs. Figure 3 shows one example SNR heat map of S-box #0 on the two different designs. All other heat maps can be found in Figs. 8 and 9 in the appendix. Each colored pixel represents the peak SNR value of the SNR trace at that measurement position for this S-box. In both maps, regions with the highest SNR are clearly distinguishable and most likely correspond to the actual physical location of the logic of S-box #0. An important observation is that the SNR values of the design with the densely placed hard-macro S-boxes are - on average - by a factor of 2 smaller than the ones from the unconstrained placement. The average peak SNR of the S-boxes on the dense placement is 0.87, compared to 1.61 on the unconstrained placement. In the case of dense placement, where SNR values are generally smaller, there are multiple positions which exhibit a relatively high SNR. As described, we simply evaluated all such locations for the corresponding S-box in the attack instead of choosing just one, which increased the measurement time of the attack.

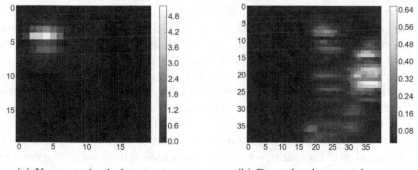

(a) Unconstrained placement.   (b) Dense hard-macro placement.

Fig. 3. SNR heat maps for S-box #0 with different placements.

Table 1. Estimated key ranks after the attacks.

S-box placement	Data complexity	Est. key rank
Unconstrained	2	2^{20}
Dense	2	2^{48}
Unconstrained	16	1
Dense	16	1

For the profiling phase, we used a maximum of 65,000 traces per position for the unconstrained design and 650,000 traces for the dense design in an effort to compensate for the lower SNR. During the attack, up to 100,000 traces were used per S-box. Table 1 summarizes the results of the attacks using all available traces. With a data complexity of 2^4 during the attack, security is completely broken and all key bytes are successfully recovered, regardless of the placement. This is a result which is similar to the findings of Belaid et al. [2].

As expected, a data complexity of 2 leads to better results. Several subkeys are not ranked first and consequentially, a higher key rank of 2^{20} remains for the unconstrained placement case with data complexity 2. However, as an important result, this is an obvious insufficient level of security.

The dense design improves the security significantly and provides a higher security level of 2^{48} compared to 2^{20}. In both cases, the achieved security level is insufficient, which is the main contribution of our investigations. This means that a minimum data complexity of 2 together with parallel S-box inputs is not suited to achieve meaningful leakage-resilient constructions, at least under the present circumstances of a 45 nm feature size FPGA implementation.

While the security level is established to be insufficient, an interesting question is, whether more profiling traces would further improve the attack, or whether the lower bound is reached. We repeated the attack with different numbers of profiling traces while using all available attack traces. The results for both designs are shown in Fig. 4. It can be noted that the gain of using more traces for profiling diminishes and the key ranks seem to approach a lower bound at about 2^{20} for the unconstrained, and about 2^{48} for the dense design. We conclude that increasing the number of profiling traces even further seems useless and that the efficiency of the attack is in fact limited by the leakage-resilience, and not by insufficient profiling due to the lower SNR. In other words, we expect that other uncorrelated noise sources are averaged out sufficiently.

In a similar manner, we investigated the number of traces required for the attack. In a real-world scenario, adversaries may have full access to one device for profiling, but limited access to the attacked device. Figure 5 shows the influence of the number of attack traces on the key rank when using templates built from the maximum number of available profiling traces. As an interesting observation, we report that the key rank seems to reach its lower bound after only about 100 attack traces, which is a surprisingly low number.

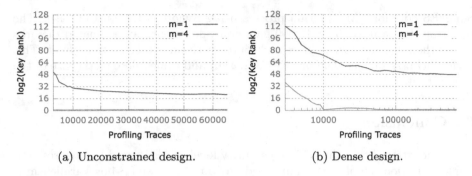

Fig. 4. Key rank evolution with varying number of profiling traces and maximum number of attack traces.

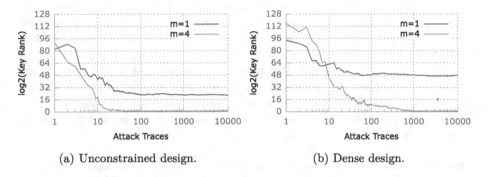

Fig. 5. Key rank evolution with varying number of attack traces and maximum number of profiling traces.

To verify the efficiency of the leakage-resilient construction against regular power attacks, we also conducted a template attack where we measured the global power consumption over a resistor in the power line with a differential probe. For increased SNR, all capacities were removed from the board. Despite using $1,000,000$ profiling traces, the attack fails to result in any significant key rank reduction. Interestingly, the correct subkeys were not even ranked highly but instead were distributed evenly across the subkey list. This is far from optimal, where correct subkeys would be ranked in the first 16 positions in all subkey lists and leave only the permutation complexity for the enumeration of the whole key. For the case of unlimited data complexity, we report that an univariate CPA using the Hamming distance leakage model already succeeds with 20.000 traces. Even though this aspect was not the focus of our research, this discrepancy is an encouraging result when adversaries are limited to global (power) attacks.

Given that our analysis is reflective of one technology, namely 45 nm FPGAs, it remains unclear, how our results affect other and smaller technologies such as ASICs or upcoming 16 nm FPGA devices. In our case study, the die area occupied by the AES is about $0.5\,\text{mm}^2$ and relatively large compared to the probe diameter of 100 μm. For a rough comparison to an ASIC design, we synthesized our AES core for UMC's 55 nm process using Synopsys Design Compiler. The resulting design uses about 10.000 gate equivalents with an estimated die area of less than $0.02\,\text{mm}^2$ when place and route overhead is taken into account. This is significantly smaller than our FPGA design and comes close to the size of the probe itself.

6   Conclusion

We demonstrated that the achieved security level of AES-based leakage resilient implementations employing minimum data complexity and S-box parallelism is insufficient in the localized EM scenario, at least in cases similar to our FPGA with 45 nm feature size. In particular, we were able to isolate the leakage of individual S-boxes and attack them separately using LDA-based, profiled, multivariate attacks, thus, circumventing the "equally leaking" and "correlated algorithmic noise" assumptions. We were able to completely recover the correct key for all designs with data complexity 2^4. A data complexity of 2 proved to be more resilient, but we were still able to reduce the key rank to 2^{20} and 2^{48} for the unconstrained and dense placement, respectively. Finally, it remains as an open question whether a denser placement and smaller feature sizes on ASIC will suffice to reach acceptable security levels against localized EM attacks. In this regard, we advise further analysis.

Acknowledgements. The work presented in this contribution was supported by the German Federal Ministry of Education and Research in the project *ALESSIO* through grant number 16KIS0629.

A   Floorplanning

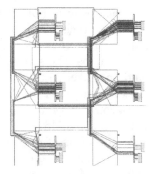

Fig. 6. Layout of one S-box in the Xilinx IDE.

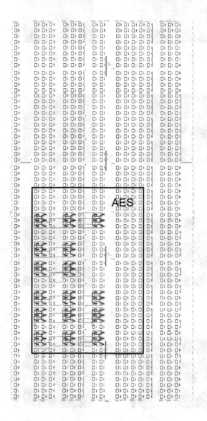

Fig. 7. Position of 16 S-boxes on the floorplan of the Xilinx Spartan 6 FPGA. The entire AES is placed within the black box.

B   SNR Heat Maps for All S-Boxes

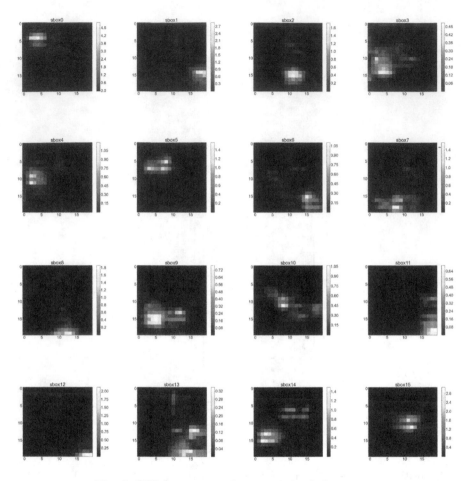

Fig. 8. SNR heat maps of unconstrained placement.

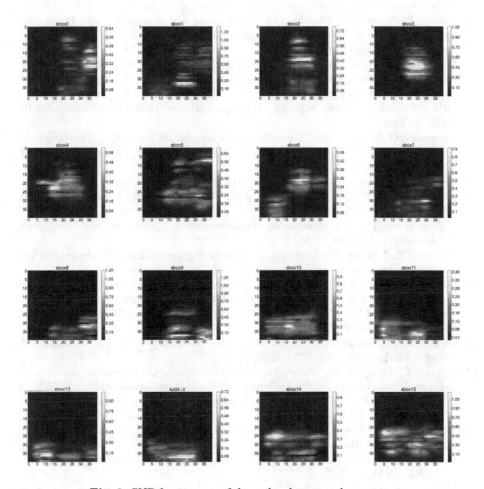

Fig. 9. SNR heat maps of dense hard-macro placement.

References

1. Archambeau, C., Peeters, E., Standaert, F.-X., Quisquater, J.-J.: Template attacks in principal subspaces. In: Goubin, L., Matsui, M. (eds.) CHES 2006. LNCS, vol. 4249, pp. 1–14. Springer, Heidelberg (2006). doi:10.1007/11894063_1
2. Belaïd, S., De Santis, F., Heyszl, J., Mangard, S., Medwed, M., Schmidt, J.M., Standaert, F.X., Tillich, S.: Towards fresh re-keying with leakage-resilient PRFs: cipher design principles and analysis. J. Cryptograph. Eng. **4**(3), 157–171 (2014)
3. Belaïd, S., Grosso, V., Standaert, F.X.: Masking and leakage-resilient primitives: one, the other(s) or both? Crypt. Commun. **7**(1), 163–184 (2015)
4. Bruneau, N., Guilley, S., Heuser, A., Marion, D., Rioul, O.: Less is more. In: Güneysu, T., Handschuh, H. (eds.) CHES 2015. LNCS, vol. 9293, pp. 22–41. Springer, Heidelberg (2015). doi:10.1007/978-3-662-48324-4_2
5. Canright, D.: A very compact S-box for AES. In: Rao, J.R., Sunar, B. (eds.) CHES 2005. LNCS, vol. 3659, pp. 441–455. Springer, Heidelberg (2005). doi:10.1007/11545262_32
6. Choudary, O., Kuhn, M.G.: Efficient template attacks. In: Francillon, A., Rohatgi, P. (eds.) CARDIS 2013. LNCS, vol. 8419, pp. 253–270. Springer, Cham (2014). doi:10.1007/978-3-319-08302-5_17
7. Choudary, O., Kuhn, M.G.: Template attacks on different devices. In: Prouff, E. (ed.) COSADE 2014. LNCS, vol. 8622, pp. 179–198. Springer, Cham (2014). doi:10.1007/978-3-319-10175-0_13
8. Fisher, R.A.: The use of multiple measurements in taxonomic problems. Ann. Eugenics **7**(7), 179–188 (1936)
9. Gammel, B., Fischer, W., Mangard, S.: Generating a session key for authentication and secure data transfer. US Patent 2014016955, 7 November 2013
10. Glowacz, C., Grosso, V., Poussier, R., Schüth, J., Standaert, F.-X.: Simpler and more efficient rank estimation for side-channel security assessment. In: Leander, G. (ed.) FSE 2015. LNCS, vol. 9054, pp. 117–129. Springer, Heidelberg (2015). doi:10.1007/978-3-662-48116-5_6
11. Goldreich, O., Goldwasser, S., Micali, S.: How to construct random functions. J. ACM (JACM) **33**(4), 792–807 (1986)
12. Heyszl, J., Mangard, S., Heinz, B., Stumpf, F., Sigl, G.: Localized electromagnetic analysis of cryptographic implementations. In: Dunkelman, O. (ed.) CT-RSA 2012. LNCS, vol. 7178, pp. 231–244. Springer, Heidelberg (2012). doi:10.1007/978-3-642-27954-6_15
13. Kirschbaum, M.: Power analysis resistant logic styles - design, implementation, and evaluation. Ph.D. thesis (2011)
14. Kocher, P.C.: Leak-resistant cryptographic indexed key update, US Patent 6,539,092, 25 March 2003
15. Mangard, S., Oswald, E., Popp, T.: Power Analysis Attacks. Springer, New York (2008)
16. Medwed, M., Standaert, F.-X., Großschädl, J., Regazzoni, F.: Fresh re-keying: security against side-channel and fault attacks for low-cost devices. In: Bernstein, D.J., Lange, T. (eds.) AFRICACRYPT 2010. LNCS, vol. 6055, pp. 279–296. Springer, Heidelberg (2010). doi:10.1007/978-3-642-12678-9_17
17. Medwed, M., Standaert, F.-X., Joux, A.: Towards super-exponential side-channel security with efficient leakage-resilient PRFs. In: Prouff, E., Schaumont, P. (eds.) CHES 2012. LNCS, vol. 7428, pp. 193–212. Springer, Heidelberg (2012). doi:10.1007/978-3-642-33027-8_12

18. Medwed, M., Standaert, F.-X., Nikov, V., Feldhofer, M.: Unknown-input attacks in the parallel setting: improving the security of the CHES 2012 leakage-resilient PRF. In: Cheon, J.H., Takagi, T. (eds.) ASIACRYPT 2016. LNCS, vol. 10031, pp. 602–623. Springer, Heidelberg (2016). doi:10.1007/978-3-662-53887-6_22
19. Petit, C., Standaert, F.X., Pereira, O., Malkin, T.G., Yung, M.: A block cipher based pseudo random number generator secure against side-channel key recovery. In: Proceedings of the 2008 ACM Symposium on Information, Computer and Communications Security, pp. 56–65. ACM (2008)
20. Pietrzak, K.: A leakage-resilient mode of operation. In: Joux, A. (ed.) EURO-CRYPT 2009. LNCS, vol. 5479, pp. 462–482. Springer, Heidelberg (2009). doi:10.1007/978-3-642-01001-9_27
21. Standaert, F.-X., Archambeau, C.: Using subspace-based template attacks to compare and combine power and electromagnetic information leakages. In: Oswald, E., Rohatgi, P. (eds.) CHES 2008. LNCS, vol. 5154, pp. 411–425. Springer, Heidelberg (2008). doi:10.1007/978-3-540-85053-3_26
22. Standaert, F.-X., Pereira, O., Yu, Y.: Leakage-resilient symmetric cryptography under empirically verifiable assumptions. In: Canetti, R., Garay, J.A. (eds.) CRYPTO 2013. LNCS, vol. 8042, pp. 335–352. Springer, Heidelberg (2013). doi:10.1007/978-3-642-40041-4_19
23. Standaert, F.X., Pereira, O., Yu, Y., Quisquater, J.J., Yung, M., Oswald, E.: Leakage resilient cryptography in practice. In: Sadeghi, A.R., Naccache, D. (eds.) Towards Hardware-Intrinsic Security. Information Security and Cryptography, pp. 99–134. Springer, Heidelberg (2010). doi:10.1007/978-3-642-14452-3_5
24. Taha, M.M.I., Schaumont, P.: Key updating for leakage resiliency with application to AES modes of operation. IEEE Trans. Inf. Forensics Secur. 10(3), 519–528 (2015)

Toward More Efficient DPA-Resistant AES Hardware Architecture Based on Threshold Implementation

Rei Ueno$^{(\boxtimes)}$, Naofumi Homma, and Takafumi Aoki

Tohoku University, Aramaki Aza Aoba 6–6–05, Aoba-ku, Sendai-shi 980-8579, Japan
ueno@aoki.ecei.tohoku.ac.jp, homma@riec.tohoku.ac.jp

Abstract. This paper presents a highly efficient AES hardware architecture resistant to differential power analyses (DPAs) on the basis of threshold implementation (TI). In contrast to other conventional masking schemes, the major feature of TI is to guarantee DPA-resistance under d-probing condition at the resister-transfer level (RTL). On the other hand, TI utilizes pipelining techniques between the non-linear functions to avoid propagating glitches, which would lead to non-negligible overheads of circuit area and latency. In this paper, we first propose a compact first-order TI-based AES S-box which has a major effect on the performance and DPA-resistance of AES hardware. The proposed S-box exploits a state-of-the-art TI construction with $d + 1$ shares in addition to the algebraic characteristics of AES S-box. We then propose an efficient AES hardware architecture suitable with the above TI-based S-box. The architectural advantage is given by register-retiming and tower-field arithmetic techniques. The performance of the proposed AES hardware was evaluated in comparison with that of conventional best ones. The logic synthesis result suggests that the proposed AES hardware architecture achieves more compact and 11–21% lower-latency than the conventional ones, which indicates that the proposed architecture can perform encryption based on TI with the lowest-energy. We also confirm the DPA-resistance of the proposed AES hardware by the Test Vector Leakage Assessment (TVLA) methodology with its FPGA implementation.

Keywords: Side-channel attacks · AES · Hardware implementation · Threshold implementation · DPA

1 Introduction

Cryptography has been widely used in many systems with secure communications, authentication, and digital signatures. According to the rapid growth of Internet of Things (IoT) applications, many cryptographic algorithms are being required in resource-constraint devices such as smart cards and sensors with limited chip area and power. On the other hand, various implementation attacks are attracting considerable attention because of the increasing applications of

© Springer International Publishing AG 2017
S. Guilley (Ed.): COSADE 2017, LNCS 10348, pp. 50–64, 2017.
DOI: 10.1007/978-3-319-64647-3_4

cryptographic hardware to IoT devices. There is definitely a high demand for efficient tamper-resistant cryptographic hardware securing IoT applications.

Side-channel attack, which is one of the most powerful implementation attacks, retrieves the secret key from operating cryptographic hardware by exploiting side-channel information such as power consumption, EM radiation, and operation timing [9]. Differential power analysis (DPA) is a typical side-channel attack on symmetric ciphers (e.g., AES) which analyzes the relation between side-channel information and the calculated values of cryptographic operations with a statistical means. Since modern ciphers commonly employ GF arithmetic for their components [15], we should design GF arithmetic circuits resistant to DPAs for DPA-resistant cryptographic hardware. Until now, many countermeasures have been developed to defeat DPAs. Masking is considered as one of the predominant countermeasures, which eliminates DPA-leaks using randomness.

Threshold implementation (TI) was presented as a provably-secure counter-measure against DPAs, including advanced DPAs exploiting power consumption caused by glitches [10,11]. While many conventional countermeasures require to use specific tools and/or libraries (e.g., symmetric layout) [21], the usage of TI makes it possible to design DPA-resistant hardware with the standard design tools including standard cells and automatic layout. In recent years, some related works on TI have been reported. They include its extension to higher-order DPAs [2,17], DPA-resistant cryptographic hardware designs based on TI [3,7,12,16], and TI-friendly cryptography where TI can be efficiently applied to the S-box [5].

In this paper, we first present a compact AES S-box based on TI and then propose a more efficient DPA-resistant AES hardware architecture. The proposed TI-based AES S-box is designed with a combination of the state-of-the-art TI construction [17] and the algebraic characteristics of AES S-box. The proposed architecture employs a byte-serial architecture commonly used for TI-based AES in order to tolerate the overheads of circuit area and random number generation [3,7,12]. In such architectures, the latency overhead caused by the pipeline registers of TI is also non-negligible. The conventional works perform SubBytes, ShiftRows, and MixColumns at serial timings despite pipelined SubBytes, which indicates that an extra latency occurs in every round due to the pipelining, and results in the loss of energy. In contrast, the proposed AES hardware architecture exploits a new register-retiming technique to perform the above operations in a partially parallel manner with a modest increase of circuit area. In addition, the proposed architecture performs all the operations over tower field for a further reduction of latency (i.e., pipeline-stage). Furthermore, our architecture saves the cost of TI applied to the key scheduling unit according to the report of [16] that it has no DPA-leaks. With the results of logic synthesis, we confirm that the proposed method has a smaller area and 11–21% lower latency than conventional architectures. In addition, we evaluate the DPA-leakage of the proposed architecture implemented on an FPGA. The t-test result shows that there is no obvious first-order DPA-leak from the proposed architecture within 500,000 traces. While this paper focuses on the first-order security, the concept of the proposed architecture can be applied to the higher-order security.

The rest of this paper is organized as follows: Sect. 2 briefly describes TI. Section 3 proposes a more compact first-order TI-based AES S-box and an efficient byte-serial AES hardware architecture equipped with the S-box. Section 4 provides the performance of the proposed AES hardware architecture with logic synthesis results and evaluates its DPA leakage with an FPGA implementation. Section 5 contains our conclusion.

2   Threshold Implementaion

TI is a state-of-the-art masking countermeasure against DPAs [14,17]. The utilization of TI makes it possible to design any kind of arithmetic circuits over $GF(2^m)$ resistant even to advanced DPAs that utilize power glitches included in observed power consumption [10,11].

The working principle of TI is to represent a secret value $a \in GF(2^m)$ with $a_0 + a_1 + \cdots + a_i + \cdots + a_{s-1}$ by means of (s, s) threshold scheme [20], where $a_i \in GF(2^m)$ $(0 \le i \le s - 1)$ is initially given by a random mask. Each element a_i is called a share. According to [17], any circuit satisfying the following three properties is secure under the dth-order DPAs. Let a and $a_0, a_1, \ldots, a_i, \ldots, a_{s-1}$ be the input and the shares of a, respectively. Let c and $c_0, c_1, \ldots, c_j, \ldots, c_{s'-1}$ be the output and the shares of c, respectively.

(i) *Correctness*

The first property implies that the sum of shares is equal to the original value at the input and output of the circuit, namely, $a = a_0 + a_1 + \cdots + a_i + \cdots + a_{s-1}$ and $c = c_0 + c_1 + \cdots + c_j + \cdots + c_{s'-1}$, where a and c are the input and output of the original function, respectively, and a_i and c_j are the input and output shares, respectively. This property indicates that the shared circuit correctly performs the original (i.e., nonshared) function.

(ii) *dth-order non-completeness*

The second property implies that the sum of chosen d output shares are independent of at least one input share. The number of input shares (i.e., s) required to meet the dth-order non-completeness is dependent on d and the algebraic degree of the circuit function, namely, the number of serially connected two-input AND gates in the combinational circuit. (Therefore, the number of input shares can be reduced by pipelining the circuit because the number of serially connected two-input AND gates are reduced in the circuits [14].)

(iii) *Uniformity*

The third property indicates that the input and output values of combinational circuits are uniformly distributed. See [3] for details.

While Properties (i) and (ii) can be realized for any GF function, some functions cannot satisfy Property (iii) under the constraints of properties (i) and (ii). However, the uniformity criterion can be satisfied by the addition of fresh mask(s), which is called a remasking scheme, to the non-uniform outputs [14]. If a circuit does not meet all the above properties, a glitch propagation between non-linear functions may leak the secret value [10]. A remasking is always necessary in the

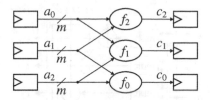

Fig. 1. Overview of circuit of function $t = 2$ meeting first-order non-completeness.

case of more than first-order security even if the output shares are uniformly distributed [17].

There were two known methods to construct circuits satisfying dth-order non-completeness. The difference of the two is the numbers of input shares. The first construction method was proposed in [14], where the number of input shares is given by $td+1$, where t is the algebraic degree of circuit function, and the number of output shares is given by $s' = \binom{s}{t}$ (e.g., $s' = s$ when $d = 1$). For example, Fig. 1 shows an overview of a circuit satisfying the first-order non-completeness when $t = 2$ (e.g., an two-input AND gate and a multiplier), where f_j indicates the function of the circuit that computes c_j. The number of input shares is 3 ($= 1 \times 2 + 1$), and each f_j has 2 inputs. Given that c_j is independent of a_j in Fig. 1, the circuit is thought to be secure under first-order DPAs from the viewpoint of first-order non-completeness. Note that, in this case, the numbers of input and output shares are the same (i.e., $s = s' = 3$); however, they become different when $d \geq 2$.

It was shown that the above TI with $td + 1$ input shares was useful especially for designing hardware architectures of lightweight ciphers [16] such as PRESENT [4] and LED [8]. This is because such ciphers employ an S-box whose algebraic degree is at most three, which leads to an efficient TI construction with simple one-stage pipelining where the number of input and output are the same. In addition, all the output shares can be satisfied with the uniformity property in the TI-based S-box in PRESENT and LED, which means that any on-the-fly random number generation is not required during one block encryption.

The second construction method was recently proposed in [17], where the number of input shares is given by $d + 1$. To construct TI-based S-box of higher algebraic degree, such as in AES, a multi-stage pipelining and an on-the-fly random number generation are required because it is known that there is no TI construction satisfying the uniformity property [3,12]. The TI with $d + 1$ input shares is more useful for designing compact and efficient hardware architecture for such a cipher. Actually, in [7], a more compact TI-based AES hardware was designed with $d + 1$ input shares. For example, Fig. 2 shows a first-order TI-based $GF(2^m)$ multiplier with 2 ($= d+1$) input shares, where a_0, a_1, b_0, and b_1 are the input shares, c_0, c_1, c_2, and c_3 are the output shares, "mult" denotes a nonshared $GF(2^m)$ multiplier, and r_0, r_1, and r_2 are fresh masks for remasking. The multiplier performs $c = a \times b$ under the first-order non-completeness, where $a = a_0 + a_1$, $b = b_0 + b_1$, and $c = c_0 + c_1 + c_2 + c_3$. More precisely, each output

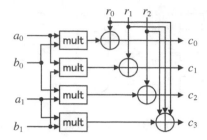

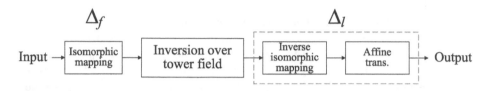

Fig. 2. First-order TI-based $GF(2^m)$ multiplier with two input shares.

Fig. 3. AES S-box based on tower-field arithmetic.

share c_j is independent of either a_0 or a_1 (b_0 or b_1), which means the multiplier meets the first-order non-completeness. The number of output share for the TI is given by at least $(d + 1)^t$.

3   Proposed Architecture

3.1   First-Order TI-Based AES S-Box

An AES S-box consists of $GF(2^8)$ inversion and $GF(2)$ affine transformation. While the inversion is performed by a polynomial basis (PB) based $GF(2^8)$ with an irreducible polynomial $x^8 + x^4 + x^3 + x + 1$, the use of tower-field arithmetic is useful for designing compact and efficient inversion circuits over $GF(2^8)$ [6,18]. Figure 3 shows the computation flow of AES S-box utilizing tower-field inversion. The input is initially mapped into a tower field. After the inversion over the tower field, the inverse mapping and affine transformation are applied to the output. Figure 4 shows a typical block diagram of tower-field inversion circuits, which consists of three stages [22]. The algebraic degrees of Stages 1, 2, and 3 are given by two, three, and two, respectively.

Until now, some TI-based inversion (i.e., S-box) circuits were proposed in the literature [3,7,12]. The major difference of the conventional circuits is the numbers of pipeline-stages and input shares. The inversion circuit in [12] was based on the TI with $td + 1$ input shares and a four-stage pipeline architecture inserting pipeline registers inside Stage 2 in addition to boundaries between Stages 1, 2, and 3. While the inversion circuit in [3] also employed the TI with $td + 1$ input shares, it was based on a three pipeline-stage architecture where

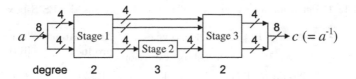

Fig. 4. Tower-field $GF(2^8)$ inversion circuit.

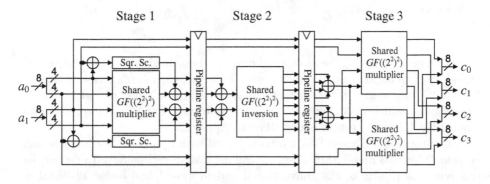

Fig. 5. Proposed first-order TI-based tower-field inversion circuit.

Stage 2 was given as a combinational circuit with algebraic degree three. Since the tower-field inversion is efficiently decomposed to three stages in terms of the algebraic expression [13], such three-stage pipeline architecture also makes the TI-based inversion circuit more efficient. On the other hand, a more compact TI-based inversion circuit with four-stage pipelining was designed in [7] on the basis of TI with $d + 1$ input shares. Note here that two more pipeline stages between the inversion and each mapping (i.e., Δ_f and Δ_l) are inserted for TI-based inversion circuits in order to satisfy dth-order non-completeness.

This paper presents a further compact and efficient TI-based inversion circuit design based on a combination of TI with $d + 1$ input shares and the above algebraic characteristics of tower-field inversion. More precisely, the proposed inversion circuit exploits a three-stage pipeline technique similar to [3], and the input of each stage is given by two shares. While the possibility of such a design was mentioned in [7], there was neither description nor evaluation in the literature. Figure 5 illustrates the proposed first-order TI-based S-box which performs $(c_0 + c_1 + c_2 + c_3) = (a_0 + a_1)^{-1}$ using three clock cycles where Stages 1, 2, and 3 correspond to those of Fig. 4, respectively. Note here that paths with fresh masks for remasking are omitted for simplicity, but Stages 1, 2, and 3 require 12-, 28-, and 24-bit random numbers for remasking [7], respectively. The block "Shared $GF((2^2)^2)$" multiplier denotes the TI-based multiplier shown in Fig. 2. The block "Sqr. Sc." performs squaring and scaling operations over $GF((2^2)^2)$ in a nonshared manner. Finally, the block "Shared $GF((2^2)^2)$ inversion" performs $GF((2^2)^2)$ inversion under the first-order non-completeness by a combinational

Table 1. Performance evaluation of first-order TI-based AES S-boxes

	Area [GE]		Clock cycles		Area-Latency product	Randomness [bit]
	compile	_ultra	S-box	Inversion		
Moradi et al. [12]	No data	4,244	5	4	21,220	44
Bilgin et al. [3]	2,835	2,224	4	3	8,896	32
Cnudde et al. [7]	1,977	1,872	6	4	11,232	54
This work	1,425	1,342	5	3	6,710	64

circuit. As stated in [7], the number of output shares is given by $(d + 1)^t$, which would have an impact on the circuit area. However, each subfunction of Shared $GF((2^2)^2)$ inversion can be efficiently factored and implemented using OR (or NOR) gates, which makes Stage 2 smaller. An example of logical expression for Shared $GF((2^2)^2)$ inversion is described in Appendix.

The proposed TI-based S-box was evaluated with Synopsys Design Compiler version D-2010.03 and TSMC 65-nm standard CMOS technology. Table 1 shows the synthesis results of the conventional and proposed first-order TI-based S-boxes, where Area denotes circuit area estimated by two-input NAND equivalent gate size (GE: Gate Equivalents), Clock cycles denotes the number of clock cycles required to perform S-box and inversion operations, Area-Latency product denotes the product of Area and Clock cycles (of S-box), and Randomness denotes the number of random bits required in a clock cycle. The columns compile and _ultra in Area were obtained by the commands compile and compile_ultra, respectively. For comparison, the values of conventional methods were derived from a table in [7]. We can confirm that our S-box achieved the smallest area without latency overhead while more randomness is consumed. In other words, our S-box is especially effective if random number generation is not critical.

3.2 Proposed Byte-Serial AES Hardware Architecture

Figure 6 shows the proposed byte-serial AES hardware architecture with the above 1st-order TI-based inversion circuit. The proposed architecture basically has an eight-bit datapath. In Fig. 6, the arrow without bit-width information denotes an eight-bit data flow. The blocks "State array" and "Key array" denote register arrays to store the intermediate values and round keys, respectively. The block "TI-based inversion" denotes the TI-based inversion circuits. Note that paths of random numbers for remasking are omitted. The blocks "Δ_f" and "Δ^{-1}" perform the isomorphic mapping and inverse mapping, respectively. Note that the output of Δ_f should be stored into the register "R" for satisfying non-completeness in the following inversion. Two State arrays are required to store $(d + 1)$-shared intermediate value. On the other hand, since it is known that the key scheduling function has no DPA-leaks [16], we do not apply TI to the Key array. SubWord in the key scheduling function is performed by a nonshared S-box in [6]. Here, the S-box should be gated using AND gates to

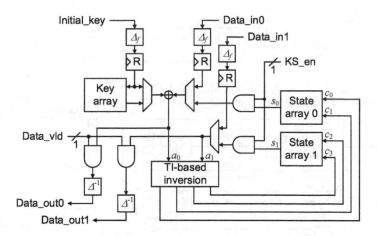

Fig. 6. Proposed byte-serial AES hardware architecture.

reduce dynamic power consumption, where the gating is controlled by one-bit signal "KS_en." Note that though the SubWord can also be performed using TI-based inversion, we use a distinct non-pipelined S-box to suppress the latency due to the pipelined Inversion. Note also that the proposed architecture would be designed with a higher-order TI-based inversion circuit in the similar manner.

Our architecture performs all operations (i.e., AddRoundkey, SubBytes, Shift-Rows, MixColumns, and key scheduling) over the tower field to reduce the number of pipeline stages (i.e., latency). Therefore, the isomorphic mappings are performed only at the input and output of the circuit. In addition, a new register-retiming technique, where the affine transformation in SubBytes is performed in State array, is introduced to further reduce the latency of the pipeline architecture. Consequently, the latency for two clock cycles are reduced by the above architectural design.

Figure 7 shows the internal structure of State array, which mainly consists of eight-bit registers and logic circuits for affine transformation and MixColumns. (Key array is omitted because it can be implemented in the same manner as [12].) We applied the above register-retiming techniques to our State array. The proposed State array is different from that of [12] because of the following three features: (i) the SubBytes of the last byte and ShiftRows are simultaneously performed in one clock cycle, (ii) MixColumns of the second and third colomns and the next round SubBytes are executed in parallel, and (iii) the SubBytes of the last four bytes and MixColumns are simultaneously performed. While the conventional State array has distinct paths only for ShiftRows, our State array performs ShiftRows and one-byte shift simultaneously using a unified path indicated in gray by allowed lines in Fig. 7 thanks to the feature (i). The output of S2 is given to TI-based inversion instead of S0 according to the feature (ii). Finally, by the feature (iii), affine transformation is performed at "Aff" in parallel during the byte shift (for the 0th–11th bytes) or MixColumns (for 12th–15th

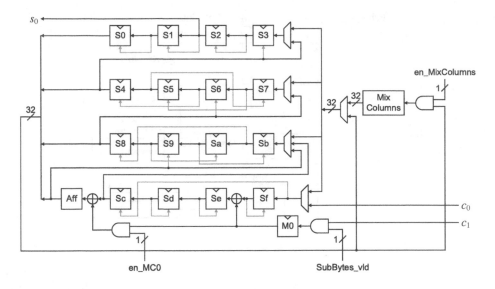

Fig. 7. State array.

bytes). It is possible to unify the affine transformation and the MixColumns in the same manner as [23]; however, we do not apply the unification technique to the proposed architecture it does not contribute to the increase of efficiency (i.e., the product of circuit area and latency). Note here that MixColumns should be gated as well as S-box for SubWord.

Figure 8 shows the timing diagrams of (a) conventional [12] and (b) proposed byte-serial AES hardware architectures, where "SubBytes k," "Inversion k," "Aff k," "SR," "MC l," and "KS l" denote the ith SubBytes, ith byte inversion, ith byte affine transformation, ShiftRows, lth column MixColumns, and lth byte SubWord in key scheduling, respectively. The blocks in gray denote operations of the previous or next round executed in parallel to the round of interest. From Fig. 8, we can confirm that our architecture achieves 20 clock cycles for one round operation while the conventional one requires 25 clock cycles because of the effect of the above resister-retiming and tower-field arithmetic techniques.

4   Evaluation

4.1   Performance Evaluation

To conduct a performance evaluation, we synthesized the proposed hardware architecture with Synopsys Design Compiler and TSMC 65 nm standard CMOS as above. Table 2 shows the synthesis result of the proposed architecture in Fig. 6. For comparison, Table 2 also shows those of the conventional ones derived in the same manner as Table 1. Note that Area of This work includes the area required for all the components in Fig. 6 and a control unit implemented with a 10-bit shift

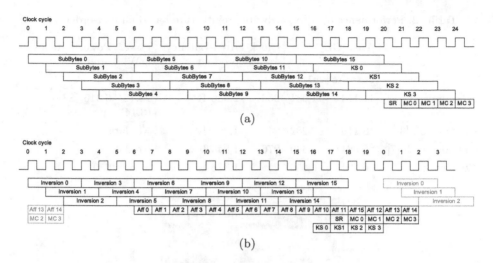

Fig. 8. Timing diagrams of (a) conventional [12] and (b) proposed byte-serial AES hardware architectures.

register and a five-bit counter. From Table 2, we confirmed that our architecture achieved 11–21% lower latency than the conventional ones. Though additional path selectors for register-retiming and MixColumns over tower fields would have an influence on the circuit area [6], our architecture achieved the smallest circuit area because of the proposed S-box and nonshared Key array. This also indicates that the circuit area can be further reduced by performing ShiftRows in a distinct clock cycle and/or replacing the tower-field MixColumns with AES-field one in exchange for increasing 10 clock cycles. Table 2 also shows the estimated power consumption based on gate-level timing simulation, where Power-Latency product indicates the product of Power and Clock cycles. The values of the conventional work was calculated using a table in [12]. The scaled values of Power and Power-Latency product in the parentheses are derived by dividing the original ones by the square of process rate (i.e., $(180/65)^2$). (The architecture in [12] was synthesized with 180 nm standard CMOS.) Note that it is quite difficult to compare power consumption estimation of hardware architectures in a fair manner, which heavily depends on the used technology and estimation method. However, the results roughly indicate that the lower latency would directly lead to lower energy of one block encryption. Thus, we confirmed the effectiveness of the proposed method.

4.2   Experimental Evaluation of DPA-Leakage

The DPA-resistance capability of our S-box was evaluated with an experiment using an FPGA implementation.

Figure 9 shows the experimental setup consisting of a Side-Channel Attack Standard Evaluation Board (SASEBO-G) [1] and an oscilloscope Tektronix

Table 2. Performance of AES hardware architecture based on first-order TI

	Area [GE]		Clock cycles	Area-Latency product	Power [μW]	Power-Latency product
	compile	_ultra				
Moradi et al. [12]	11,114	11,031	266	2,956 K	24.12 (3.14)	6,415 (835)
Bilgin et al. [3]	8,119	7,282	246	1,997 K	No data	
Cnudde et al. [7]	6,681	6,340	276	1,844 K	No data	
This work	6,321	6,053	219	1,376 K	3.06	670

Fig. 9. Experimental setup.

DPO7254. The proposed AES hardware architecture with the proposed S-box was implemented on an FPGA (Xilinx Virtex II Pro) on the SASEBO-G, and the power variation was sampled with the sampling rate of 1 GS/s.

We evaluated the resistance and vulnerability of the AES hardware architecture by Test Vector Leakage Assessment (TVLA) based on Welch's t-test (a.k.a. non-specific t-test) [19]. The TVLA examines t-values which indicate the existence of dth-order DPA-leakage exploitable by the attackers.

Figures 10(a) and 11(a) show examples of power traces at around the nineth with and without a pseudo random number generator (PRNG) implemented on the FPGA, respectively. When the PRNG is turned on, the TI works. We can find the small spikes between the big spikes in Fig. 11(a) because AES and PRNG are asynchronously active. Thus, the PRNG would not have a significant impact on the following TVLA result.

Figures 10 and 11 show the (b) first-order and (c) second-order TVLA results. We used 10,000 and 500,000 traces for Figs. 10 and 11, respectively. It is known that the absolute t-value of more than 4.5 indicates a high confidence in the existence of exploitable DPA-leakage. The results suggest that our design is resistant to the first-order DPAs under the condition of 500,000 traces by means

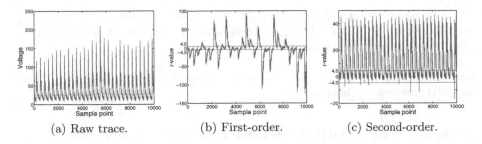

(a) Raw trace.   (b) First-order.   (c) Second-order.

Fig. 10. Measurement and TVLA results without PRNG.

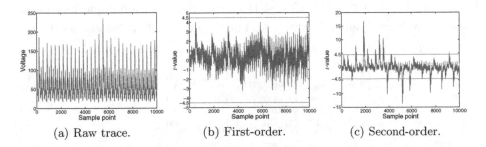

(a) Raw trace.   (b) First-order.   (c) Second-order.

Fig. 11. Measurement and TVLA results with PRNG.

of the first-order TI. On the other hand, we can see the second-order leakage in both Figs. 10 and 11 due to the limitation of the first-order TI. Thus, we could validate the DPA resistance of the proposed hardware architecture in the experimental condition with 500,000 traces.

5 Conclusion

This paper presented an efficient DPA-resistant AES hardware architecture based on the 1st-order TI. We first described the most compact first-order TI-based S-box design by combining the TI with $d + 1$ input shares and the algebraic characteristics of S-box. We then proposed a more efficient AES hardware architecture based on register-retiming and tower-field arithmetic in addition to the proposed S-box. The logic synthesis result showed that the proposed architecture achieved 11–21% lower latency and smaller area than the conventional ones, which would lead to the lowest-energy encryption secure against first-order DPAs. The DPA-resistance was validated through an experimental evaluation based on TVLA with 500,000 traces.

Our architecture can also be easily extended to higher-order TI-based S-boxes. On the other hand, the proposed S-box is not necessarily useful for compact implementation for the case of higher orders because the number of Stage 2 outputs increases by the cubic of TI-order, that is, by $(d + 1)^3$. A further evaluation of the proposed architecture and S-box for higher-order security would be required in the future. It is also demanded to consider the (partially) uniform sharing of TI-based $GF((2^2)^2)$ inversion for a further efficient first-order DPA-resistant implementation.

Acknowledgments. This research has been supported by JSPS KAKENHI Grants No. 16K12436 and No. 16J05711.

Appendix: First-Order TI-Based $GF((2^2)^2)$ Inversion with $d + 1$ Input Shares

Let $a^{(n)}$ and $a_i^{(n)}$ $(0 \le n \le 3, 0 \le i \le 1)$ be the nth bit of input and its shares, respectively. Let $c^{(n)}$ and $c_j^{(n)}$ $(0 \le j \le 7)$ be the nth bit of output and its shares, respectively. Here, the least-significant bits correspond to $a_i^{(0)}$ and $c_j^{(0)}$.

$$c_0^{(0)} = a_0^{(1)} a_0^{(2)} a_0^{(3)} + a_0^{(1)} a_0^{(3)} + a_0^{(2)}, \tag{1}$$

$$c_0^{(1)} = a_0^{(0)} a_0^{(2)} a_0^{(3)} + a_0^{(0)} a_0^{(3)} + a_0^{(2)}, \tag{2}$$

$$c_0^{(2)} = a_0^{(0)} a_0^{(1)} a_0^{(3)} + a_0^{(1)} a_0^{(3)} + a_0^{(0)}, \tag{3}$$

$$c_0^{(3)} = a_0^{(0)} a_0^{(1)} a_0^{(2)} + a_0^{(1)} a_0^{(2)} + a_0^{(0)}, \tag{4}$$

$$c_1^{(0)} = a_0^{(1)} a_0^{(2)} a_1^{(3)} + a_0^{(1)} a_0^{(2)} + a_0^{(0)} a_1^{(3)}, \tag{5}$$

$$c_1^{(1)} = a_0^{(0)} a_0^{(2)} a_1^{(3)} + a_0^{(0)} a_1^{(3)}, \tag{6}$$

$$c_1^{(2)} = a_0^{(0)} a_0^{(1)} a_1^{(3)} + a_0^{(1)} a_1^{(3)}, \tag{7}$$

$$c_1^{(3)} = a_0^{(0)} a_0^{(1)} a_1^{(2)} + a_0^{(1)} a_1^{(2)}, \tag{8}$$

$$c_2^{(0)} = a_0^{(1)} a_1^{(2)} a_0^{(3)} + a_0^{(0)} a_1^{(2)} + a_0^{(0)} a_0^{(3)}, \tag{9}$$

$$c_2^{(1)} = a_0^{(0)} a_1^{(2)} a_0^{(3)} + a_1^{(1)} a_0^{(3)} + a_0^{(3)}, \tag{10}$$

$$c_2^{(2)} = a_0^{(0)} a_1^{(1)} a_0^{(3)} + a_1^{(1)} a_0^{(2)} + a_0^{(0)} a_0^{(2)}, \tag{11}$$

$$c_2^{(3)} = a_0^{(0)} a_1^{(1)} a_0^{(2)} + a_1^{(1)} a_1^{(3)} + a_1^{(1)}, \tag{12}$$

$$c_3^{(0)} = a_0^{(1)} a_1^{(2)} a_1^{(3)} + a_0^{(0)} a_1^{(3)}, \tag{13}$$

$$c_3^{(1)} = a_0^{(0)} a_1^{(2)} a_1^{(3)} + a_0^{(1)} a_1^{(3)}, \tag{14}$$

$$c_3^{(2)} = a_0^{(0)} a_1^{(1)} a_1^{(3)} + a_0^{(0)} a_1^{(2)} + a_1^{(1)} a_1^{(2)}, \tag{15}$$

$$c_3^{(3)} = a_0^{(0)} a_1^{(1)} a_1^{(2)} + a_1^{(1)} a_0^{(3)}, \tag{16}$$

$$c_4^{(0)} = a_1^{(1)} a_0^{(2)} a_0^{(3)} + a_1^{(1)} a_0^{(3)}, \tag{17}$$

$$c_4^{(1)} = a_1^{(0)} a_0^{(2)} a_0^{(3)} + a_1^{(1)} a_0^{(3)}, \tag{18}$$

$$c_4^{(2)} = a_1^{(0)} a_0^{(1)} a_0^{(3)} + a_1^{(0)} a_0^{(2)} + a_0^{(1)} a_0^{(2)}, \tag{19}$$

$$c_4^{(3)} = a_1^{(0)} a_0^{(1)} a_0^{(2)} + a_0^{(1)} a_1^{(3)}, \tag{20}$$

$$c_5^{(0)} = a_1^{(1)} a_0^{(2)} a_1^{(3)} + a_1^{(0)} a_0^{(2)}, \tag{21}$$

$$c_5^{(1)} = a_1^{(0)} a_0^{(2)} a_1^{(3)} + a_1^{(1)} a_1^{(3)} + a_1^{(3)}, \tag{22}$$

$$c_5^{(2)} = a_1^{(0)} a_0^{(1)} a_1^{(3)} + a_0^{(1)} a_1^{(2)} + a_1^{(0)} a_1^{(2)}, \tag{23}$$

$$c_5^{(3)} = a_1^{(0)} a_0^{(1)} a_1^{(2)} + a_0^{(1)} a_0^{(3)} + a_0^{(3)}, \tag{24}$$

$$c_6^{(0)} = a_1^{(1)} a_1^{(2)} a_0^{(3)} + a_1^{(0)} a_1^{(2)} + a_1^{(0)} a_0^{(3)}, \tag{25}$$

$$c_6^{(1)} = a_1^{(0)} a_1^{(2)} a_0^{(3)} + a_1^{(1)} a_0^{(3)}, \tag{26}$$

$$c_6^{(2)} = a_1^{(0)} a_1^{(1)} a_0^{(3)} + a_1^{(1)} a_0^{(3)}, \tag{27}$$

$$c_6^{(3)} = a_1^{(0)} a_1^{(1)} a_0^{(2)} + a_1^{(1)} a_0^{(2)}, \tag{28}$$

$$c_7^{(0)} = a_1^{(1)} a_1^{(2)} a_1^{(3)} + a_1^{(1)} a_1^{(3)} + a_1^{(2)}, \tag{29}$$

$$c_7^{(1)} = a_1^{(0)} a_1^{(2)} a_1^{(3)} + a_1^{(0)} a_1^{(3)} + a_1^{(2)}, \tag{30}$$

$$c_7^{(2)} = a_1^{(0)} a_1^{(1)} a_1^{(3)} + a_1^{(1)} a_1^{(3)} + a_1^{(0)}, \tag{31}$$

$$c_7^{(3)} = a_1^{(0)} a_1^{(1)} a_1^{(2)} + a_1^{(1)} a_1^{(2)} + a_1^{(0)}. \tag{32}$$

References

1. Side-channel attack standard evaluation board (SASEBO). http://www.rcis.aist.go.jp/special/SASEBO
2. Bilgin, B., Gierlichs, B., Nikova, S., Nikov, V., Rijmen, V.: Higher-order threshold implementations. In: Sarkar, P., Iwata, T. (eds.) ASIACRYPT 2014. LNCS, vol. 8874, pp. 326–343. Springer, Heidelberg (2014). doi:10.1007/978-3-662-45608-8_18
3. Bilgin, B., Gierlichs, B., Nikova, S., Nikov, V., Rijmen, V.: Trade-offs for threshold implementations illustrated on AES. IEEE Trans. Comput. Aided Des. Integr. Syst. **34**(7), 1188–1200 (2015)
4. Bogdanov, A., Knudsen, L.R., Leander, G., Paar, C., Poschmann, A., Robshaw, M.J.B., Seurin, Y., Vikkelsoe, C.: PRESENT: an ultra-lightweight block cipher. In: Paillier, P., Verbauwhede, I. (eds.) CHES 2007. LNCS, vol. 4727, pp. 450–466. Springer, Heidelberg (2007). doi:10.1007/978-3-540-74735-2_31
5. Boss, E., Grosso, V., Güneysu, T., Leander, G., Moradi, A., Schneider, T.: Strong 8-bit Sboxes with efficient masking in hardware. In: Gierlichs, B., Poschmann, A.Y. (eds.) CHES 2016. LNCS, vol. 9813, pp. 171–193. Springer, Heidelberg (2016). doi:10.1007/978-3-662-53140-2_9
6. Canright, D.: A very compact S-Box for AES. In: Rao, J.R., Sunar, B. (eds.) CHES 2005. LNCS, vol. 3659, pp. 441–455. Springer, Heidelberg (2005). doi:10.1007/11545262_32

7. De Cnudde, T., Reparaz, O., Bilgin, B., Nikova, S., Nikov, V., Rijmen, V.: Masking AES with $d + 1$ shares in hardware. In: Gierlichs, B., Poschmann, A.Y. (eds.) CHES 2016. LNCS, vol. 9813, pp. 194–212. Springer, Heidelberg (2016). doi:10.1007/978-3-662-53140-2_10

8. Guo, J., Peyrin, T., Poschmann, A., Robshaw, M.: The LED block cipher. In: Preneel, B., Takagi, T. (eds.) CHES 2011. LNCS, vol. 6917, pp. 326–341. Springer, Heidelberg (2011). doi:10.1007/978-3-642-23951-9_22

9. Kocher, P., Jaffe, J., Jun, B.: Differential power analysis. In: Wiener, M. (ed.) CRYPTO 1999. LNCS, vol. 1666, pp. 388–397. Springer, Heidelberg (1999). doi:10.1007/3-540-48405-1_25

10. Mangard, S., Pramstaller, N., Oswald, E.: Successfully attacking masked AES hardware implementations. In: Rao, J.R., Sunar, B. (eds.) CHES 2005. LNCS, vol. 3659, pp. 157–171. Springer, Heidelberg (2005). doi:10.1007/11545262_12

11. Moradi, A., Mischke, O., Eisenbarth, T.: Correlation-enhanced power analysis collision attack. In: Mangard, S., Standaert, F.-X. (eds.) CHES 2010. LNCS, vol. 6225, pp. 125–139. Springer, Heidelberg (2010). doi:10.1007/978-3-642-15031-9_9

12. Moradi, A., Poschmann, A., Ling, S., Paar, C., Wang, H.: Pushing the limits: a very compact and a threshold implementation of AES. In: Paterson, K.G. (ed.) EUROCRYPT 2011. LNCS, vol. 6632, pp. 69–88. Springer, Heidelberg (2011). doi:10.1007/978-3-642-20465-4_6

13. Morioka, S., Satoh, A.: An optimized S-Box circuit architecture for low power AES design. In: Kaliski, B.S., Koç, K., Paar, C. (eds.) CHES 2002. LNCS, vol. 2523, pp. 172–186. Springer, Heidelberg (2003). doi:10.1007/3-540-36400-5_14

14. Nikova, S., Rijmen, V., Schläffer, M.: Secure hardware implementation of nonlinear functions in the presence of glitches. J. Cryptol. 24, 292–321 (2011)

15. Nyberg, K.: Differentially uniform mappings for cryptography. In: Helleseth, T. (ed.) EUROCRYPT 1993. LNCS, vol. 765, pp. 55–64. Springer, Heidelberg (1994). doi:10.1007/3-540-48285-7_6

16. Poschmann, A., Moradi, A., Khoo, K., Lim, C.W., Wang, H., Ling, S.: Side-channel resistant crypto for less than 2,300 GE. J. Cryptol. 24, 322–334 (2011)

17. Reparaz, O., Bilgin, B., Nikova, S., Gierlichs, B., Verbauwhede, I.: Consolidating masking schemes. In: Gennaro, R., Robshaw, M. (eds.) CRYPTO 2015. LNCS, vol. 9215, pp. 764–783. Springer, Heidelberg (2015). doi:10.1007/978-3-662-47989-6_37

18. Satoh, A., Morioka, S., Takano, K., Munetoh, S.: A compact Rijndael hardware architecture with S-Box optimization. In: Boyd, C. (ed.) ASIACRYPT 2001. LNCS, vol. 2248, pp. 239–254. Springer, Heidelberg (2001). doi:10.1007/3-540-45682-1_15

19. Schneider, T., Moradi, A.: Leakage assessment methodology. In: Güneysu, T., Handschuh, H. (eds.) CHES 2015. LNCS, vol. 9293, pp. 495–513. Springer, Heidelberg (2015). doi:10.1007/978-3-662-48324-4_25

20. Shamir, A.: How to share a secret. Commun. ACM 22, 612–613 (1979)

21. Tiri, K., Verbauwhede, I.: A logic level design methodology for a secure DPA resistant ASIC or FPGA implementation. In: Design, Automation and Test in Europe Conference and Exhibition (DATE), vol. 1, pp. 246–251 (2004)

22. Ueno, R., Homma, N., Sugawara, Y., Nogami, Y., Aoki, T.: Highly efficient $GF(2^8)$ inversion circuit based on redundant GF arithmetic and its application to AES design. In: Güneysu, T., Handschuh, H. (eds.) CHES 2015. LNCS, vol. 9293, pp. 63–80. Springer, Heidelberg (2015). doi:10.1007/978-3-662-48324-4_4

23. Ueno, R., Morioka, S., Homma, N., Aoki, T.: A high throughput/gate AES hardware architecture by compressing encryption and decryption datapaths. In: Gierlichs, B., Poschmann, A.Y. (eds.) CHES 2016. LNCS, vol. 9813, pp. 538–558. Springer, Heidelberg (2016). doi:10.1007/978-3-662-53140-2_26

Enhanced Elliptic Curve Scalar Multiplication Secure Against Side Channel Attacks and Safe Errors

Jeremy Dubeuf[1]([✉]), David Hely[2]([✉]), and Vincent Beroulle[2]([✉])

[1] Security Excellence Lab, Maxim Integrated, San Jose, USA
jeremy.dubeuf@maximintegrated.com
[2] University of Grenoble Alpes, LCIS, Grenoble, France
{david.hely,vincent.beroulle}@lcis.grenoble-inp.fr

Abstract. Elliptic curve cryptography (ECC) is involved in many secure schemes. Such schemes involve the elliptic curve scalar operation which is particularly security sensitive. Many algorithms of this operation have been proposed including security countermeasures. This paper discusses the security issues of such algorithms when running on a device that can be physically accessed. Leveraging these issues, new simple attack schemes to recover scalar bit information are presented and a new detailed attack based on C safe-error, probability and lattice is described against an Elliptic Curve Digital Signature Algorithm (ECDSA) using the Montgomery ladder algorithm. This new attack shows that Montgomery ladder can be sensitive to C safe-errors under some conditions. Finally, new secure elliptic curve scalar operation algorithms are presented with solutions to the discussed issues and guidance for their secure implementations.

Keywords: Elliptic curve cryptography · Point multiplication algorithms · Security issues · Forced bit · Safe-error · Dummy operands · Smartcard

1 Introduction

Physical attacks are commonly performed against cryptographic devices and have been proven to be a very efficient way to defeat strong cryptographic schemes. It is then necessary to consider such attacks when implementing cryptographic functions. Elliptic Curve Cryptography (ECC) is now used in most systems requiring asymmetric cryptography and thus faces a wide range of attacks. Some of these attacks are well known as they have already been performed against other cryptosystems and have just been adapted to ECC. Among them, we can mention Simple Power Analysis (SPA), Differential/Correlation Power Analysis (DPA/CPA) [23] or template attacks [9], which are part of side channel analysis. Active attacks are also very efficient against ECC such as safe-error [31] or Differential Fault Attack (DFA) [6]. Due to the mathematical properties of ECC, new kinds of attacks have been introduced such as weak curve attack

© Springer International Publishing AG 2017
S. Guilley (Ed.): COSADE 2017, LNCS 10348, pp. 65–82, 2017.
DOI: 10.1007/978-3-319-64647-3_5

and sign change among other. [12,13] survey different attacks on ECC and countermeasures. This work first discusses current problems encountered in elliptic curve scalar multiplication algorithms that are not covered in [13] and then proposes solutions to secure ECC. Several security issues are risen in Sect. 2, and new attacks are described based on them. Section 3 provides new elliptic curve scalar multiplication algorithms that partially solves the aforementioned problems thanks to their structure. Section 4 provides guidance for an efficient implementation and attempts to solve all the described problems by using the new algorithms alongside the well-known existing countermeasures. Finally, Sect. 5 concludes the paper.

2   Security Issues in Current Elliptic Curve Scalar Multiplication Algorithms

As explained in [12], the algorithm used to compute the scalar elliptic curve operation has to be selected and implemented extremely carefully when used with secrets. Even a partial leakage could help an attacker to fully recover secrets. In [2], authors demonstrated that the old Bleichenbacher attack presented in 2001 in [7] can be conducted with only one bit of leakage against an ECDSA scheme based on a 160-bit curve. The authors questioned the feasibility of such an attack on a 256-bit curve. In any case, [2] clearly demonstrates that every leakage source, even the smallest, should be removed from the elliptic curve scalar point multiplication algorithms. In this section, we describe different security issues in current elliptic curve scalar point multiplication algorithms that are not considered in recent academic surveys. We also describe new attacks leveraging these security issues that can be used to recover partial information on the scalar. A new C safe-error and lattice based attack against an ECDSA signature that uses the Montgomery ladder to compute the scalar is also presented.

2.1   MSB Given Away

In elliptic curve arithmetic, the neutral element is a very special point that usually generates particular cases in the computation. Basic elliptic curve scalar point multiplication algorithms usually initialize a working register to this neutral element [19] thus leading to particular computations until the register gets updated. This leads to leaking the scalar length through side channels such as power consumption or timing with disastrous consequences as in [8]. Indeed, curves represented with a Weierstrass form over F_p do not provide any working elliptic curve point addition and point double for the infinity point [19]. In order to deal with this special point during an elliptic curve point addition, a verification is usually performed on the operands and if the neutral element is detected, then the second operand is returned. If the neutral element is detected this approach leads to a shortened computation time and also to a different power consumption [12]. In order to avoid this leakage, it is a common practice to modify the register initialization and reduce the number of loop iteration or to directly force the scalar MSB to 1. Coron's algorithm initially presented in [10] and described below as Algorithm 1 is a good example of the problem.

Algorithm 1. Coron always Double-and-add

Input: $k = (k_{t-1}, ..., k_1, k_0)_2, P \in E(F_q)$
Output: $k.P$

1: $Q[0] \leftarrow P$
2: **for** $i = t - 2$ to 0 **do**
3:     $Q[0] \leftarrow 2Q[0]$
4:     $Q[1] \leftarrow Q[0] + P$
5:     $Q[0] \leftarrow Q[k_i]$
6: **end for**
7: **return** $(Q[0])$

Due to the initialization, in line 1, and the for loop, the MSB k_{t-1} is set to 1. Thus, a naive implementation of this algorithm either results in having a loop dependent of the length of the secret scalar (i.e. reduced to the first non-null bit) or to give away 1 bit. Giving away a single bit can help to avoid side channel leakage however this practice mathematically biases the system and thus may offer new attack possibilities to any systems. Indeed, as example in the case of a remote access, where side channel information are not observable, due to this introduced bias the system may be at risk. While public research does not provide a practical attack on curves with more than 160-bits with one bit bias, [2] discourages such a practice as the risk does not seem acceptable. [33] also noted that Bleichenbacher in [7] originally targeted an even smaller bias in the DSS in which the DSA nonces were generated by randomly picking up a number and applying a modulus reduction. Even if the risk is unclear and may seems acceptable regarding the current public research, a good practice should be to avoid to force any scalar bit nor to leak the length.

2.2 Local Dummy Operations

In [12], the authors only focus on side channel leakages. However, real world implementations may also face up other attacks such as fault injections. Some elliptic scalar point multiplication algorithms are designed to resist against Simple Power Analysis (SPA) by adding dummies operations. These operations aim at providing a constant operation without being useful from an arithmetic point of view. Algorithm 1 is a good example of such practice. In [32], authors have presented the C safe-error attack against the dummy operation inside an RSA exponentiation. The basic idea behind this attack is to inject a fault during a dummy operation and see if it is propagated or not. From Algorithm 1 it is obvious that a fault on $Q[1] = Q[0] + P$ will not be propagated if $k_i = 0$ as the faulty result will never be reused. Most common algorithms are known to be sensitive against the C safe-error attack and require specific countermeasures. Usually only the Montgomery ladder presented in Algorithm 2 is referenced as resistant against this attack by construction. In [24], the authors claim this resistance as there are no dummy operations in the Montgomery ladder. Also, they described the M safe-error attack against the Montgomery ladder that targets a memory value instead of a computation and then they provide a modified Montgomery ladder 'protected' against side-channel, M safe-error and also C safe-error per the previous claim.

Algorithm 2. Montgomery scalar operation

Input: $k = (1, k_{t-2}, ..., k_1, k_0)_2, P \in E(F_q)$
Output: $k.P$

1: $R_0 \leftarrow P$
2: $R_1 \leftarrow 2P$
3: **for** $i = t - 2$ to 0 **do**
4:     $R_{1-k_i} \leftarrow R_0 + R_1$
5:     $R_{k_i} \leftarrow 2R_{k_i}$
6: **end for**
7: **return** (R_0)

One missing information is that, from Algorithm 2, if $k_i = 0$ the operation $R_{1-k_i} \leftarrow R_0 + R_1$ may become a dummy operation. It is clearly visible that for $k = (1, 0, ..., 0, 0)_2$ the register R_1 is never involved in the result. This observation can be used in order to break an ECDSA. Indeed, if a nonce is randomly selected and then used as the scalar k in Algorithm 2, an attacker may inject faults into $R_{1-k_i} \leftarrow R_0 + R_1$ for some k LSBs, deduce if $k_{LSB} = 0$ and then mount a lattice attack [27]. As we did not find any public record of such an attack, here is a short example of the principle. An attacker looking for 70 signatures with 9 known bits in order to mount a lattice attack against an ECDSA based on NIST P-256 will need to fault $R_0 + R_1$ during evaluation of the nine LSBs for $2^9 \cdot 70 = 35840$ signatures generation. 2^9 signatures are needed to find one signature with the nine nonce LSB equal to zero. A valid signature means that the faults did not propagate to the result and thus that the nine LSB were set to 0. This attack scheme is realistic as each elliptic curve operation requires time, any fault on the targeted operation can be used and the algorithm is highly homogeneous facilitating the fault injection. As required faults are in a sequential order and physically can be injected in the same location, a single LED based laser is enough to inject such faults. Our number in term of required signatures (35840) is realistic in order to break an ECDSA based on NIST P-256 with a basic, non optimized lattice attack and does not take into account the information of the scalar MSB bit directly given by the algorithm. The MSB leakage, in this case, can easily be taken in account inside the lattice attack. Indeed, from [28], the s part of the ECDSA signature is equal to:

$$s_i = k_i^{-1}(H_i(m) + dr_i)$$
$$\Leftrightarrow k_i - dr_i/s_i - H_i(m)/s_i \equiv 0 \mod n$$
$$\Leftrightarrow k_{i_{MSB}} + k_{i_{unknow}} \cdot 2^x + k_{i_{LSBs}} - dr_i/s_i - H_i(m)/s_i \equiv 0 \mod n$$

Where $H_i(m)$ is the hash result of the message m, d is the private key, k_i the random nonce, r_i the r part of the signature and i represents the ith signature.

With x depending on the number of LSB set to 0.

By considering a 256 bit curve, $k_{i_{MSB}} == 1$ and $k_{i_{LSBs}} == 0$:

$$\Leftrightarrow k_{i_{unknow}} \cdot 2^x - dr_i/s_i - H_i/s_i + 2^{255} \equiv 0 \mod n$$
$$\Leftrightarrow k_{i_{unknow}} - dr_i/(s_i \cdot 2^x) - H_i/(s_i \cdot 2^x) + 2^{255-x} \equiv 0 \mod n$$

Thus, if the MSB bit is given away, attackers can easily insert this knowledge inside the equations. As one bit of is provided, attackers will seek one bit less and then the expected consequence is to divide by almost two the total number of required signatures to ask to the system and thus the attack time. Indeed, in our example, with $2^8 \cdot 70 = 17920$ signatures, attackers will obtain 70 of them with 9 known bits (8 LSBs + 1 MSB). The M safe-error countermeasure presented in [24] does not affect this attack as it simply changes the operand order depending on the secret bit. We can then conclude that the Montgomery ladder, as opposed to a common thought ([11,13,24,30]...), is in some case sensitive to C safe-error attacks. It also demonstrates that safe-error attacks can be used even with an ephemeral scalar value in the ECDSA case. A good practice could be to systematically verify that $R_0 - R_1 = P$ prior any result exposure. However, this will not be enough in the case of transient faults where fault exist over the scalar loop and disappear prior the verification.

2.3   Unused Memory Values

Similarly, errors can be inserted on a memory saved value which is no longer used. As an example, in [25] authors presented an algorithm based on the binary expansion and the randomized initial point (RIP, initially presented in [22]). The presented algorithm aims to be resistant against Simple Power Analysis (SPA), Differential Power analysis (DPA) [10], Refined Power Analysis (RPA) [17] and Zero-value Point Attacks (ZPA) [1]. It is described as Algorithm 3.

Algorithm 3. Binary Expansion with RIP (BRIP)

Input: $k = (k_{t-1}, ..., k_1, k_0)_2, P \in E(F_q)$
Output: $k.P$

1: $R \leftarrow randompoint()$
2: $T \leftarrow P - R$
3: $Q \leftarrow R$
4: **for** $i = t - 1$ to 0 **do**
5:     $Q \leftarrow 2Q$
6:     **if** k_i **then**
7:         $Q \leftarrow Q + T$
8:     **else**
9:         $Q \leftarrow Q - R$
10:     **end if**
11: **end for**
12: **return** $(Q - R)$

For $k = (k_{t-1}, ..., k_1, k_0)_2 = 0$, T is never used, therefore, an attacker can fault T inside the memory prior evaluation of the scalar LSBs and thus detect if they are equal to zero similarly to Sect. 2.2. This attack works with both transient and permanent faults. In the case of transient faults, the result is even worse as attackers could target any bit of the scalar. Usually, this problem arises with most of pre-calculated algorithms as pre-computed values are used based on the scalar.

2.4   The Infinity Point and Dummy Operands

Despite the fact that the infinity point (i.e. the neutral element) usually generates particular cases in the computation, some scalar point multiplication algorithms in even the most modern implementations still use it. Algorithm 4 is an example of such algorithms. It uses pre-calculated points in order to speed up the computation and a simple loop to sequentially add the correct points. The scalar k is represented with both positive and negative coefficients in order to reduce the total number of pre-calculated points and thus the memory and implementation cost.

Algorithm 4. Scalar operation with pre-computation

Input: $k = (r_{\lceil t/4 \rceil}, ..., r_1, r_0)_{24}$, $r_i.16^a.P \in E(F_q)$, with $0 \le a \le \lceil t/4 \rceil$ for any $r_i \in \{-8, -7, -6, -5, -4, -3, -2, -1, 0, 1, 2, 3, 4, 5, 6, 7\}$
Output: $k.P$

1: $Q \leftarrow \infty$
2: **for** $i = \lceil t/4 \rceil$ to 0 **do**
3:     $Q \leftarrow Q + r_i.16^i.P$
4: **end for**
5: **return** (Q)

As defined above, this algorithm uses the infinity point for the initialization and also as a pre-calculated point. It can then be expected that, in most cases, attackers may recover all $r_i == 0$ through side channel analysis. In [5] the authors describe a similar algorithm used with an Edwards curve. An interesting property of Edwards curves is the fact that the elliptic curve point addition formula is a unified and complete addition law that works for both point addition, point doubling and also with the neutral element without any special condition. However, the neutral element is still special. Indeed, the neutral element $(0, 1)$ under the twisted Edwards addition law presented in [4] generates special arithmetic cases. The given law is the following:

$$(x_1, y_1) + (x_2, y_2) = (\frac{x_1 y_2 + x_2 y_1}{1 + d x_1 x_2 y_1 y_2}, \frac{y_1 y_2 + x_1 x_2}{1 - d x_1 x_2 y_1 y_2}) \tag{1}$$

With the neutral element, we obtain:

$$(x_1, y_1) + (0, 1) = (\frac{x_1 \cdot 1 + 0 \cdot y_1}{1 + d x_1 \cdot 0 \cdot y_1 \cdot 1}, \frac{y_1 \cdot 1 + x_1 \cdot 0}{1 - d x_1 \cdot 0 \cdot y_1 \cdot 1}) = (x_1, y_1) \tag{2}$$

In order to avoid side channel attacks, the system should perform $x_1 y_2 \mod p$, $x_1 \cdot 1 \mod p$ and $0 \cdot y_1 \mod p$ similarly. And also $A/1 \mod p$ similarly to $A/B \mod p$ for any A and B. This means, without any noticeable timing difference due to a simplification nor without a different power consumption that could be due to carry propagation, modular reduction or other. Even considering the design is safe against side channel analysis, other problems appear if fault injection is considered. Indeed, in Eq. (2), a fault can be injected in y_1 operand when $0 \cdot y_1$ is computed. Faults can also be injected in dx_1 operands or computation without affecting the result. Such faults will not be propagated to the result in the case where the neutral element is used. Thus, due to this special point the design may face up safe-error attacks. Table 1 provides the identity element representation in different systems. From this table, it can be expected that the neutral element, due to the zero and one values, will generate special arithmetic cases in most systems and thus should be avoided.

Table 1. Representation of the neutral element in different systems

System	∞
Weierstrass, affine coordinates	None
Weierstrass, projective coordinates	(0:1:0)
Weierstrass, Chudnovsky coordinates	(1:1:0:0:0)
Edwards, affine coordinates	(0,1)
Edwards, homogeneous projective coordinates	(0:1:1)

2.5 Unnecessary Scalar Manipulation

In order to speedup the computation, some elliptic curve scalar point multiplication algorithms require the scalar in a non-conventional representation. This is the case in Algorithm 4. The Non-Adjacent Form (NAF) and the Join sparse form [19] are also often used. As the scalar may be used elsewhere in the system, this ends up with different representation, manipulation and use case of a same secret. This may extend the attack surface and introduce new security risks. For example, from [28], an implementation of ECDSA signature requires the elliptic curve scalar operation $k \cdot P$ with a random k for the r part of the signature. It also requires the computation of $k^{-1}(H(m) + dr) \mod n$ with the same k for the s part. If the scalar $k \cdot P$ is computed with Algorithm 4, it is highly possible that the system uses a conversion function $\phi(k) = k'$ to convert the random nonce k from a binary representation to k', the signed representation required in the algorithm. In this case, the designer has to ensure that $\phi(k)$ is implemented with side channel and fault countermeasures. Indeed, the carry attack against the scalar blinding [15] is a good demonstration that any scalar manipulation can leak information through side channel. The fault countermeasure is also important in order to guarantee that $k = k'$ independently of the representation otherwise the system may be vulnerable to a lattice-based fault attack. As example, if k' is used to compute r, and k to compute s, and if, due to a fault, a bit flip such that $k' = k \oplus 2^i$ then the signature (r, s) generated by the system will be invalid. However, an attacker can easily recover the injected fault by recovering $Q = k'P$ from r and then try all $Q \pm 2^i \cdot P$ to find a new r value that allows the verification to succeed. From this, attackers can then deduce the value of the flipped bit of k. By inserting a bigger fault (i.e. on 8 bits) more bits can be recovered and then used inside a lattice attack to recover the private key.

3   Our Secure Scalar Point Multiplication

If we consider current attacks and the aforementioned problems, from a security point of view a secure scalar point multiplication algorithm should use an operation flow independent of the scalar. Indeed such a deterministic flow can easily be online checked with a control flow integrity (CFI) system. For the same reason, it should also avoid the use of any dummy operation, even local. The infinity point should not be used and no scalar bit should be given away. The algorithm should also avoid using any specific representation of the scalar. From an implementation point of view, the algorithm should use a minimum number of working registers and allows acceleration through pre-calculation. Plenty of elliptic curve scalar multiplication algorithms already exist, however, as far as we now, non published algorithms meet all these requirements. In this section we describe a solution that meets all these criteria.

3.1   Scalar Operation

Our proposition presented in Algorithm 5 is based on the classical double-and-add algorithm and aims at correcting the different security issues.

Algorithm 5. scalar operation

Input: $E(F_q), k = (k_{t-1}, ..., k_1, k_0)_2, P \in E(F_q)$
Output: $2k.P$
  1: $Q \leftarrow P$
  2: **for** $i = t - 1$ to 0 **do**
  3:      $Q \leftarrow 2.Q$
  4:      $Q \leftarrow Q + (-1)^{\overline{k_i}}.P$ //add or subtract P, depending on k_i
  5: **end for**
  6: $Q \leftarrow Q - P$
  7: **return** (Q)

The idea behind our algorithm is to first initialize Q to a point $E \in E(F_q)$ instead of the infinity point as in the standard double-and-add algorithm. This will be transformed to $2E$ during the loop due to the point doubling. In order to maintain the point E over the loop indexes and reject the doubling effect, E should be subtracted from Q. This leads to two possibilities, if $k_i = 1$, $Q \leftarrow Q + (P - E)$ otherwise, $Q \leftarrow Q - E$. At the end of the loop, E is subtracted from Q in order to remove the initialization. By carefully selecting $E = \frac{P}{2}$, the two solutions become, if $k_i = 1$, $Q \leftarrow Q + \frac{P}{2}$ otherwise, $Q \leftarrow Q - \frac{P}{2}$. By using P instead of $\frac{P}{2}$, this leads to Algorithm 5 that compute $2k \cdot P$. This proposed algorithm removes most problems encountered in the basic double-and-add. It ensures that no infinity point is used without forcing any bit of the scalar. The operation flow is homogeneous thanks to the E subtraction and branches condition are removed. It is to be noted that Algorithm 5 avoids dummy operations that can be detected through safe-error attacks, does not force any specific treatment on any bits and also minimizes the number of required registers to prevent faults on unused memory values. Thanks to the use of either $+P$ or $-P$, the algorithm is also immune to Address bit DPA (ADPA) [21] as the addresses flow is homogeneous and independent of the scalar. The number of working registers is the same than a standard double-and-right, left-to-right algorithm. The data dependent leakage is however not removed as operands values depends on the scalar. It is also to be noted that Algorithm 5 adds a new horizontal threat. Indeed, at the beginning and until a scalar bit equal to one is encountered; the computation of $Q \leftarrow Q - P$ nullifies the double operation. Thus, the following elliptic curve double operation will double the same point as the previous operation. This will lead attackers to detect the scalar length through collision correlation analysis [3]. However existing solutions can be used and are discussed in Sect. 4.

If the platform does not provide indistinguishable field addition/subtraction, the elliptic curve point addition/subtraction can be computed with Algorithm 6.

This algorithm is similar to standard ECC mixed Affine/Jacobian point addition excepted from lines 4 to 9. These lines compute the intermediate result for both ECC point addition and ECC point subtraction and save them randomly in $T_3[0]$ and $T_3[1]$ depending on r. Line 6 selects which intermediate result is to be used regarding the asked operation and the random bit r. This randomness aims at avoiding fault injection inside the memory (M safe-error) as attackers will not know which value $-T_2$ or T_2 is faulted. Lines 8 and 9 aim at preventing C safe-error attacks when $-T_2$ is computed.

Algorithm 5 improves the security level compared to the algorithm presented in [18] that uses the zeroless signed-digit expansion. In [18], the authors note

---

Algorithm 6. ECC Jacobian-affine point addition/subtraction

Input: $P = (X_1 : Y_1 : Z_1)$ in Jacobian and $Q = (x_2, y_2)$ in affine $\in E(F_q)$, b operation selection, r a random bit
Output: if $b = 0$, $P - Q$ else $P + Q$

1: $T_1 \leftarrow Z_1^2$	10: $T_2 \leftarrow T_2 - Y_1$	19: $X_3 \leftarrow T_2^2$
2: $T_2 \leftarrow T_1 \cdot Z_1$	11: **if** $T_1 == 0$ **then**	20: $X_3 \leftarrow X_3 - T_1$
3: $T_1 \leftarrow T_1 \cdot x_2$	12: **return** (error)	21: $X_3 \leftarrow X_3 - T_3[1]$
4: $T_3[r] \leftarrow -T_2$	13: **end if**	22: $T_3[0] \leftarrow T_3[0] - X_3$
5: $T_3[\bar{r}] \leftarrow T_2$	14: $Z_3 \leftarrow Z_1 \cdot T_1$	23: $T_3[0] \leftarrow T_3[0] \cdot T_2$
6: $T_2 \leftarrow T_3[b \oplus r] \cdot y_2$	15: $T_3[0] \leftarrow T_1^2$	24: $T_3[1] \leftarrow T_3[1] \cdot Y_1$
7: $T_1 \leftarrow T_1 - X_1$	16: $T_3[1] \leftarrow T_3[0] \cdot T_1$	25: $Y_3 \leftarrow T_3[0] - T_3[1]$
8: $T_2 \leftarrow T_2 + T_3[0]$	17: $T_3[0] \leftarrow T_3[0] \cdot X_1$	
9: $T_2 \leftarrow T_2 + T_3[1]$	18: $T_1 \leftarrow T_3[0] + T_3[0]$	

that any group of w bits $00...01$ can be replaced with a group of w signed digits $1\bar{1}\bar{1}...\bar{1}$ with $\bar{1} = -1$. This remark leads to a zeroless signed representation of the scalar k. An on the fly conversion is possible by initializing Q to the base point P and then by performing a for loop for all bits (except the LSB) that contains a point double $Q \leftarrow 2Q$ and two possibilities, $Q \leftarrow Q + P$ if $k_i = 1$ or $Q \leftarrow Q - P$ if $k_i = 0$. Nevertheless, the zeroless signed representation works only for an odd scalar k thus forcing the LSB to one. Another downside of [18] is that an on the fly conversion with pre-calculated points seems harder to implement. Also, as mentioned in Sect. 2.5, recoding the scalar value in a signed representation prior using the elliptic curve scalar operation may extends attack possibilities. Thus, our Algorithm 5 improves the security as no scalar extra manipulation is required and no scalar bit is constrained to any value while allowing performance/cost flexibility thanks to pre-calculation possibilities as presented in the next section. Compared to Algorithms 3 and 5 losts some randomness provided by the random point. However, this allows preventing M safe-errors and ADPA as all registers are used during each loop independently of the secret. Randomness can be recovered as described in Sect. 4 without effecting the M-safe resistance property of the algorithm.

3.2   Scalar Operation with Pre-calculation

When performance is needed, point precalculation is usually used. The number of pre-computed points can be used to determine a trade-off between implementation cost and performance. In this section we present how our approach can be applied to the fixed-base comb method. This leads to similar performance with half of the required pre-calculated points.Compared to [20], in our case the scalar k does not need any modified representation simplifying the implementation. In order to achieve that, the same reasoning than Algorithm 5 is applied. The working register Q is initialized to a carefully selected point instead of the neutral element. The introduced error is partially corrected over the for loop and then fully removed after the loop. By carefully selecting the initialization point, the number of pre-calculated points is divided by two and the neutral element is removed from the list. Algorithm 7 describes this approach. Each pre-calculated point is used either in an elliptic curve point addition or subtraction. This leads to a better performances/cost ratio if pre-computation is allowed in the system. It is to be noted that the initialization point is one of the pre-calculated ones and the neutral element is never used. As described, Algorithm 7 needs to pre-compute 2^{w-1} points and uses $\lceil \frac{t}{w} \rceil$ elliptic curve point double operations and $\lceil \frac{t}{w} \rceil + 1$ elliptic curve point addition operations.

The 'represent k as:' does not necessarily mean that the scalar representation change in the register. Indeed, an hardware implementation can easily use the standard binary representation of k and evaluate the required bit without transforming the scalar.

Algorithm 7. Modified Comb method

Input: $E(F_q), k = (k_{t-1}, ..., k_1, k_0)_2, P \in E(F_q)$, window width $w, d = \lceil \frac{t}{w} \rceil$

Output: $2k.P$

Pre-computation: compute $2 \cdot [1, a_{w-2}, ...a_1, a_0].P - [1_{w-1}, ...1_1, 1_0].P$ for all possible binary values of $a_{w-2}, ...a_1, a_0$, with $[1, a_{w-2}, ...a_1, a_0].P = 2^{(w-1)d}P + ... + a_1 2^d P + a_0 P$.

Represent k as: $\begin{pmatrix} k_{d-1}^0 & \cdots & k_1^0 & k_0^0 \\ \vdots & \ddots & \ddots & \vdots \\ k_{d-1}^{w-1} & \cdots & k_1^{w-1} & k_0^{w-1} \end{pmatrix}$ //if necessary, pad 0s as k MSBs.

1: $Q \leftarrow [1_{w-1}, ...1_1, 1_0].P$ //represents the highest pre-calculated point.

2: **for** $i = d - 1$ to 0 **do**

3:    $Q \leftarrow 2.Q$

4:    $Q \leftarrow Q + (-1)^{\overline{k_i^{w-1}}} . \overline{[k_i^{w-1} \oplus [k_i^{w-2}, ...k_i^1, k_i^0]].P}$

5: **end for**

6: $Q \leftarrow Q - [1_{w-1}, ...1_1, 1_0].P$

7: **return** (Q)

3.3   The Non-standard $2k \cdot P$ Result and Exceptional Cases

Standard ECC schemes usually require the computation of $k \cdot P$, instead of $2k \cdot P$ as the presented algorithms. Different simple solutions exist. First, the less recommended solution would be to compute $c = k \cdot 2^{-1} \mod n$ prior using our algorithms with c. This solution is not the best one as it extends the number of manipulation of the secret k thus extending the attack surface (k is a secret but also c in this case). Another solution would be to save $\frac{P}{2}$ in the system instead of the fixed base point P during the system development. However, this solution cannot be applied if algorithms are intended to be used with an elliptic curve point coming from outside the system (e.g. a Diffie-Hellman). Finally, a last solution would be to consider during the secret generation that $k \cdot 2^{-1} \mod n$ is generated and saved in memory instead of k and take it into account if needed. For example, during an ECDSA signature, c can be generated from a TRNG. Then for the r part of the signature, by using c with our algorithms, $k \cdot P$ will be returned. The s part can be modified into: $s = c^{-1}(H(m) + d \cdot r) \cdot 2 \mod n$ leading to the good signature. This solution seems to be the best as it can be applied to most ECC schemes without extending the attack surface nor requiring special computation such as point halving. It is also to be noted that few exceptional cases exist within our algorithms depending on the scalar. Indeed, as example, if n represents the ECC point order used within Algorithm 5, then if $k = 0$, $k = n - 1$, $k = n - 2$, $k = n - 3$, $k = \frac{n-1}{2}$ and $k = \frac{n-1}{2} - 1$, special computation such as $P + P$, $P - P$, $-P + P$, $-P - P$ or $2 \cdot \infty$ will happen leading to special power consumption or fault exposure as explained in Sect. 2.4. However, the probability to encounter these cases is really low (around $\frac{5}{2^{256}}$ for NIST P256) and thus attacker cannot expect to observe such behavior.

4 Secure Implementation Strategy

Usually, systems requiring ECC implement different schemes such as ECDSA, ECDH, ECIES. These schemes face up different threat models as they use the elliptic scalar operation in different scenario. In an ECDH, the scalar is a private key while the base point is a given public key. Thus chosen point attacks such as RPA [1]/ZPA [17] or attacks requiring multiple executions with the same scalar such as DPA can be used to attack the device and fully recover the secret. However, during an ECDSA signature, the elliptic scalar operation is used with a refreshed nonce and the curve base point. Thus, these attacks cannot be used. Nevertheless, due to Lattice attack a partial leakage of the scalar is enough to recover the secret [12]. The work presented in [13] surveys different active and passive attacks, their prerequisites and countermeasures. Resource-restricted systems such as embedded system or smart-card need clever implementations in order to provide both functionalities and security with acceptable performance. In this section we present an efficient implementation strategy based on previous algorithms that can be used in different scenarios. Three use cases are considered. Firstly the elliptic curve scalar operation $k \cdot P$ with P the curve base point which can be hardcoded. Secondly the sum of two elliptic curve scalar operation $k \cdot P + v \cdot G$ with two different points P and G that can be different from the curve base point. Finally, these algorithms will lead to a security enhanced elliptic curve scalar operation $k \cdot G$ that considers any point G. It is to be noted that the different cases differ only from slight algorithmic modifications. Thus, a single implementation can be used over these use-cases in order to save memory or silicon as only the input and configuration will change.

4.1 Scalar with a Fixed Base Point: $k \cdot P$

The $k \cdot P$ operation can be implemented using ECC point precomputation technique as described in Algorithm 7. In Algorithm 8, we use a window width parameter of two, allowing to speed-up the computation with a factor of $\times 2$. It requires only two ECC points on the curve: $2^{t/2}P + P$ and $2^{t/2}P - P$.

In this algorithm, RandomAfftoJac() represents the common affine to Jacobian random representation conversion and provides countermeasure against data dependent leakage. Shuffleregisters() is a function that randomly reassign $P[0]$ and $P[1]$ in order to avoid address dependent leakages. Lines 11–13 aim to involve $P[r]$ in the result prior verification for integrity check. Without these two operations, an attacker can fault $P[r]$ prior the last l bits of the scalar and use the consequences to detect l consecutive 1 inside the LSBs (M-safe-error attacks). This countermeasure does not consider transient faults as only the value at the end is verified.

This algorithm requires $t/2 + 5$ ECC point additions and $t/2$ ECC point doubling operations while requiring 2 precomputed points. The classical comb method would require 3 ECC points to perform in $t/2$ ECC point doubling and ECC point additions. As the base point is fixed, precomputation cost is neglected.

Algorithm 8. kP operation

Input: $k = (k_{t-1}, ..., k_1, k_0)_2$, P and $2^{t/2}P \in E(F_q)$
Output: $2k.P$

1:   $r \leftarrow randombit()$
2:   $P[r] \leftarrow 2^{t/2}P - P$ //in Affine coordinates
3:   $P[\bar{r}] \leftarrow 2^{t/2}P + P$ //in Affine coordinates
4:   $Q \leftarrow RandomAfftoJac(P[1])$ //use random affine to Jacobian conversion
5:   **for** $i = t/2 - 1$ to 0 **do**
6:       $Q \leftarrow 2Q$
7:       $Q \leftarrow Q + (-1)^{\overline{k_{i+t/2}}} P[r \oplus \overline{(k_i \oplus k_{i+t/2})}]$
8:       $r \leftarrow randombit()$ //refresh the random r
9:       $shuffleregisters(P[0],P[1],r)$ //shuffle P[0] and P[1] according to r
10: **end for**
11: $Q \leftarrow Q + P[r]$ //add $P[r]$ for system integrity
12: $Q \leftarrow Q - P[\bar{r}]$
13: $Q \leftarrow Q - P[r]$ //remove $P[r]$
14: $Q \leftarrow JactoAff(Q)$ //use Jacobian to affine conversion
15: $Verify(Q)$
16: **return** Q

4.2   Sum of Two Scalar: $k \cdot P + v \cdot G$

ECDSA verifications for example require the computation of $k \cdot P + v \cdot G$ with always the same known ECC point P and G another ECC point representing a public key. Usually precomputation is not convenient to use in this case as the point G is usually not predictable. The best known solution to speed-up the computation is Shamir's trick [19] that aims to simultaneously compute both scalar. In our implementation, we use Shamir's trick combined with our previous presented solution. The result is described as Algorithm 9 and allows a speed-up with a factor of ×2 compared to two distinct classical ECC scalar operations. During the for loop, two bits of scalars are considered, one from k and another from v ending-up with four different cases. With the classic Shamir's trick, either $Q+\infty$, $Q+P$, $Q+G$ or $Q+(P+G)$ are considered. In our case, the four different cases become either $Q-(P+Q)$, $Q-(G-P)$, $Q+(G-P)$ or $Q+(G+P)$. This algorithm does not aim to improve the security level of the implementation as all manipulated values are public. Instead, it aims to end-up with a very similar implementation than the $k \cdot P$ computation. The descriptions of Algorithms 8 and 9 are almost identical. The differences are the initialization of $P[r]$ and $P[\bar{r}]$, the loop length, the bits used to select the ECC point addition/subtraction and the precomputed value to use. By doing so, implementation cost can be reduced as a parameter can be used to configure the algorithm depending on the requested operation.

    This algorithm requires $t + 5$ ECC point additions and t ECC point doubling. The classic Shamir's trick would get slightly better performances, in average t ECC point doubling and $0.75t$.

Algorithm 9. $k \cdot P + v \cdot G$ operation

Input: $k = (k_{t-1}, ..., k_1, k_0)_2, v = (v_{t-1}, ..., v_1, v_0)_2$, P and $G \in E(F_q)$
Output: $2kP + 2vG$

1: $r \leftarrow randombit()$
2: $P[r] \leftarrow G - P$ //in Affine coordinates
3: $P[\overline{r}] \leftarrow G + P$ //in Affine coordinates
4: $Q \leftarrow RandomAfftoJac(P[1])$ //use random affine to Jacobian conversion
5: **for** $i = t - 1$ to 0 **do**
6:      $Q \leftarrow 2Q$
7:      $Q \leftarrow Q + (-1)^{\overline{v_i}} P[r \oplus \overline{(k_i \oplus v_i)}]$
8:      $r \leftarrow randombit()$ //refresh the random r
9:      $shuffleregisters(P[0],P[1],r)$ //shuffle P[0] and P[1] according to r
10: **end for**
11: $Q \leftarrow Q + P[r]$ //add $P[r]$ for system integrity
12: $Q \leftarrow Q - P[\overline{r}]$
13: $Q \leftarrow Q - P[r]$ //remove $P[r]$
14: $Q \leftarrow JactoAff(Q)$ //use Jacobian to affine conversion
15: $Verify(Q)$
16: **return** Q

4.3 Scalar with Any Base Point: $k \cdot G$

The $k \cdot G$ operation is required for example in a Diffie-Hellman key agreements. It differs from the $k \cdot P$ operation as the ECC point used is not predictable meaning that usual precomputation techniques cannot be used efficiently. Another difference is from a security perspective as this operation is vulnerable to chosen point attacks. As example, RPA [1]/ZPA [17] can defeat the random affine to Jacobian point representation conversion. These attacks rely on the fact that some special points such as $(0, Y, Z), (X, 0, Z)$ remains in the same form whatever the value of Z and thus the random number used during the conversion. In this attack scenario, attackers input a chosen value Q that will be transformed to the special point at a targeted stage depending on the scalar value. E.g. in Algorithm 8, if P is chosen such that $(2^{t/2} + 1)P = (0, Y, Z)/3$ then if $k_{t-1} = k_{t/2-1} = 1$, the value $(0, Y, Z)$ will appear in the system at the end of the first loop iteration. This can be detected through a CPA by attackers allowing them to recover two bits. By iteratively performing the attack, attackers end-up with all the scalar value. To prevent this attack, we can use the $k \cdot P + v \cdot G$ algorithm as presented previously with a random ECC point G, $v = 0$ and $P = Q$.

From an arithmetic point of view, $k \cdot P + 0 \cdot G$ is equal to $k \cdot P$, however inside Algorithm 9, computations are based on $G - P$ and $G + P$. Thus with a random G, the algorithm computes the good result from random points. Attackers that input their own special point P will not be able to predict the values of $G - P$ and $G + P$. With this use of Algorithm 9, included countermeasures become mandatory. In the given example, the ECC point G is based on a 32 bits random number that can be computed similarly than $k \cdot P$ with a reduced loop in order to reduce the effect on the overall performances. An even more interesting use of Algorithm 9 to compute $k \cdot Q$ is with scalar splitting such that $k \cdot P = k_1 P + k_2(r \cdot P)$. This is described in Algorithm 10. Indeed, Algorithm 9 security relies on indistinguishable point addition or subtraction and also indistinguishable use of $P[0]$ or $P[1]$ during each loop iteration. By using scalar splitting, attackers will be forced to look for both information at the same time. In Algorithm 10, $2 \times t + 32$ bits of secret are used instead of t. If an attacker is able to distinguish elliptic curve point addition from point doubling, he obtains at most t bits of secret meaning that $t + 32$ bits remain. Similarly, if an attacker is able to recover which precomputed point is used, he will obtain at most $t + 32$ bits and thus t bits will remain. The use of r aims to remove the probability dependence between bits that exist and can be used in the simple additive splitting as explained in [26]. The reduction to a 32 bit loop should be carefully implemented by the designer. Indeed, if using an input parameter to configure the loop size, he has to ensure that this entry cannot be faulted in order to reduce the computation of a full size scalar to only 32 bits. Nevertheless, a CFI should easily be able to detect any iteration errors within the loop.

Algorithm 10. kG operation computed as $k \cdot G = k_1 G + k_2(r \cdot G)$

Input: $k = (k_{t-1}, ..., k_1, k_0)_2$, $Q \in E(F_q)$
Output: $2k \cdot G$
1: $r \leftarrow random([0, 2^{32}])$
2: $v \leftarrow random([0, \#E(F_q) - 1])$
3: $k \leftarrow k - v$
4: $v \leftarrow v \cdot r^{-1} \mod n$
5: $Q \leftarrow AlgKP(r, G)$ //Algorithm 8 reduced for 32bits of scalar
6: $R \leftarrow AlgkPvG(k, G, v, Q)$
7: **return** R

This algorithm requires $t + 26$ ECC point additions and $t + 16$ ECC point doubling operations. The classic always double-and-add algorithm would be faster as only t ECC point doubling and additions would be required, however, Algorithm 9 uses a fully masked scalar. Scalar blinding technique could also be used alongside the classic always double-and-add algorithm. Nevertheless, a too small random used with the scalar blinding may conduct to a partially masked scalar [14] due to the particular modulus value or to bias the probability as described in [29] or also to face doubling attacks due to the birthday paradox [16]. From [14], a NIST-P256 implementation would require a random on 64 bits leading to $t + 64$ ECC point doubling and additions. From this perspective, Algorithm 9 is around 15% faster.

5   Conclusion

In this paper we have illustrated different wide spread problems present in elliptic curve scalar point multiplication algorithms. We pointed out that some algorithms force a scalar bit value thus reducing their resistance against both purely mathematical attack and combination of side channel and lattice. A novel C safe-error attack against an ECDSA scheme using the Montgomery ladder has been presented. This lattice-fault based attack demonstrates that the Montgomery ladder, as opposed to a common thought, is also vulnerable to C safe-error due to local dummy operations. Problems related to safe-error against unused memory values was also presented. We recalled that the neutral element is a special point generating special arithmetic cases, even with Edwards curves despite the fact of providing a complete addition law. And we also discussed the need of a special representation (e.g. signed form) of the scalar during the computation of the scalar as it may extend the attack surface and the system complexity. We then presented a new suggestion of elliptic curve scalar multiplication algorithm that aims at improving the security level regarding the previously mentioned issues. This algorithm uses the traditional binary representation avoiding scalar recoding. It has a constant operation flow without any dummy operation. It ensures that the neutral element is not used and avoids any specific treatment on the scalar bits and provides M safe-error and ADPA immunity. We also provided an algorithm similar to the fixed-base comb method however with a reduced requirement in terms of number of precomputed points. This achievement was obtained by reusing precomputed points in different cases similarly to [18] however without having to recode the scalar in a signed representation avoiding the insertion of new vulnerabilities. In the last part of this paper we also provided an efficient implementation strategy that can be used for smart-cards when different ECC schemes are required by providing a unified algorithm.

References

1. Akishita, T., Takagi, T.: Zero-value point attacks on elliptic curve cryptosystem. In: Boyd, C., Mao, W. (eds.) ISC 2003. LNCS, vol. 2851, pp. 218–233. Springer, Heidelberg (2003). doi:10.1007/10958513_17
2. Aranha, D.F., Fouque, P.-A., Gérard, B., Kammerer, J.-G., Tibouchi, M., Zapalowicz, J.-C.: GLV/GLS decomposition, power analysis, and attacks on ECDSA signatures with single-bit nonce bias. In: Sarkar, P., Iwata, T. (eds.) ASIACRYPT 2014. LNCS, vol. 8873, pp. 262–281. Springer, Heidelberg (2014). doi:10.1007/978-3-662-45611-8_14
3. Bauer, A., Jaulmes, E., Prouff, E., Wild, J.: Horizontal collision correlation attack on elliptic curves. In: Lange, T., Lauter, K., Lisoněk, P. (eds.) SAC 2013. LNCS, vol. 8282, pp. 553–570. Springer, Heidelberg (2014). doi:10.1007/978-3-662-43414-7_28
4. Bernstein, D.J., Birkner, P., Joye, M., Lange, T., Peters, C.: Twisted Edwards curves. Cryptology ePrint archive, report 2008/013 (2008). http://eprint.iacr.org/
5. Bernstein, D.J., Duif, N., Lange, T., Schwabe, P., Yang, B.: High-speed high-security signatures. IACR cryptology ePrint archive 2011, 368 (2011). http://eprint.iacr.org/2011/368
6. Biehl, I., Meyer, B., Müller, V.: Differential fault attacks on elliptic curve cryptosystems. In: Bellare, M. (ed.) CRYPTO 2000. LNCS, vol. 1880, pp. 131–146. Springer, Heidelberg (2000). doi:10.1007/3-540-44598-6_8
7. Bleichenbacher, D.: On the generation of DSS one-time keys. Preprint (2001)
8. Brumley, B.B., Tuveri, N.: Remote timing attacks are still practical. In: Atluri, V., Diaz, C. (eds.) ESORICS 2011. LNCS, vol. 6879, pp. 355–371. Springer, Heidelberg (2011). doi:10.1007/978-3-642-23822-2_20

9. Chari, S., Rao, J.R., Rohatgi, P.: Template attacks. In: Kaliski, B.S., Koç, K., Paar, C. (eds.) CHES 2002. LNCS, vol. 2523, pp. 13–28. Springer, Heidelberg (2003). doi:10.1007/3-540-36400-5_3
10. Coron, J.S.: Resistance against differential power analysis for elliptic curve cryptosystems (1999)
11. Danger, J.L., Guilley, S., Hoogvorst, P., Murdica, C., Naccache, D.: A synthesis of side-channel attacks on elliptic curve cryptography in smart-cards. J. Cryptogr. Eng. 3(4), 241–265 (2013). https://hal.inria.fr/hal-00934333
12. Dubeuf, J., Hely, D., Beroulle, V.: ECDSA passive attacks, leakage sources, and common design mistakes. ACM Trans. Des. Autom. Electron. Syst. 21(2), 3101–3124 (2016). http://doi.acm.org/10.1145/2820611
13. Fan, J., Verbauwhede, I.: An updated survey on secure ecc implementations: attacks, countermeasures and cost. In: Naccache, D. (ed.) Cryptography and Security: From Theory to Applications. LNCS, vol. 6805, pp. 265–282. Springer, Heidelberg (2012). doi:10.1007/978-3-642-28368-0_18
14. Feix, B., Roussellet, M., Venelli, A.: Side-channel analysis on blinded regular scalar multiplications. Cryptology ePrint archive, report 2014/191 (2014). http://eprint.iacr.org/
15. Fouque, P.-A., Réal, D., Valette, F., Drissi, M.: The carry leakage on the randomized exponent countermeasure. In: Oswald, E., Rohatgi, P. (eds.) CHES 2008. LNCS, vol. 5154, pp. 198–213. Springer, Heidelberg (2008). doi:10.1007/978-3-540-85053-3_13
16. Fouque, P.-A., Valette, F.: The doubling attack – why upwards is better than downwards. In: Walter, C.D., Koç, Ç.K., Paar, C. (eds.) CHES 2003. LNCS, vol. 2779, pp. 269–280. Springer, Heidelberg (2003). doi:10.1007/978-3-540-45238-6_22
17. Goubin, L.: A refined power-analysis attack on elliptic curve cryptosystems. In: Desmedt, Y.G. (ed.) PKC 2003. LNCS, vol. 2567, pp. 199–211. Springer, Heidelberg (2003). doi:10.1007/3-540-36288-6_15
18. Goundar, R.R., Joye, M., Miyaji, A., Rivain, M., Venelli, A.: Scalar multiplication on Weierstraß elliptic curves from Co-Z arithmetic. J. Cryptogr. Eng. 1(2), 161–176 (2011). http://dblp.uni-trier.de/db/journals/jce/jce1.html#GoundarJMRV11
19. Hankerson, D., Menezes, A.J., Vanstone, S.: Guide to Elliptic Curve Cryptography. Springer, New York (2003)
20. Hedabou, M., Pinel, P., Bénéteau, L.: Countermeasures for preventing comb method against SCA attacks. In: Deng, R.H., Bao, F., Pang, H.H., Zhou, J. (eds.) ISPEC 2005. LNCS, vol. 3439, pp. 85–96. Springer, Heidelberg (2005). doi:10.1007/978-3-540-31979-5_8
21. Itoh, K., Izu, T., Takenaka, M.: Address-bit differential power analysis of cryptographic schemes OK-ECDH and OK-ECDSA. In: Kaliski, B.S., Koç, K., Paar, C. (eds.) CHES 2002. LNCS, vol. 2523, pp. 129–143. Springer, Heidelberg (2003). doi:10.1007/3-540-36400-5_11
22. Itoh, K., Izu, T., Takenaka, M.: Efficient countermeasures against power analysis for elliptic curve cryptosystems. In: Smart Card Research and Advanced Applications VI, IFIP 18th World Computer Congress, TC8/WG8.8 & TC11/WG11.2 Sixth International Conference on Smart Card Research and Advanced Applications (CARDIS), 22–27 August 2004, Toulouse, France, pp. 99–113 (2004). http://dx.doi.org/10.1007/1-4020-8147-2_7
23. Joye, M., Tymen, C.: Protections against differential analysis for elliptic curve cryptography — an algebraic approach —. In: Koç, Ç.K., Naccache, D., Paar, C. (eds.) CHES 2001. LNCS, vol. 2162, pp. 377–390. Springer, Heidelberg (2001). doi:10.1007/3-540-44709-1_31

24. Joye, M., Yen, S.-M.: The montgomery powering ladder. In: Kaliski, B.S., Koç, K., Paar, C. (eds.) CHES 2002. LNCS, vol. 2523, pp. 291–302. Springer, Heidelberg (2003). doi:10.1007/3-540-36400-5_22

25. Mamiya, H., Miyaji, A., Morimoto, H.: Efficient countermeasures against RPA, DPA, and SPA. In: Joye, M., Quisquater, J.-J. (eds.) CHES 2004. LNCS, vol. 3156, pp. 343–356. Springer, Heidelberg (2004). doi:10.1007/978-3-540-28632-5_25

26. Muller, F., Valette, F.: High-order attacks against the exponent splitting protection. In: Public Key Cryptography, pp. 315–329 (2006)

27. Nguyen, P.Q., Shparlinski, I.: The insecurity of the digital signature algorithm with partially known nonces. J. Cryptol. 15(3), 151–176 (2002). http://dx.doi.org/10.1007/s00145-002-0021-3

28. NIST: Digital Signature Standard (DSS), FIPS PUB 186 (2013). http://nvlpubs.nist.gov/nistpubs/FIPS/NIST.FIPS.186-4.pdf

29. Okeya, K., Sakurai, K.: Power analysis breaks elliptic curve cryptosystems even secure against the timing attack. In: Roy, B., Okamoto, E. (eds.) INDOCRYPT 2000. LNCS, vol. 1977, pp. 178–190. Springer, Heidelberg (2000). doi:10.1007/3-540-44495-5_16

30. Rondepierre, F.: Revisiting atomic patterns for scalar multiplications on elliptic curves. In: Francillon, A., Rohatgi, P. (eds.) CARDIS 2013. LNCS, vol. 8419, pp. 171–186. Springer, Cham (2014). doi:10.1007/978-3-319-08302-5_12

31. Yen, S.M., Joye, M.: Checking before output may not be enough against fault-based cryptanalysis. IEEE Trans. Comput. 49(9), 967–970 (2000). http://dx.doi.org/10.1109/12.869328

32. Sung-Ming, Y., Kim, S., Lim, S., Moon, S.: A countermeasure against one physical cryptanalysis may benefit another attack. In: Kim, K. (ed.) ICISC 2001. LNCS, vol. 2288, pp. 414–427. Springer, Heidelberg (2002). doi:10.1007/3-540-45861-1_31

33. Zapalowicz, J.C.: Security of the pseudorandom number generators and implementations of public key signature schemes. Theses, Université Rennes 1 (2014). https://tel.archives-ouvertes.fr/tel-01135998

SafeDRP: Yet Another Way Toward Power-Equalized Designs in FPGA

Maik Ender[✉], Alexander Wild, and Amir Moradi

Horst Görtz Institute for IT-Security, Ruhr-Universität Bochum, Bochum, Germany
{maik.ender,alexander.wild,amir.moradi}@rub.de

Abstract. Side-channel analysis attacks, particularly power analysis attacks, have become one of the major threats, that hardware designers have to deal with. To defeat them, the majority of the known concepts are based on either masking, hiding, or rekeying (or a combination of them). This work deals with a hiding scheme, more precisely a power-equalization technique which is ideally supposed to make the amount of power consumption of the device independent of its processed data. We propose and practically evaluate a novel construction dedicated to Xilinx FPGAs, which rules out the state of the art with respect to the achieved security level and the resource overhead.

1 Introduction

Unintended communication channels of a cryptographic device may leak information about the processed data. These channels – also known as side channels – can be used to recover a secret key from the device. Targeting the power consumption as a side channel, in the literature the corresponding attack is known as Power Analysis (PA) attacks. As a powerful and low-cost attack vector, it makes use of statistical dependencies between the processed data and the power consumption of the cryptographic device. Amongst the known countermeasures, those which are known as *hiding* techniques mainly try to reduce the exploitability of the leakages [12]. To this end, one solution for hardware platforms is to equalize the power consumption, and hence decorrelate the leakage from the processed data. Such countermeasures usually follow the Dual-Rail Precharge (DRP) concept, where the source of the data-dependent power consumption, i.e., transition in transistors, is addressed. Among several such schemes, we can refer to SABL [27], WDDL [28], DRSL [7], MDPL [21], and iMDPL [20], which have particularly been designed for Application-Specific Integrated Circuits (ASICs). However, due to the predefined structure and limited routing resources, they cannot be easily employed on Field Programmable Gate Arrays (FPGAs). It is noteworthy that several other works, e.g., [4,8–11,13,14,18,22,32], have already tried to adopt the concept of DRP schemes to FPGAs. Such schemes reduce the vulnerability against PA attacks, but almost all of them suffer from at least one of the known major pitfalls: the early propagation effect, glitches, and imbalanced routings. To the best of our knowledge, GliFreD [17] is of the rare FPGA-based

© Springer International Publishing AG 2017
S. Guilley (Ed.): COSADE 2017, LNCS 10348, pp. 83–101, 2017.
DOI: 10.1007/978-3-319-64647-3_6

solutions addressing all pitfalls by a massive utilization of Flip Flops (FFs). Note that a more optimized variant of GliFreD has been introduced in [30].

In this work, as an FPGA-based power-equalizing solution, we introduce Safe-DRP which also avoids the three aforementioned major problems. The selling point of SafeDRP is its low resource overhead. More precisely, it significantly reduces the number of utilized FFs with the price of a more complicated control logic. We investigate its effectiveness by a case study, i.e., an Advanced Encryption Standard (AES) encryption engine on a Kintex-7 FPGA. Our investigations include resource overhead as well as practical side-channel analysis evaluations. We further introduce a process on how to convert an unprotected circuit into its SafeDRP-variant.

It is noteworthy that power-equalization schemes in general cannot fully avoid the exploitability of leakages due to e.g., imperfection of balanced routings. Instead, in case of effective solutions, they can reduce the leakage interpreted by e.g., lower Signal-to-Noise ratio (SNR) leading to harder attacks with respect to the number of required side-channel measurements. Therefore, a combination of such a solution with a sound masking scheme, e.g., Threshold Implementation (TI) [19], is known to provide a high level of practical security (e.g. [17]). Hence, a suitable construction would be to implement a masked design under the concept of SafeDRP. However, in order to solely examine the effectiveness of SafeDRP, we considered an ordinary (not masked) design as the case study. Since the concept of SafeDRP and GliFreD are relatively similar, the same security achievements as in [17] are expected if TI and SafeDRP are merged.

2 Background

As the name says, the Dual-Rail Precharge (DRP) logic is a combination of Dual-Rail (DR) logic and precharge logic. DR logic makes use of a differential encoding of the signals, where every signal a is encoded to $(a, \bar{a})$. Hence, a logic Hi is encoded to $(1, 0)$ and a logic Lo to $(0, 1)$. In general, a DR gate expects differentially-encoded input signals, and also produces a differentially-encoded output. Therefore, a DR gate has a double number of input and output signals compared to a functionally-equivalent single-rail gate. Commonly, the differential encoding of DR logic forms two networks which are often noted as positive and negative network.

A circuit built upon precharge logic alternates between a precharge and an evaluation phase. During the precharge phase, the gates of the circuit propagate a predefined constant value. During the evaluation phase, the gates evaluate the data intended to be processed by the circuit. Hence, the input and output signals of a circuit built in precharge logic are set to a constant value before the given data set is processed.

The combination of both (DR and precharge) techniques forms the concept of DRP logic. Hence, DRP uses a differential encoding of their signals which alternate between a precharge and evaluation phase. During the precharge phase, the encoded signals are set to a constant value, i.e., $(0, 0)$ or $(1, 1)$. Hence, at a phase transition from precharge to evaluation exactly one of the wires changes their state, i.e., $(0,0) \rightarrow (1,0)$ or $(0,0) \rightarrow (0,1)$ (when precharge is $(0,0)$). The same holds for the transition from the evaluation phase back to precharge. This behavior can be scaled from a single gate to a full circuit, which results in a circuit with a constant number of wire transitions, independent of the data it processes. Hence, following the DRP concept, a circuit forms a promising basis to equalize the dynamic power consumption independent of the data it processes.

Beyond constant number of wire transitions, DRP schemes have to deal with three major pitfalls. The first pitfall is the *early evaluation* effect, which occurs if the scheme shows a data-dependent time of evaluation. The second is known as *glitches*, which are temporary, faulty transitions of a gate output signal. Thus, the output should change only once per phase transition. Therefore, a DRP scheme has to ensure that all input signals are stable at the gate's inputs before it evaluates. The third problem is *imbalanced routing*, which occurs especially in FPGAs due to the limited routing resources. The routing of the positive and negative networks need to be identical with respect to their length, i.e., capacitance. Otherwise, the power consumption to load coupled wires will be different and hence impede the aimed power equalization.

FPGAs are integrated circuits with reprogrammable logic cells. The cells are organized on a grid. Every cell can be individually addressed by its X/Y-coordinates in the grid. The first level of the hierarchy is the clock region. Next the tiles group different classes of distinct elements in the grid, such as the slices and specialized elements like Block RAM (BRAM) and Digital Signal Processors (DSPs). A slice holds several Look-Up-Tables (LUTs) and FFs which are the basic elements in an FPGA[1]. A LUT is the reprogrammable logic element of the FPGA, it evaluates every Boolean function based on the configuration. The LUT is realized by SRAM cells and a multiplexer tree. After each LUT a FF or rather a latch is placed which can store the LUT's output.

The connection between all elements is realized by a routing engine. The routes between the elements are reprogrammable as well. Reprogrammable switching matrices before a group of elements (e.g. two slices) connect these groups with distinct hardwired routes. Beside the routes between the elements, an FPGA is equipped with a clock network, since clock signals have usually a high fan-out and need a minimal skew to guarantee the proper instantiation of synchronous circuits. This clock network is also known as *clock tree*, which has a direct connection to every clocked element placed on the FPGA. Depending on the FPGA architecture, the clock tree is able to reach non-clocked elements as well.

Our proposed scheme is implemented and evaluated on a Xilinx 7-Series FPGA which makes use of the Vivado tool flow. The Vivado Design Suite is used to synthesis, translate, map, and placed and route a Hardware Description Language (HDL) code to generate a bitstream, which programs the FPGA. It also supports the Tool Command Language (Tcl) command line interface and

---

[1] The exact number of available resources in an FPGA highly depends on its device family and differs between the available architectures and manufacturer.

scripting language which are essential for our work. With Tcl we are able to manipulate objects within the design after each synthesis step.

3   Concept

The majority of cryptographic algorithms implemented on FPGAs are based on the standard components like LUTs, FFs and latches. Hence, the concept of SafeDRP focuses on these components. This does not automatically conclude that other dedicated hardware components cannot be instantiated in a SafeDRP-based design. For instance, the inclusion of BRAMs may be achieved by adapting a concept proposed in [3].

3.1   Controlling LUTs

In order to define a proper DRP scheme we need to address the problem of early evaluation and glitches. Therefore, we define control signals to trigger the phase transition of the LUTs. The number of used control signals, further referred as *active* signals, highly depends on the underlying hardware and the logic depth of the hardware design. Figure 1(b) shows the waveforms of four *active* signals, connected to the simple circuit given in Fig. 1(a). As shown, the *active* signals are periodic and run at the same frequency. They are phase shifted to each other, and come up with different duty cycles.

Each LUT is connected to one of the *active* signals which controls the LUT's precharge and evaluation phases. Note that this grants no restriction to the LUT's logical function but reduces the number of available inputs per LUT by one. It is because the LUT functionality must be of the form

$$f'(x) = active \land f(x), \tag{1}$$

while the LUT evaluates at $active = \mathrm{Hi}$ and is set to precharge on $active = \mathrm{Lo}$. In order to avoid glitches at the LUT output, the *active* signal is supposed to be the last arriving signal at the LUT to trigger the phase transition after all input signals are stable. Therefore, consecutive LUTs are not connected to the same *active* signal.

In detail, a combinatorial circuit is organized in LUT stages. The stage of a LUT is defined by the maximum number of LUTs the input signals have to pass to reach it. The LUTs with the same stage label are connected to the same *active* signal. The *active* signals of consecutive stages are slightly phase shifted so that they evaluate one after another. The reduced duty cycle ensures a glitch-free transition from evaluation to precharge of the stages in reverse order. A small exemplary circuit with corresponding waveforms is given in Fig. 1.

The *active* signals define the phase transitions of the LUTs, and are hence very critical signals which require a low skew. Therefore, the *active* signals are routed via the special routing network called *clock tree*. To reach the first LUT input (and the FF's reset pin), the clock tree is left in the switching matrix attached to the slice. It is noteworthy that a switch matrix is limited to move just two signals from the clock tree to the logic.

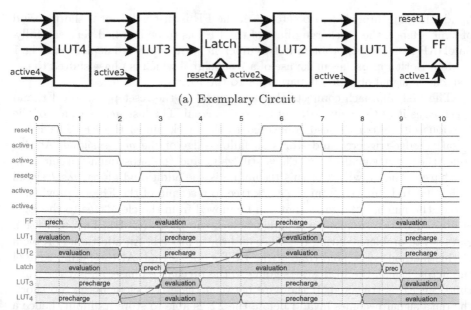

(a) Exemplary Circuit

(b) Waveform for *active* signals and phases of LUTs, FF and latch of the exemplary circuit

Fig. 1. Visualization of the SafeDRP concept.

3.2 Controlling Memory Elements

FFs and latches play an important role in sequential circuits. In a DRP scheme, a FF is mostly built on a master-slave fashion, as one of the two FFs must be in precharge while the other FF holds the data value. Alternatively, it is not mandatory to consecutively place the FFs in a design. To minimize the critical path and hence increase the performance of a circuit, one of the FF stages can be moved into the middle of the combinatorial circuit. This splits the combinatorial circuit into two parts. The control signals of the circuit parts work inverse to each other to keep the FFs alternating between precharge and evaluation. Hence, either the first part (LUTs and FFs) of the circuit is in precharge and the second part evaluates or the other way around. This can be seen in Fig. 1 as well. In general, a combinatorial circuit can be split into an even number of parts while every second part is connected to the same control signals. The number of parts used to form the circuit depends on the circuit constraints. By increasing the number of parts, the critical path will be reduced which results in a higher clock frequency but also increases the latency of the circuit. Further, the overhead with respect to the number of FFs increases as well but does not necessarily increase the area consumption of the design since the FPGA structure provides a FF right after each LUT which left unused in most cases.

The FF provides the data for the subsequent part, and can hence be precharged only when all stages of the subsequent part are in precharge. On the other hand, the data provided to be stored in the FF is just valid for a short period of time. Hence, the precharge phase of the FF has to be handled very carefully. SafeDRP handles this problem in two different ways which are depicted in Fig. 1 as well. Both approaches make use of a *reset* signal which is phase shifted to the last *active* signal of a circuit part prior to the FF.

The first approach connects the *reset* signal to the reset pin of the FF and precharges it at the rising edge of the *reset* signal. The last *active* signal of the previous circuit part is also connected to the clock pin of the FF. The FF is negative edge-triggered and stores on falling edge of the *active* signal. At this time, the input signal of the FF is still valid and the subsequent circuit is in precharge. The drawback of this technique is the additional global *reset* signal which has to be moved from the clock tree to reset pin of the FF. As previously noted, the number of signals that can be moved from the clock tree to the logic is limited to two signals per Configurable Logic Block (CLB). This limitation can lead to problems during the place and route process. An additional problem of this method is the hold time of the LUT's data signal. At the falling edge of the *active* signal, the LUT starts going to the precharge phase, and at the same time the FF is triggered. If the wire between these two elements is too short, the data signal becomes invalid before the FF is able to store its input, hence a hold time violation[2].

The second approach replaces the edge-triggered FF with a level-triggered latch. The *reset* signal is connected to the clock pin of the latch which will be transparent at *reset* = Hi. Indeed, in this case the latch is not forced to reset, but the *reset* signal plays the role of the enable signal of the latch. The *reset* signal enables the latch slightly before the LUT (that is connected to the input pin of the latch) evaluates. Hence, the latch first passes the precharge value (which is the prechagred LUT output), and then holds the output after the LUT output is evaluated. Since all latches in a circuit should be controlled by the same *reset* signal, this method requires the entire latches to be connected to the last LUT stage of the underlying circuit part. In other words, the input of all latches should be supplied by the LUTs from the same stage. Otherwise, the latch(es) might not pass a precharged value.

3.3   Duplication

Facing the problem of imbalanced routing, the method proposed in [32] is utilized to create the positive and negative networks. Therefore, each cell from the positive network (e.g. LUTs and FFs) needs to be cloned and inverted. First, a single rail circuit (so-called positive network) which follows the principles explained above (Sects. 3.1 and 3.2) is fully placed and routed at a defined area on the FPGA. Second, the positive network is copied and placed at another equivalent reserved area on the FPGA while the relative routing and placement

---

[2] This failure appears on the Kintex-7 if the FF and the LUT controlled by the same active signal are placed at the same slice.

is retained. Third, by inverting the functionality of the copied circuit, the negative network of the design is formed. Indeed, the combination of both circuits follows the DRP definition and shows a balanced routing structure. More details about the duplication process is given in Sect. 4.3.

3.4   Resources and Limits

Every LUT stage of the circuit needs one *active* signal, also an additional *reset* signal for the FF/latch is needed. Note that the invert of every *active* signal of one circuit part exists in the other part, except for *active* signals connected to the last LUT stage. Figure 2 shows that $active_2$ and $active_5$ or $active_3$ and $active_6$ are complementary. Therefore, the *active* signals of the first part can be reused in the second part, only the last LUT stage needs an individual *active* signal. Note that, the active signals are not inverted for the second part. Instead, the LUTs in the second part are configured to evaluate on $active = \mathrm{Lo}$. It is useful to have the same number of stages in both parts to share the most number of *active* signals. In total, $d + 1 + 2$ *active* signals are required, where d is the maximum logic depth of a circuit part, 1 individual *active* signal for the last LUT stage and 2 *reset* signals for the two FF/latch stages. Since the clock tree is used to distribute the *active* signals, the maximum logic depth is limited by the number of available clock trees. For example, on the 7-Series FPGAs 12 clock trees exist in every clock domain. Thus, the maximum logic depth for each circuit part is $d = 12 - 3 = 9$.

The duplication concept requires to invert the functionality of every logical element in the positive network. Some dedicated hardware components like the DSPs or multiplexer are not invertible and hence not usable in this construction. The *active* signal connected to the LUT reduces its functionality to an $(N - 1)$-to-1 LUT, i.e., in the 7-Series only 5-to-1 LUTs can be used rather than the 6-to-1 LUTs. The duplication process doubles the consumed resources as the negative network is a copy of the positive network.

The maximum frequency of this construction strongly depends on the design, i.e., the logic depth of the circuit parts which defines the number of *active* signals and the critical path of every stage. Here the critical path is the maximal signal delay between two consecutive LUT stages. The duty cycle $duty_i$ at stage i is

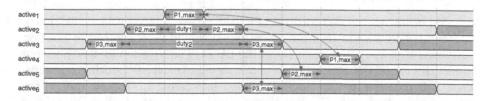

Fig. 2. The relation between the *active* signals, the composition of their duty cycles and phase shifts.

defined by its critical path[3] $p_{i,max}$ and the duty cycle of the subsequent LUT stage. An exception is the last LUT stage, where the duty cycle depends only on the critical path from the last LUT stage to the FF/latch. This results in

$$duty_i \geq \begin{cases} p_{1,max} & \text{if } i = 1 \\ 2 \cdot p_{i,max} + duty_{i-1} & \text{otherwise.} \end{cases}$$

The frequency is defined by the first LUT stage, as it has the highest duty cycle (e.g. $active_3$ or $active_6$), and a margin to reset the FF/latch. The $active$ signals are phase shifted in a way that the delay between the rising edges of $active_{i-1}$ and $active_i$ is $p_{i,max}$. The duty cycle estimation and signal alignment is visualized in Fig. 2.

In a practical instantiation, the device capabilities limit the phase shift, duty cycle and hence the frequency. For example, we can make use of the Mixed-Mode Clock Manager (MMCM) to generate the control signals (more details in Sect. 4.2), but its capabilities on the 7-Series FPGAs directly influence the discrete duty cycle steps and the minimal phase shift angle (see [31] for more detailed information).

It is worth to note that all control signals should become active only when the corresponding stage of the circuit is in the precharge phase in order to introduce no glitches. For example, a multiplexer, which selects two different signals, should only switch while both signals hold the precharge value. One can use the last LUT stage's $active$ signal as a control clock to switch the other part, e.g. in Fig. 2 the circuit connected to $active_4$ to $active_6$ can be controlled while $active_1$ is Hi.

4   Tooling

Developing a design which follows the above-illustrated concept is an intensive task. To reduce the workload and support hardware designers, this section introduces a tool flow which helps to transform arbitrary combinatorial circuits written in HDL into circuits which follows the concept of SafeDRP. Since the hardware design process highly depends on the FPGA architecture, we focus on the Xilinx 7-Series and the associated Vivado tool flow.

4.1   Circuit Mapping

As stated before, the phase transition of a LUT is determined by an $active$ signal. Following the concept of [17], the $active$ signal should be connected to the first multiplexer stage, i.e., first input pin of the LUTs. The Vivado tool flow does not support this constraint to append an $active$ signal to each LUT during synthesis. Hence, it is required to map the HDL design into 5-to-1 LUTs and append the control signals afterwards. Further, the maximal logic depth of a SafeDRP-based circuit highly depends on the number of available clock trees, which requires to add memory elements into large combinatorial circuits.

---

[3] Since we reuse the $active$ signals to control stages of all circuit parts, we should consider the highest stage delay at all circuit parts.

To synthesize, optimize, edit, and map a given HDL code we used ABC [2], an open-source tool from the Berkeley Logic Synthesis and Verification Group. The tool strongly supports the needs of our construction and is natively capable to map a given HDL design into 5-to-1 LUTs. To reduce the manual overhead of a hardware designer, we extended ABC by two functions. First, our modified version of ABC is now capable to add memory elements to a combinatorial circuit after a defined logic depth. Second, each 5-to-1 LUT of the mapped design is replaced by a 6-to-1 LUT. Based on their logic depth, the corresponding *active* signal is attached to the LUTs first input pin which drives the first multiplexer stage. Further, the LUT's *INIT* attribute is extended to fulfill Eq. 1.

4.2   Placement Restrictions

Notwithstanding, for a large design – e.g. an AES round function – the place and route algorithms of Vivado with standard settings could mostly not generate a routable SafeDRP design or rather a design with a higher leakage than expected, as the routing of non-clocking signals is challenging.

As stated before, we make use of clock trees to route *active* signals with low skew. Hence, to generate the *active* signals we use an MMCM, which is hardwired to the clock tree. The clock tree is also hardwired to each switch matrix located in the CLBs. As given before, the limited resources of the switch matrix allows to extend only two clock trees to data pins of the slices located in the CLB.

The placer is not aware of the constraint to connect the *active* signal to the LUTs' first input, and may also place two different LUT stages in one slice, which results in an unroutable design. It may also happen that the *active* signal leaves the clock tree at a different switching matrix and is routed via the routing fabric, which results in a larger skew. Hence, the placement and the routing problem need to be addressed. We consider two methods to overcome this problem. First, the placer can be restricted to use just a single LUT per slice which obviously results in a heavy area increase. Second, the placer is not restricted, but the placement has to be corrected manually. Swapping the LUTs between nearby slices can group the LUTs to from the same stage in one slice.

4.3   Design Duplication

The output of ABC forms just the positive network of SafeDRP. In order to address the problem of imbalanced routing, SafeDRP adapts the duplication concept proposed in [32]. Hence, the placed-and-routed positive network is duplicated and inverted to form the negative network. The full duplication process is split into the following sub-processes:

1. Place an additional instance of the positive network at a reserved area on the FPGA and keep its placement and routing structure.
2. Invert the second instance's LUT functionality to form the negative network.
3. Logically connect all I/O signals of the negative network to the control logic.
4. Route the I/O signals of the negative network to the control logic.

Vivado includes the Tcl shell and scripting language, which is used to manipulate objects within the design. We are using a Tcl script to perform the duplication process and no third party tool is needed to manipulate the objects within the design. First, all cells of the positive network are cloned. Our script can deal with all primitives of the FPGA, which are in general LUTs, FFs, latches, and clock buffers. Consequently, a new object is created and all properties are copied from the positive cell to the new (negative) cell. The set of negative cell hence form the negative network of our scheme. To place the cells of the negative network at a different location, a constant value is added to the X-/Y-coordinate. Note that the addition of a constant to the location coordinates does not change the relative placement structure between the cells of the negative network. In order to invert the logic of the negative network, just the LUTs' content are changed, since only the LUTs define the logical function of the design. The LUT content is adjusted to fulfill Eq. (2). The behavior of the *active* signal is maintained, as mentioned before.

$$active \wedge f_{\text{neg}}(x) = active \wedge \overline{f_{\text{pos}}(\overline{x})} \tag{2}$$

The routes of the positive network are cloned in a similar way. First, we have to make a distinction between internal nets, i.e., nets which connect only cells inside the SafeDRP-based circuit, and external nets, i.e., nets which connect the SafeDRP-based circuit with the remaining design (control logic). Internal nets contribute to the leakage and are important for the DRP logic, while external nets do not contribute to the leakage. Similar to internal signals, the routes of the *active* signal inside a clock region are kept equal during the duplication process. The remaining part of the *active* signals used to reach the clock region is handled separately and routed to the clock source. In order to route a given internal net of the positive network, a new net is created and connected to the respective negative cells. The routing information (located in the "ROUTE" property) of the positive route is copied to its negative pendant. Since this information is relative to its location and the relative placement of the negative network is equivalent to the positive network, no further changes are needed. Next, the external signals are connected to the negative network. Each connection of each external signal is connected to its negative pendant. In the last step another run of the Vivado routing algorithm is needed to route the external signals, while the already existing routes are preserved.

5   Evaluation

In a case study we examine the effectiveness of our construction by an round-based AES-128 encryption on a Kintex-7 FPGA.

The used AES core is designed to compute one full round in one clock cycle. For the S-Boxes the area optimized Canright S-Box [6] is employed. Figure 3 shows a block diagram of the final design, while TraceGetter is a custom framework which provides a communication interface between the computer,

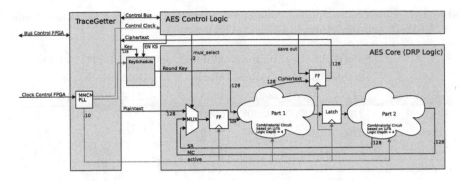

Fig. 3. The case study's design of an AES-128 on the Kintex-7.

oscilloscope and the targeted device. During the implementation, the ABC synthesizer was queried until the AES core had a logic depth of 8, then the core was split-up in the middle to get two equally-sized parts. FFs are used in front of Part 1 as the multiplexer before the FFs introduces enough delay that no hold time violation occurs. Also, the multiplexer has no *active* signal, thus the reset signal to the FF is the only *active* signal leaving the clock tree in the corresponding switch matrix. Part 1 and Part 2 are divided by means of latches, i.e., the second approach explained in Sect. 3.2. Here 382 latches are placed in front of Part 2. In 222 cases a LUT, which is connected to the last *active* signal in Part 1, exists right before the latch. The latch gets its precharge value from this LUT. If the LUT that supplies the latch's input is not connected to the last *active* signal, a pass-through LUT is inserted. This pass-through LUT is connected to the last *active* signal to provide the precharge value to the latch (see Fig. 1(b)). This is done in the 160 remaining cases ($222 + 160 = 382$). We further have used an MMCM and a Phase-Locked Loop (PLL) to generate the *active* signals.

5.1 Resource Utilization

To measure the resource overhead of our construction, we performed a normal Place and Route (PAR) without any modifications. We, in fact, replaced the synthesized AES from ABC with the unmodified Verilog sources. In addition, we implemented the same AES design under the concept of GliFreD to enable a fair comparison. Table 1 shows the resources used for the AES core in SafeDRP both before and after duplication, the same for a corresponding GliFreD variant, and the AES in plain. The resource consumption is doubled after the duplication process, as all cells are duplicated, and no further cells are added. Comparing the normally PAR design to the single-rail (SafeDRP/GliFreD) core, we get the overhead by the reduced LUT input pins.

An overall comparison to plain design reveals a factor of 2.94 more 5-LUT and 3.30 more slices, while the number of registers has an overhead of a factor of 4.98. The number of added registers ($636 - 256 = 382$) results from cutting the circuit into two parts. Compared to the improved GliFreD, the number of

Table 1. Resource consumptions, comparison between a normally PAR (Plain), Safe-DRP, GliFreD, and the improved GliFreD AES designs.

	SafeDRP		Improved GliFreD [30]		GliFreD [17]		Plain
	Doubled	Single	Doubled	Single	Doubled	Single	
5-LUT	3712	1856	3466	1733	3466	1733	1262
Register	1276	638	11360	5680	22080	11040	256
Slices	1296	648	11638	5819	15502	7751	392
Latency[a]	11		154		308		11
Pipeline	0		14		14		0
Throughput[b]	116		116		58		116

[a]Clock cycles
[b]MBit/s @ 10 MHz

registers is drastically reduced, i.e., by a factor of $\frac{5680}{638} \approx 8.9$. In order to examine the overhead of the 5-LUT design, we need to subtract the 160 pass-through 5-LUT from the 1856 5-LUT used in the single-rail design. Since these delayed 5-LUT are added in order to precharge the latches, they add no overhead for the 5-LUT design. Therefore, the overhead of the 5-LUT design compared to plain is $\frac{1856-160}{1262} = 1.343$, while the 5-LUT design can slightly reduce the number of 5-LUT compared to GliFreD (4-LUT) $(1856 - 160) = 1696 < 1733$. Including the pass-through 5-LUT, the number of 5-LUT is slightly increased by SafeDRP compared to GliFreD.

Table 1 also shows the throughput for the three considered designs. Since SafeDRP does not form any pipeline, its throughput – at the same frequency – is the same as the plain design, but it outperforms the first GliFreD variant due to the high number of pipeline stages of GliFreD. However, the GliFreD design can operate at an extremely higher frequency compared to SafeDRP, whose frequency is limited by the logic depth and the performance of the employed MMCM. Therefore, at maximum frequency the GliFreD design has a much higher throughput than its SafeDRP variant. Our design reaches up to 81 MHz, as the critical path in each stage is ≈1.1 ns. In order to increase the frequency, one could insert pipeline stages as usual or enhance the placement process, as stated before.

5.2 Measurement Setup

We used a SAKURA-X evaluation board [1], equipped with a Kintex-7 XC7K160T FPGA, which hosts the investigated AES core(s). The power consumption is measured by a PicoScope 6402B at 1.25 GS/s. Between the measurement points on the SAKURA-X and the PicoScope, we placed two Mini-Circuits ZFL-1000LN+ AC amplifiers to amplify the signal. A trigger generated by the targeted FPGA ensures well-aligned traces.

We developed three different profiles of our design to examine the effectiveness of our construction, while we activate and deactivate the different protection mechanisms between the profiles. All modifications are done on the placed-and-routed design, which keeps the placement and routing untouched and hence allows a fair comparison.

- Profile 1 is the reference profile, in which the SafeDRP design is untouched.
- In Profile 2 the duplicated negative circuit is removed to test the Dual-Rail concept.
- In Profile 3 the duplicated negative circuit is removed, the precharge of the LUTs and FFs/latches are turned off and the LUTs are always active.

It should be noted that Profile 3 is a fully unprotected AES core. The only difference to a normal AES circuit is the unused first input pin of the LUTs (by constantly keeping the *active* = Hi) and the delay LUTs in front of the latches.

We run the AES core at a frequency of 10 MHz, thus one encryption requires 1.1 μs. An exemplary trace per profile are displayed in Fig. 4, which cover the full 2,000 measured sample points. The encryption starts around point 230 and terminates around point 1670. The plaintext bytes for each encryption are randomly selected from a uniform distribution and the key is kept constant.

5.3   Side-Channel Analysis

We used different side-channel analysis methods in order to quantify leakage reduction caused by SafeDRP. We applied

- Signal-to-Noise ratio (SNR) [12] which examines how large the exploitable signal compared to the available noise is,
- Information-Theoretic (IT) metric [24] that gives an overview about the information available in the side-channel leakages with respect to the concept of information theory,
- Correlation Power Analysis (CPA) [5] with the Hamming Weight (HW) and Bit models to get an impression about the required number of traces of common key-recovery attacks, and
- Moments-Correlating DPA (MC-DPA) [16] attack to relax the necessity of a suitable power model in case of CPA, and examine the exploitable leakage through first-order leakages.
- Semi-fix vs. random Welch's t-test [23] which gives an overview about the existing detectable leakage.

Power-equalization schemes are designed to reduce the SNR and make the attacks harder. Therefore, we limit our investigations to univariate first-order analyses. Thus, every method is conducted on each sample point separately. For each design, we measured $n = 10,000,000$ traces, and all attacks and analysis methods – except the t-test – make use of the entire measured traces while focusing on the first key byte.

We start with $SNR = \frac{var(signal)}{var(noise)}$. Following the procedure given in [12], we first categorize the traces by the value of the targeted plaintext byte and estimate

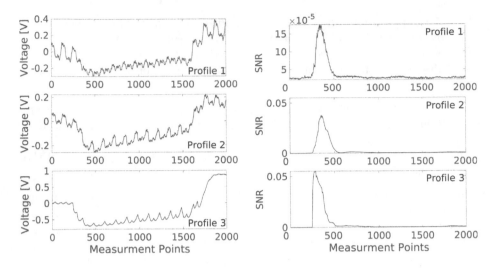

Fig. 4. Exemplary power traces.   **Fig. 5.** SNR curves.

the mean and variance for each group. Then the variance of the means states the $var(signal)$ since the average over each group represents mostly the noise-free signal depending on the underlying plaintext byte value. Further, the mean over the variance traces determines the $var(noise)$. Figure 5 shows the SNR curves based on the first plaintext byte of all profiles. No high non-sensitive peak is visible as the plaintext is applied to the circuit before the measurement and encryption start. Comparing Profile 2 and Profile 3, the precharge and evaluation of the LUTs and FFs/latches could only slightly reduce the SNR. However, when it is combined with DR, i.e., Profile 1, the SNR is reduced by a factor of $0.055/0.00018 \approx 313$.

By Information-Theoretic (IT) analysis [24] we can measures the amount of exploitable information by estimating the mutual information. Since our construction is a realization of hiding schemes and we limit our analyses to first order, in order to estimate the Mutual Information (MI) we can estimate the conditional entropy by means of Probability Density Functions (PDFs) based on Gaussian distributions. To this end, we can re-use the mean and variance traces estimated for the SNR. The resulting curves are given in Fig. 6. The results are similar to the SNR; comparing Profile 1 and Profile 3, mutual information reduced by a factor of ≈ 265.

We also conducted the commonly-used CPA attacks with different power models. We have first examined the attacks with HW of the S-Box output, which all led to unsuccessful key recovery. The reason is the underlying architecture of our case study, where the S-Box outputs are not stored in registers. Instead, the design has two FF stages (see Fig. 3). The first FF stage contains the plaintext in the first round. After the first round, the output of the MixColumns is saved

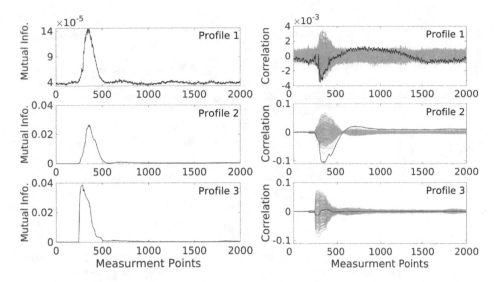

Fig. 6. Mutual information curves.   **Fig. 7.** CPA attack results, HW model.

in these FFs. In order to predict the value of these FFs, at least 4 key bytes need to be guessed. Hence, we focus on the latches after Part 1, which are easier to predict.

We should emphasize that our design somehow merged the S-Boxes with their subsequent MixColumns. However, the latches are placed before the end of the S-Box calculations. It means that every latch bit depends on only one plaintext byte and one key byte. Further, for each state byte (plaintext XOR key) between 18 and 30 latches have been instantiated, which is mainly caused by the application of ABC's synthesis algorithms and the fact that ABC synthesizes the full AES round. Figure 7 depicts the result of the CPA attacks with the HW of the intermediate values, i.e., value of the latches. In the graphics, the curve for the correct key candidate is plotted in black while that of other candidates in gray. The attack is unsuccessful for Profile 3, since the latches in this Profile never become precharged. We did not change the power model to HD in Profile 3 since the distance is hard to predict in the first round. As well a comparison with an attack in the last round to the first round would not be fair. However, we can already observe the advantage of Profile 1 over Profile 2. More precisely, due to the quadratic inverse relation between the correlation and the required number of traces [12], i.e., $\rho^2 \propto \frac{1}{n}$, we can conclude that the attack on Profile 1 needs around $\left(\frac{0.1}{0.0035}\right)^2 \approx 816$ times more traces comapared to that on Profile 2.

To conduct a successful attack on Profile 3, we considered the bit model as well, i.e., correlating the traces to a predicted certain latch bit. We have examined all latches independently; the result of the best attack (on Profile 3) is shown by Fig. 8. In this case, the advantage of each profile compared to the others can be observed. For example, the effect of precharge concept, i.e., Profile 2 versus Profile 3, can be expressed by $\left(\frac{0.2544}{0.0599}\right)^2 \approx 18$ times more traces, and the effect of

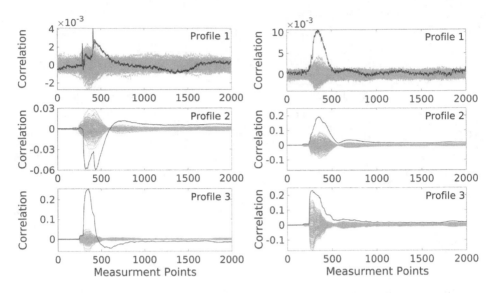

Fig. 8. CPA attack results, bit model.   **Fig. 9.** MC-DPA results.

duplication (i.e., Profile 1 versus Profile 2) can be seen by $\left(\frac{0.0599}{0.0041}\right)^2 \approx 213$ more traces to exploit the leakage.

The feasibility of such CPA attacks strongly depends on the soundness of the underlying hypothetical power model, i.e., its linear relation to the actual leakage of the device. Therefore, we applied MC-DPA attack [16] as a sophisticated scheme that relaxes such a necessity and does not require any predefined power model. We used the profiled version of MC-DPA, where the first $n/2$ traces are used to generate the profiles, and the second $n/2$ traces for the attack. Figure 9 shows the correlation curves as the result of the attack on all profiles. The attack successfully recovers the correct key for all profiles. This is indeed expected because SafeDRP– as am power-equalization scheme – can only reduce the leakage, which results in a higher number of required traces for a successful attack. With respect to the concept of MC-DPA that can exploit any first-order leakage independent of the actual leakage function of the device, we can conclude that our construction Profile 1 succeeds in reducing the exploitable leakage with respect to the number of required traces. Comparing the results, the attack on Profile 1 needs $\left(\frac{0.195}{0.0106}\right)^2 \approx 338$ and $\left(\frac{0.230}{0.0106}\right)^2 \approx 470$ more traces compared to Profile 2 and Profile 3 respectively.

We also used the Welch's t-test [23] to quantify the leakage of our proposed scheme using $1,000,000$ traces. Following the same procedure in [29], we applied the semi-fix vs. random test to discard the leakage associated to plaintext/ciphertext. Prior to each measurement, a coin is flipped thereby selecting the plaintext from either the semi-fix or the random poll. The random plaintexts are selected from a uniform distribution, while the semi-fix plaintexts have been pre-computed in such a way that half of the cipher state (i.e., 64 bits) at the

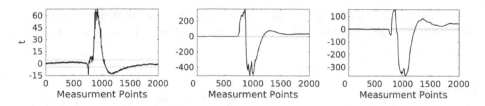

Fig. 10. Welch's semi-fix vs. random t-test using $1,000,000$ traces.

fifth round is filled by zero. Figure 10 shows the results of the t-test, where the reduction in leakage is visible again between the profiles. It is noteworthy that due to the memory effect of the employed amplifier [15], the detected leakage still appears after the fifth cipher round.

6 Conclusion

In this paper we introduced a novel equalization scheme SafeDRP and presented a general method to transform unprotected circuits into SafeDRP logic. By an AES implementation in SafeDRP as a case study we practically evaluated the effectiveness of SafeDRP to reduce the exploitable leakage resulting in hardening the key-recovery attacks.

In almost all duplication schemes, the power consumption of the positive and negative networks are still slightly different caused by process variations, different temperatures, aging of cells, and the chip internal supply voltage differences. Further, different times of evaluation are caused by e.g., the skew of the control signals[4]. Therefore, similar to the other power-equalization schemes, SafeDRP cannot completely avoid the leakage, but the practical results showed its success to extremely reduce such leakages. As a side note, for a complete practical protection such power-equalization techniques, e.g., SafeDRP, should be combined with proper masking schemes, e.g., the case shown in [17].

Apart from its high level of security, the advantage of SafeDRP over the known and similar schemes is its low overhead. Compared to the GliFreD scheme, SafeDRP reaches a slightly higher factor of reduced leakage. GliFreD reduces the SNR and MI by a factor of ≈ 100 (SafeDRP ≈ 300). The number of required traces and the factor for the CPA and MC-DPA attacks is mostly the same for both schemes. This leads to the following assumption: GliFreD's analysis is done on a 45 nm FPGA, while this case study is done on a 28 nm FPGA. Thus the smaller manufacturing process most likely causes the reduced leakage. Nevertheless, the tested SafeDRP-based AES design uses fewer resources than a GliFreD-based design. With the cost of a more complicated control logic and 7% more LUTs SafeDRP requires only 11% of the FFs which are essential in corresponding GliFreD design.

---

[4] Such features are already used for identification purposes [25] as well as randomness generation [26].

Future research might be the adaptation of a different duplication strategy, as the separate true and false DRP cores could exploit some leakage in a localized EM attack.

References

1. Side-channel attack user reference architecture. http://satoh.cs.uec.ac.jp/SAKURA/index.html
2. Berkeley logic synthesis, verification group, ABC: a system for sequential synthesis and verification, release ae0be2deffef. http://www.eecs.berkeley.edu/alanmi/abc/
3. Bhasin, S., Danger, J., Guilley, S., He, W.: Exploiting FPGA block memories for protected cryptographic implementations. TRETS **8**(3), 16 (2015)
4. Bhasin, S., Guilley, S., Flament, F., Selmane, N., Danger, J.: Countering early evaluation: an approach towards robust dual-rail precharge logic. In: WESS 2010, p. 6. ACM (2010)
5. Brier, E., Clavier, C., Olivier, F.: Correlation power analysis with a leakage model. In: Joye, M., Quisquater, J.-J. (eds.) CHES 2004. LNCS, vol. 3156, pp. 16–29. Springer, Heidelberg (2004). doi:10.1007/978-3-540-28632-5_2
6. Canright, D.: A very compact S-box for AES. In: Rao, J.R., Sunar, B. (eds.) CHES 2005. LNCS, vol. 3659, pp. 441–455. Springer, Heidelberg (2005). doi:10.1007/11545262_32
7. Chen, Z., Zhou, Y.: Dual-rail random switching logic: a countermeasure to reduce side channel leakage. In: Goubin, L., Matsui, M. (eds.) CHES 2006. LNCS, vol. 4249, pp. 242–254. Springer, Heidelberg (2006). doi:10.1007/11894063_20
8. He, W., de la Torre, E., Riesgo, T.: A precharge-absorbed DPL logic for reducing early propagation effects on FPGA implementations. In: ReConFig 2011, pp. 217–222. IEEE Computer Society (2011)
9. He, W., Otero, A., de la Torre, E., Riesgo, T.: Automatic generation of identical routing pairs for FPGA implemented DPL logic. In: ReConFig 2012, pp. 1–6. IEEE Computer Society (2012)
10. Kaps, J., Velegalati, R.: DPA resistant AES on FPGA using partial DDL. In: FCCM 2010, pp. 273–280. IEEE Computer Society (2010)
11. Lomné, V., Maurine, P., Torres, L., Robert, M., Soares, R., Calazans, N.: Evaluation on FPGA of triple rail logic robustness against DPA and DEMA. In: DATE 2009, pp. 634–639. IEEE Computer Society (2009)
12. Mangard, S., Oswald, E., Popp, T.: Power Analysis Attacks: Revealing the Secrets of Smart Cards. Springer, New York (2007)
13. McEvoy, R.P., Murphy, C.C., Marnane, W.P., Tunstall, M., Isolated, W.: A hiding countermeasure for differential power analysis on FPGAs. TRETS **2**(1), 3:1–3:23 (2009)
14. Moradi, A., Immler, V.: Early propagation and imbalanced routing, how to diminish in FPGAs. In: Batina, L., Robshaw, M. (eds.) CHES 2014. LNCS, vol. 8731, pp. 598–615. Springer, Heidelberg (2014). doi:10.1007/978-3-662-44709-3_33
15. Moradi, A., Mischke, O.: On the simplicity of converting leakages from multivariate to univariate. In: Bertoni, G., Coron, J.-S. (eds.) CHES 2013. LNCS, vol. 8086, pp. 1–20. Springer, Heidelberg (2013). doi:10.1007/978-3-642-40349-1_1
16. Moradi, A., Standaert, F.-X.: Moments-correlating DPA. In: Workshop on Theory of Implementation Security, TIS 2016, pp. 5–15. ACM (2016)

17. Moradi, A., Wild, A.: Assessment of hiding the higher-order leakages in hardware. In: Güneysu, T., Handschuh, H. (eds.) CHES 2015. LNCS, vol. 9293, pp. 453–474. Springer, Heidelberg (2015). doi:10.1007/978-3-662-48324-4_23

18. Nassar, M., Bhasin, S., Danger, J., Duc, G., Guilley, S.: BCDL: a high speed balanced DPL for FPGA with global precharge and no early evaluation. In: DATE 2010, pp. 849–854. IEEE Computer Society (2010)

19. Nikova, S., Rijmen, V., Schläffer, M.: Secure hardware implementation of nonlinear functions in the presence of glitches. J. Cryptol. **24**(2), 292–321 (2011)

20. Popp, T., Kirschbaum, M., Zefferer, T., Mangard, S.: Evaluation of the masked logic style MDPL on a prototype chip. In: Paillier, P., Verbauwhede, I. (eds.) CHES 2007. LNCS, vol. 4727, pp. 81–94. Springer, Heidelberg (2007). doi:10.1007/978-3-540-74735-2_6

21. Popp, T., Mangard, S.: Masked dual-rail pre-charge logic: DPA-resistance without routing constraints. In: Rao, J.R., Sunar, B. (eds.) CHES 2005. LNCS, vol. 3659, pp. 172–186. Springer, Heidelberg (2005). doi:10.1007/11545262_13

22. Sauvage, L., Nassar, M., Guilley, S., Flament, F., Danger, J., Mathieu, Y.: DPL on Stratix II FPGA: what to expect? In: ReConFig 2009, pp. 243–248. IEEE Computer Society (2009)

23. Schneider, T., Moradi, A.: Leakage assessment methodology. In: Güneysu, T., Handschuh, H. (eds.) CHES 2015. LNCS, vol. 9293, pp. 495–513. Springer, Heidelberg (2015). doi:10.1007/978-3-662-48324-4_25

24. Standaert, F.-X., Malkin, T.G., Yung, M.: A unified framework for the analysis of side-channel key recovery attacks. In: Joux, A. (ed.) EUROCRYPT 2009. LNCS, vol. 5479, pp. 443–461. Springer, Heidelberg (2009). doi:10.1007/978-3-642-01001-9_26

25. Suh, G.E., Devadas, S.: Physical unclonable functions for device authentication and secret key generation. In: Proceedings of the 44th Design Automation Conference - DAC 2007, San Diego, CA, USA, 4–8 June 2007, pp. 9–14. IEEE Computer Society (2007)

26. Sunar, B., Martin, W.J., Stinson, D.R.: A provably secure true random number generator with built in tolerance to active attacks. IEEE Trans. Comput. **56**(1), 109–119 (2007)

27. Tiri, K., Akmal, M., Verbauwhede, I.: A dynamic and differential CMOS logic with signal independent power consumption to withstand differential power analysis on smart cards. ESSCIRC **2002**, 403–406 (2002)

28. Tiri, K., Verbauwhede, I.: A logic level design methodology for a secure DPA resistant ASIC or FPGA implementation. In: DATE 2004, pp. 246–251. IEEE Computer Society (2004)

29. Wild, A., Moradi, A., Güneysu, T.: Evaluating the duplication of dual-rail precharge logics on FPGAs. In: Mangard, S., Poschmann, A.Y. (eds.) COSADE 2014. LNCS, vol. 9064, pp. 81–94. Springer, Cham (2015). doi:10.1007/978-3-319-21476-4_6

30. Wild, A., Moradi, A., Guneysu, T.: GliFreD: Glitch-Free Duplication - towards power-equalized circuits on FPGAs. IEEE Trans. Comput. (2017). http://doi.ieeecomputersociety.org/10.1109/TC.2017.2651829

31. Xilinx: UG472 7 series FPGAs clocking resources user guide, June 2015

32. Yu, P., Schaumont, P.: Secure FPGA circuits using controlled placement and routing. In: CODES+ISSS 2007, pp. 45–50 (2007)

On the Construction of Side-Channel Attack Resilient S-boxes

Liran Lerman[1], Nikita Veshchikov[1(✉)], Stjepan Picek[2],
and Olivier Markowitch[1]

[1] Quality and Security of Information Systems, Département d'Informatique,
Université libre de Bruxelles, Brussels, Belgium
nveshchi@ulb.ac.be
[2] KU Leuven ESAT/COSIC and IMEC, Kasteelpark Arenberg 10,
3001 Leuven-heverlee, Belgium

Abstract. Side-channel attacks exploit physical characteristics of implementations of cryptographic algorithms in order to extract sensitive information such as the secret key. These physical attacks are among the most powerful attacks against real-world crypto-systems. In recent years, there has been a number of proposals how to increase the resilience of ciphers against side-channel attacks. One class of proposals concentrates on the intrinsic resilience of ciphers and more precisely their S-boxes. A number of properties has been proposed such as the transparency order, the confusion coefficient and the modified transparency order. Although results with those properties confirm that they are (to some extent) related with the S-box resilience, there is still much to be investigated. There, the biggest drawback stems from the fact that even S-boxes with the best possible values of those properties have only slightly improved side-channel resistance. In this paper, we propose to construct small sized S-boxes based on the results of the measurements of the actual physical attacks. More precisely, we model our S-boxes to be as resilient as possible against non-profiled and profiled physical attacks. Our results highlight that we can design 4×4 and 5×5 S-boxes that possess increased resistance against various real-world attacks.

Keywords: S-box construction · Lightweight cryptography · Genetic algorithms · Side-channel analysis · Correlation power analysis · Template attacks

1 Introduction

The pervasive presence of interconnected lightweight devices has lead to a massive interest in security features provided among others by cryptography. For decades, designers estimated the security level of a cryptographic algorithm independently of its implementation in a cryptographic device. However, since the publication on implementation attacks in the mid-nineties, the physical attacks have become an active research area by analysing physical leakages measured on

© Springer International Publishing AG 2017
S. Guilley (Ed.): COSADE 2017, LNCS 10348, pp. 102–119, 2017.
DOI: 10.1007/978-3-319-64647-3_7

the target cryptographic device [1]. The rationale is that there is a relationship between the manipulated data (e.g., the secret key), the executed operations and the physical properties observed during the execution of the cryptographic algorithm by a device. A side-channel attack (SCA) represents a process that exploits leakages in order to extract sensitive information such as the key. This paper analyses the non-linear part (called S-boxes) of ciphers, which is often targeted by implementation attacks. Note that other functions could be analysed, which constitutes an interesting future work.

Three categories of countermeasures against physical attacks exist: masking, hiding and leakage resiliency. Masking blinds sensitive operations (manipulating key-related information) using random numbers and hiding minimises the signal-to-noise ratio in the leakage by shuffling operations or adding a noise generator. Leakage resiliency regularly updates the secret key in order to prevent the aggregation of information from several leakages. The extreme constraints of Radio-Frequency Identification based on chip (in short RFID tags) as well as the hostile environments in which the RFID tags are manipulated raise the need of lightweight countermeasures against side-channel attacks minimising the power consumption, the clock cycles, and the used random numbers.

In 2014, Picek et al. generated S-boxes of various sizes providing high resistance to physical attacks without the need of extra random numbers (like masking or shuffling) during the execution of cryptographic primitives [2]. More precisely, they used genetic programming and genetic algorithms to evolve S-boxes minimising the transparency order metric (that relates to the side-channel resistance of the S-boxes). The main advantage of these approaches (compared to exhaustive search) lies in the execution time of the research: exhaustive search generates 2^{2^n} different $n \times n$ S-boxes[1] while genetic algorithms optimise this research in an automatic way. At the same year, Picek et al. obtained two S-boxes of sizes 4×4 and 8×8 by exploiting genetic algorithms optimising the confusion coefficient (representing another metric related to the side-channel resistance of the S-boxes) [3]. Finally, Picek et al. built a 4×4 S-box using genetic algorithms optimising an improved transparency order [4].

Our Contributions. The success probability (also known as success rate) represents the probability of an adversary to extract the sensitive information from physical leakages measured on the target device. This (security) metric provides the strength of a strategy against an implementation. Surprisingly, all the previous works generated S-boxes by optimising the properties (e.g., confusion coefficient) related with the side-channel resilience, but up to now no one explored whether it is possible to design S-boxes with the success rate as a metric, i.e., by obtaining it already in the design phase of the S-boxes and not only a posteriori.

In this paper, we shed new insights on the generation of S-boxes by focusing on a security metric that is directly related to the strength of a side-channel adversary. More precisely, we provide several S-boxes minimising the success probability of two well-known side-channel attacks called (non-profiled) correlation power analysis and (profiled) template attacks. Correlation power analysis represents the

---

[1] $(2^n)!$ if we only consider permutations.

state-of-the-art when considering non-profiled attacks while template attacks are the most powerful physical attacks from an information theoretic point of view.

Furthermore, we present the first 5×5 S-boxes minimising the security metric that can be directly exploited in cryptographic primitives.

Differing from previous works, we also consider S-boxes where their inverse has good resilience against side-channel attacks. This approach is of high importance since the attacker can concentrate either on the first round and the plaintext or the last round and the ciphertext (in which case the inverse of the S-box is targeted) during the side-channel attack phase.

Finally, to depict the increased resilience of our new S-boxes, we also design S-boxes that provide the worst resilience regarding the considered physical attacks as well as the considered devices. Following the kleptography concept [5], these results highlight that malicious designers of cryptographic primitives can weaken a target device (or family of devices) by carefully selecting an S-box (that still has good cryptographic properties) for the cipher.

We provide all the new (4×4 and 5×5) S-boxes in Table 1 and in Table 2 taking into account respectively non-profiled attacks and profiled attacks. Our results can be of major value (1) for industry that wants to (easily and quickly) increase the protection of the executed cryptographic primitive according to the considered device, and (2) for the scientific community that can pursue research on lightweight countermeasures with different optimisation goals.

Cautionary Note. This paper relates to the protection of one low-cost party in a communication protocol that involves (for example) an RFID tag and an RFID reader. More precisely, we assume that an RFID tag (having strong cost constraints) requires lightweight countermeasures (provided in this paper) while the RFID readers (implementing the same cryptographic primitive or its inverse) can be protected with more expensive means such as masking and shuffling. Indeed, in this paper we provide S-boxes having good resilience against side-channel attacks when we implement these S-boxes in a specific device (such as an RFID tag) while this protection could be undermined (by a physical attack) when they are implemented in other devices (such as RFID readers) having different physical characteristics. Note however that our approach can be generalised to protect several devices at the same time, which constitutes a future work.

Outline. This paper is organised as follows. Section 2 starts with the basic notions of relevant cryptographic properties and side-channel attacks. Moreover, this section discusses about the evaluation procedure from the side-channel attacks perspective. Next, Sect. 3 presents our search strategy as well as the obtained results. Finally, Sect. 4 provides conclusions of the paper and gives several directions for future works.

2 Background

2.1 Cryptographic Properties of S-boxes

Let $\mathbb{F}_2^n$ be the vector space that contains all the n-bit binary vectors. Let F be a substitution box (denoted S-box). S-boxes provide the confusion property in cryptographic primitives by substituting values from $\mathbb{F}_2^n$ to $\mathbb{F}_2^m$ (denoted as an S-box $n \times m$ as well as an (n, m)-function). The S-box can be seen as a vector of m Boolean functions $[\mathsf{F}_1, \mathsf{F}_2, ..., \mathsf{F}_m]$ where each Boolean function represents a mapping from $\mathbb{F}_2^n$ to $\mathbb{F}_2$.

We denote the inner product of two vectors $a = (a_0, a_1, \ldots, a_{n-1})$ and $b = (b_0, b_1, \ldots, b_{n-1})$ as $a \cdot b$ (which is equal to $a \cdot b = \oplus_{i=1}^n a_i b_i$). The Hamming weight of a vector $a \in \mathbb{F}_2^n$ (denoted $\mathsf{HW}(a)$) represents the number of non-zero positions in the vector.

The **nonlinearity** N_F of an (n, m)-function F is equal to the minimum non-linearity of all non-zero linear combinations $v \cdot \mathsf{F}$, with $v \neq 0$, of its coordinate functions F_i, i.e.:

$$N_F = 2^{n-1} - \frac{1}{2} \max_{\substack{a \in \mathbb{F}_2^n \\ v \in \mathbb{F}_2^{m*}}} \|W_\mathsf{F}(a, v)\|, \tag{1}$$

where $\|x\|$ symbolises the absolute value of x, and $W_\mathsf{F}(a, v)$ represents the Walsh-Hadamard transform of F that is equal to:

$$W_\mathsf{F}(a, v) = \sum_{x \in \mathbb{F}_2^m} (-1)^{v \cdot \mathsf{F}(x) + a \cdot x}, \quad a \in \mathbb{F}_2^n, v \in \mathbb{F}_2^m. \tag{2}$$

The nonlinearity N_F of any (n, n)-function F must satisfy the inequality:

$$N_F \leq 2^{n-1} - 2^{\frac{n-1}{2}}. \tag{3}$$

Let F be a function from $\mathbb{F}_2^n$ into $\mathbb{F}_2^n$ and $a, b \in \mathbb{F}_2^n$. We denote:

$$\mathsf{D}(a, b) = \{x \in \mathbb{F}_2^n : \mathsf{F}(x + a) + \mathsf{F}(x) = b\}. \tag{4}$$

$\delta(a, b)$ denotes the cardinality of $\mathsf{D}(a, b)$ and

$$\delta_F = \max_{a \neq 0, b} \delta(a, b). \tag{5}$$

Almost Bent (AB) functions contain an equality in Eq. (3) while when a function is differentially 2-uniform, it is called Almost Perfect Nonlinear (APN) function. Every AB function is also APN, but the other direction does not hold in general. AB functions exist only in an odd number of variables, while APN functions also exist for an even number of variables. Furthermore, the maximal algebraic degree of AB functions equals $(n + 1)/2$ while for the inverse APN equals $n - 1$. We refer to the following papers [6,7] for the interested readers about the theory of Boolean functions and S-boxes.

Leander and Poschmann define optimal 4-bit S-boxes as being bijective, with the minimal possible linearity (or, maximal possible nonlinearity) and with a minimal differential uniformity. For optimal 4×4 S-boxes, both N_F and the differential uniformity are equal to 4 [8]. PRESENT considers S-boxes from the 16 classes suggested by Leander and Poschmann [8], but some lightweight ciphers use 4×4 S-boxes with different cryptographic conditions. For instance, the authors of the PRINCE cipher impose several additional criteria on the 4×4 S-box and therefore they accept only 8 out of the 16 classes [9].

When considering 5×5 S-boxes, the cryptographic properties one can obtain differ with regards to the choice of the S-box. As a first example, we consider the Keccak S-box for which both the nonlinearity and differential uniformity are equal to 8 [10]. Note that those values are relatively far from the optimal ones. Furthermore, the algebraic degree of Keccak is low, and it actually equals the minimal possible algebraic degree for a nonlinear function. However, the Keccak S-box has an extremely efficient hardware implementation. The S-box used in Ascon [11] is an affine transformation of the Keccak S-box in order to remove the fixed points and to increase the differential branch number value. On the other hand, the PRIMATEs S-box [12] is based on an almost bent permutation, which means it has a nonlinearity equal to 12 and a differential uniformity equal to 2, while the algebraic degree equals 2.

2.2   Side-Channel Attacks

We assume that the adversary wants to retrieve the secret key used when the cryptographic device (that executes a known encryption algorithm) encrypts known plaintexts and provides known ciphertexts. In order to find the key, the adversary targets a set of key-related information (called the *target intermediate values*) with a *divide-and-conquer approach*. The divide-and-conquer strategy extracts information on separate parts of the key (e.g., the adversary extracts each byte of the key independently) and then combines the results in order to get the full secret key. In the following, we systematically use the term key to denote the target of our attacks, though, in fact, we address one byte at a time.

During the execution of the encryption algorithm, the cryptographic device processes a function F (e.g., the S-box of the block-cipher AES)

$$\mathsf{F} \colon \mathcal{P} \times \mathcal{K} \to \mathcal{Y} \tag{6}$$
$$y_i = \mathsf{F}_k(p),$$

that outputs the target intermediate value y_i and where $k \in \mathcal{K}$ is a key-related information (e.g., one byte of the secret key), $p \in \mathcal{P}$ represents information known by the adversary (e.g., one byte of the plaintext), and i is a number related to k and p.

Physical Characteristics. Let ${}^j T_i$ be the j-th leakage (also known as trace) measured when the device manipulates the target value y_i. In the following, we represent each leakage with a vector of real values (of length n_s) measured at different instants on the analysed device. We denote ${}^j_t T_i$ the j-th leakage (associated to the target value y_i) measured at time t such that:

$$ {}^j_t T_i = {}_t\mathsf{L}\left(\mathsf{F}_k\left(p\right)\right) + {}^j_t \epsilon_i, \tag{7}$$

where ${}^j_t \epsilon_i \in \mathbb{R}$ is the noise of the trace ${}^j_t T_i$ following for example a Gaussian distribution with zero mean, and ${}_t\mathsf{L}$ is the (deterministic) leakage function at time t. The function $_t\mathsf{L}$ can be *linear* (e.g., the weighted sum of each bit of the input value) or *nonlinear* (e.g., the weighted sum of products of bits of the input value). Evaluators often model linear leakage functions as the Hamming weight of the manipulated value y_i for software implementations. A *side-channel attack* is a process during which an attacker analyses leakages measured on a target device in order to extract information on the secret value. Several side-channel attacks exist such as the Correlation Power Analysis (CPA) [13] and Template Attack (TA) [14]. We refer to the work of Chakraborty et al. [15] for a detailed description of the improved transparency order metric and to the work of Fei et al. introducing the confusion coefficient that evaluates the resistance of S-boxes against side-channel attacks in a theoretical point of view [16,17].

Correlation Power Analysis. CPA recover the secret key from a cryptographic device by selecting the key that maximises the dependence between the actual leakage and the estimated leakage based on the assumed secret key. More precisely, CPA selects the secret key $\widehat{k}$ such that:

$$\widehat{k} \in \arg\max_{k \in \mathcal{K}} \left\| \rho\left(\widehat{\mathcal{T}}_{(k)}, \mathcal{T}\right)\right\|, \tag{8}$$

where $\rho(\mathcal{X}, \mathcal{Y})$ represents the Pearson's correlation between 2 lists $\mathcal{X}$ and $\mathcal{Y}$, and:

- $\mathcal{T} = \left[{}^1T, ..., {}^{N_a}T\right]$ represents a list of N_a traces measured when the target device manipulates the S-box (where ${}^iT$ denotes the i-th measurement on the target device and N_a is the number of attack traces), and
- $\widehat{\mathcal{T}}_{(k)} = \left[\widehat{\mathsf{L}}(\mathsf{F}(k \oplus p_{[1]})), ..., \widehat{\mathsf{L}}(\mathsf{F}(k \oplus p_{[N_a]}))\right]$ refers to a list of estimated leakages (with a leakage model $\widehat{\mathsf{L}}$) parametrised with the output of the S-box combining (with the exclusive-or operation denoted $\oplus$) an estimated key k and known plaintext $p_{[i]}$ associated to ${}^iT$.

Template Attacks. (Gaussian) Template attacks assume that $\Pr\left[{}^jT_i \mid y_i\right]$ follows a Gaussian distribution $\mathcal{N}(\widehat{\mu}_i, \widehat{\Sigma}_i)$ for each value y_i where $\widehat{\mu}_i \in \mathbb{R}^{n_s}$ and $\widehat{\Sigma}_i \in \mathbb{R}^{n_s \times n_s}$ are respectively the sample mean and the sample covariance matrix of the traces associated to y_i. In what follows we assume that the noise is independent of y_i in unprotected contexts. This property allows to estimate the same physical noise (represented by Σ) for all the target values.

During the attack step, the adversary classifies the list $\left[{}^1T, ..., {}^{N_a}T\right]$ by using:

$$\widehat{k} \in \arg\max_{k \in \mathcal{K}} \prod_{j=1}^{N_a} \Pr\left[{}^jT \mid k, p_j\right] \times \Pr\left[k, p_j\right], \tag{9}$$

$$\approx \arg\max_{k \in \mathcal{K}} \prod_{j=1}^{N_a} \widehat{\Pr}\left[{}^jT \mid y_i = \mathsf{F}_k(p_j); \widehat{\theta}_i\right] \times \widehat{\Pr}\left[y_i = \mathsf{F}_k(p_{[j]})\right], \tag{10}$$

where $\widehat{\theta}_i$ denotes the two parameters $\{\widehat{\mu}_i, \widehat{\Sigma}_i\}$.

The designers of cryptographic devices measure the resistance of an implementation against a physical attack by using (among others) the first order Success Rate (SR) [18]. The success rate (also known as the success probability) represents the probability that the physical attack extracts the secret key.

3   Experiments

Tables 1 and 2 display all the generated 4×4 and 5×5 S-boxes taking into account respectively the correlation power analysis and the template attacks. We note that all presented 4×4 S-boxes also have maximal possible algebraic degree that is equal to 3. For the 5×5 size, algebraic degree varies from 2 to 4 where we note that for all optimal S-boxes it equals 2 (since optimal 5-bit S-boxes are actually AB functions, meaning that the algebraic degree is upper bounded with $\frac{n+1}{2}$ that equals to 3). Note that our new 5×5 S-boxes have better nonlinearity and differential uniformity values than Keccak or Ascon, but we can easily adapt our strategy to output S-boxes with any combinations of values.

Table 1. Properties of evolved S-boxes when considering correlation power analysis. Values of S-boxes are given in hexadecimal format. Strategy F represents S-boxes optimised for the success rate. Strategy $F + I$ represents S-boxes optimised in the forward direction as well as their inverse. Strategy K represents S-boxes optimised for the kleptography concept.

Size	Name	N_F	δ_F	Strategy	S-box
4×4	Evolved$_{SR1}$	4	4	F	2,4,8,0,F,B,7,D,6,5,E,3,1,9,C,A
	Evolved$_{SR2}$	4	4	$F+I$	F,E,0,A,1,8,9,B,7,6,4,C,5,2,3,D
	Evolved$_K$	4	4	K	0,F,1,9,B,5,8,2,E,3,C,6,D,4,A,7
5×5	Evolved$_{SR1}$	8	6	F	1E,07,15,02,0E,09,19,04,17,12,0B,08,1C,0A,1D,06 0C,1B,05,0D,00,14,18,1F,10,13,11,1A,01,16,03,0F
	Evolved$_{SR2}$	8	6	$F+I$	15,02,1F,0A,19,11,1B,12,08,0E,0C,07,06,0F,10,16 13,00,17,09,1D,18,0D,03,04,1A,14,1C,05,1E,01,0B
	Evolved$_{SR3}$	10	6	$F+I$	1D,15,03,02,1C,0A,0C,09,11,10,1F,0D,18,14,19,16 06,12,0F,17,01,04,13,1B,0B,07,0E,05,1A,1E,00,08
	Evolved$_{SR4}$	10	4	$F+I$	0A,1C,01,13,04,08,12,10,06,05,03,0D,02,18,09,00 0F,1B,1A,11,14,1D,0B,0E,16,07,15,19,0C,17,1E,1F
	Evolved$_{SR5}$	8	6	F	04,17,1C,18,07,00,12,19,0E,14,10,15,06,13,1F,08 1A,11,0C,0B,05,1E,0F,01,02,1D,1B,09,0D,03,0A,16
	Evolved$_{SR6}$	8	4	F	09,05,1E,1C,0D,16,14,06,07,1D,01,10,03,02,13,1F 1B,15,08,18,04,00,0F,1A,0A,12,0B,0E,19,17,11,0C
	Evolved$_{SR7}$	10	4	F	1B,13,17,16,0B,0F,0D,1A,03,06,01,09,02,14,08,11 10,12,00,0A,1F,18,05,0C,1D,1C,04,07,0E,1E,15,19
	Evolved$_{SR8}$	12	2	F	00,0E,1C,16,19,01,0D,11,13,08,02,1D,1A,17,03,0A 07,0B,10,18,04,1E,1B,05,15,0C,0F,12,06,09,14,1F
	Evolved$_{SR9}$	12	2	$F+I$	00,07,0E,0B,1C,10,16,18,19,04,01,1E,0D,1B,11,05 13,15,08,0C,02,0F,1D,12,1A,06,17,09,03,14,0A,1F
	Evolved$_K$	8	4	K	15,07,06,03,18,0E,04,01,0C,05,0A,16,1F,1D,19,13 12,0F,11,1B,09,1A,17,10,08,0B,00,14,02,1C,1E,0D

Table 2. Properties of S-boxes Evolved for ATMega 328 microcontroller using template attacks. Values of S-boxes are given in hexadecimal format. Strategy F represents S-boxes optimised for the success rate. Strategy $F + I$ represents S-boxes optimised in the forward direction as well as their inverse.

Size	Name	N_F	δ_F	Strategy	S-box
4×4	EvolvedTA$_{SR1}$	4	4	F	0,5,7,C,A,6,2,4,9,8,B,F,D,E,1,3
	EvolvedTA$_{SR2}$	4	4	$F+I$	4,2,D,E,B,1,6,5,7,8,3,A,F,0,C,9
	EvolvedTA$_{SR3}$	4	4	F	9,4,5,D,3,0,1,F,B,2,C,7,E,8,A,6
	EvolvedTA$_{SR4}$	4	4	$F+I$	4,0,6,7,1,2,A,F,5,3,C,E,D,9,B,8
5×5	EvolvedTA$_{SR1}$	8	6	F	1F,15,01,0C,14,1D,12,00,1A,09,08,17,05,0E,0B,0d 04,18,1B,0A,13,11,06,1E,10,19,16,02,0F,07,03,1c
	EvolvedTA$_{SR2}$	10	6	$F+I$	07,14,1D,11,12,02,06,13,19,0F,09,0C,1C,15,0A,08 01,0B,1F,0D,03,17,1E,05,04,1B,0E,00,1A,18,10,16
	EvolvedTA$_{SR3}$	12	2	F	1F,01,02,1A,04,1B,15,0C,08,1E,17,07,0B,1C,18,09 10,0D,1D,06,0F,13,0E,14,16,03,19,0A,11,05,12,00
	EvolvedTA$_{SR4}$	12	2	$F+I$	1F,01,02,1E,04,0E,1D,15,08,13,1C,05,1B,14,0B,06 10,0F,07,1A,19,12,0A,03,17,0D,09,11,16,18,0C,00
	EvolvedTA$_{SR5}$	10	4	F	01,13,11,16,0F,03,08,10,0B,1A,02,1B,0C,1E,17,12 19,0D,14,00,05,04,18,07,1F,1D,06,15,0A,1C,0E,09
	EvolvedTA$_{SR6}$	8	4	$F+I$	1F,15,18,05,01,06,08,0B,12,02,17,0D,03,1C,04,16 0F,1B,09,13,0C,11,1E,1A,00,19,0A,07,1D,14,0E,10

In order to compare our generated S-boxes we used the following existing S-boxes:

- 4×4 S-boxes: Evolved$_{CC}$ [3], Evolved$_{TO}$ [4], Klein [19], PRESENT [20] and PRINCE [9];
- 5×5 S-boxes: ASCON [11], Keccak (Ketje, Keyak) [21] and PRIMATE [22].

The S-boxes Evolved$_{CC}$ and Evolved$_{TO}$ were also generated using genetic algorithms while taking into account theoretical metrics (i.e., the confusion coefficient and the modified transparency order) in order to estimate their resistance against side-channel attacks.

3.1 Search Strategy

We use a genetic algorithm (GA) exploiting simple variation operators and solution encodings. We follow this line of research in an effort to make our search process as fast as possible as well as to make comparison with previous works as fair as possible. We encode solutions as lists of values between 0 and $2^n - 1$ where n is the size of the S-box. Note that this representation (i.e., permutation encoding) is highly efficient since this ensures that solutions are bijections (which is a necessary condition we enforce on our S-boxes).

We use the tournament selection mechanism in order to avoid the need to tune the crossover rate parameter. We work with the 3-tournament selection which is the option that offers the fastest convergence [23]. This mechanism selects three solutions randomly and discards the worst solution. Then, from the remaining two solutions, the crossover operator creates a new offspring. For variation operators, we use the Toggle mutation and the Order crossover. In the Toggle mutation we randomly select two values and swap them. The Order crossover (OX) works by first randomly selecting two crossover points and copying everything between those two points from the first parent to the offspring. Then, starting from the second crossover point in the second parent, the unused numbers are copied in the order they appear in that parent [23]. The initial population is created uniformly at random and the population size equals 100. As a termination criterion, we use the number of evaluations without improvement, which we set here to 100 generations.

In our experiments, we maximise the nonlinearity while minimising the differential uniformity as well as the success probability (denoted SR), hence the subtraction from 2^n value and 1, respectively:

$$\text{fitness} = N_F + (2^n - \delta_F) + (1 - \text{SR}). \tag{11}$$

We give equal weights to both N_F and δ_F since our experiments show there is no statistically significant difference in those two cases[2].

Regarding the complexity of our search strategy, on average one generation (100 individuals) needs around 1 s to evolve. In that estimation we include the cost of the evaluation of the cryptographic properties, but not the cost of the evaluation of the attack strategy. We note that although here we work with GA, our methodology is not exclusive for that algorithm, but it could work with any other heuristics that supports the permutation encoding. Naturally, it is to be expected that in such case one could also need to change the fitness function and the stopping criterion. For further details about genetic algorithms, we refer readers to the work of Eiben and Smith [23].

3.2   Results for Correlation Power Analysis

We generated synthetic leakages by considering that the leakage function equals to the Hamming weight and the leakage model (of the adversary) also equals to the Hamming weight (i.e., the adversary has a perfect knowledge on how the device leaks information). We use the same level of noise of 0.5 variance representing a signal-to-noise ratio (1) of 2.13 when considering 4×4 S-boxes, and (2) of 2.58 when considering 5×5 S-boxes. It is worth to note that the order of the (generated) S-boxes sorted by the resistance against SCA are not influenced by the signal-to-noise ratio.

---

[2] Note that in general case of a fitness function it is possible to sacrifice one parameter in order to boost another. However, in our case it is impossible to sacrifice the nonlinearity N_F in order to improve the success rate due to the fact that $N_F \in \mathbb{N}$ and $\text{SR} \in \mathbb{R}$ and $0 < \text{SR} < 1$. In other words the minimal step in values of N_F is 1, while 1 is the maximum increase that the SR can get, thus, the whole fitness will decrease if N_F decreases while boosting the SR.

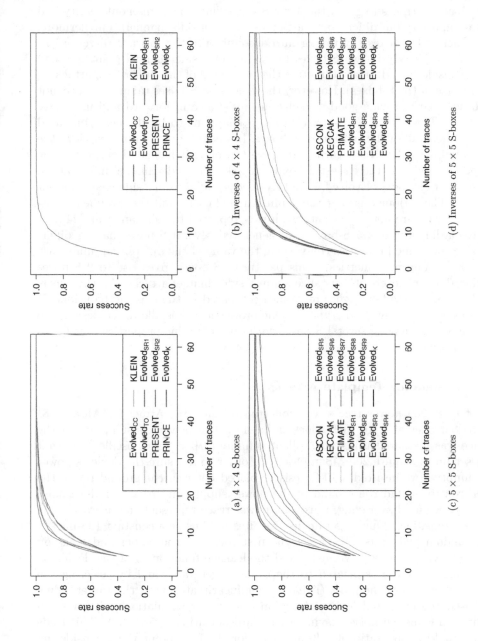

Fig. 1. Success rate of correlation power analysis on 4×4 and 5×5 S-boxes.

Figure 1 provides the success rate of CPA on the (three) new generated S-boxes as well as their inverses. The first observation is that the nonlinearity of an S-box and its delta uniformity are not the (only) metrics impacting side-channel attacks (e.g., all the 4 × 4 S-boxes have the same nonlinearity and delta uniformity but differ from the point of view of side-channel). Furthermore, the generated S-boxes (by taking into account only the forward direction) as well as the already known S-boxes are weak (in a side channel point of view) when considering adversary targeting the last round of the cipher (i.e., attacking the inverse of the S-boxes). However, the generated S-boxes taking into account such adversary provide good side-channel resistance in forward and in inverse direction. The new 4 × 4 S-box Evolved$_{SR2}$ happens to be the best generated S-box among all of the considered S-boxes. In a kleptography point of view, the generated 4 × 4 Evolved$_K$ turns out to be the best: it has good cryptographic properties and it is the easiest S-box to attack using side-channel information. Note that the S-boxes Evolved$_{CC}$ and Evolved$_{TO}$ differ from a side-channel point of view. The rationale is that the confusion coefficient and the modified transparency order are not equivalent, as already reported by Lerman et al. [24].

Regarding the 5 × 5 S-boxes, we generated several S-boxes having different cryptographic properties (by varying the value of the differential uniformity and the nonlinearity metrics). This palette of S-boxes gives rise to 9 S-boxes having different levels of resistance against side-channel attacks. All the generated S-boxes provide a higher resistance compared to the existing (considered) S-boxes while having good cryptographic properties. This allows the designer to choose S-boxes among several S-boxes with cryptographic properties that fit his requirements.

3.3 Results for Template Attacks

A set of 80 000 power traces was collected on an 8-bit Atmel (ATMega 328) microcontroller at a 16 MHz clock frequency. The power consumption of the device was measured using an Agillent Infiniium 9 000 Series oscilloscope that was set up to acquire 200 MSamples/s. In order to measure the device's power consumption we inserted a 10 Ω resistor placed between the ground pin of the microcontroller and the ground of the power supply. In order to reduce noise in traces we used averaging, thus each power trace represents an average of 64 single acquisitions. Our target device executes AES using a constant 128-bit key and random plaintexts. We target the first round of the cipher and focus on the first byte of the key. We extracted the leakage function $_tL$ of the device by averaging all traces associated to the same target value and by selecting the 8 instants that are the most (linearly) nonlinearityated with the target value. We used the extracted leakage function during our simulations with a small additional Gaussian noise[3] having a standard deviation of 5×10^{-6}. This leads to a signal-to-noise ratio of 0.40 and 0.37 for the best point when considering

---

[3] A small amount of noise is necessary in order to avoid numerical issues during template attacks.

respectively an 4×4 S-box and an 5×5 S-box. It is worth to note that we do not claim in this paper that this profiled attack represents the optimal physical attack against the analysed implementation. Other profiled attacks could provide higher success rates [25]. In other words, our purpose here is to provide S-boxes resilient against *chosen* profiled attacks.

Figure 2 shows the success rate of template attacks on the considered S-boxes. We can notice that Fig. 2a and b show results similar to the results that we obtain in the previous section: when we consider well-known S-boxes or newly generated S-boxes (while considering only the forward strategy) the corresponding inverse S-boxes show them weaker against side-channel attacks. The 4×4 EvolvedTA$_{SR2}$ S-box that was generated by taking into account the S-box and its inverse gives the best result: it is as good as PRESENT S-box in terms of its inverse and it is one of the best among well known 4×4 S-boxes (in the forward direction) with the exception of 4×4 EvolvedTA$_{SR1}$ that was designed to be good in the forward direction (but not as an inverse). In terms of 5×5 S-boxes, 5×5 EvolvedTA$_{SR5}$ provides the best result in forward direction. In the inverse direction, EvolvedTA$_{SR6}$ outperforms all the known S-boxes. Note that it is still difficult to create resilient S-boxes while having good cryptographic properties and being better than existing S-boxes in both forward and inverse directions. However, we deem that we can still create a more resilient 5×5 S-box against template attacks since 5×5 S-boxes provide a large set of possible solutions.

3.4 Discussion

The previous sections report the improvement of the success probability of physical attacks on 4×4 and 5×5 S-boxes. Our results highlight that the improvement is more significant for the 5×5 S-boxes than for the 4×4 S-boxes. The reason relies on the fact that 5×5 S-boxes have a wider range of obtainable values for the success rate property when compared with 4×4 S-boxes.

Figure 3 provides the success rate on each S-box targeted by an adversary exploiting the plaintext (by attacking the forward S-box used in the first round of the cryptographic primitive) and the ciphertext (by attacking the inverse S-box used in the last round of the primitive). Plots on these figure correspond to the maximum of the two attacks (between the attack on an S-box and on its inverse). The results highlight the usefulness of our approach by providing new S-boxes outperforming well known S-boxes in several contexts. More precisely, the 4×4 Evolved$_{SR2}$ and the 5×5 Evolved$_{SR2}$ S-boxes provide the best results against correlation power analysis while the 4×4 EvolvedTA$_{SR4}$ and the 5×5 EvolvedTA$_{SR6}$ S-box provide the best results against template attacks.

Note also that all our results report the success rate of adversaries targeting *one* nibble of the key. It is worth to note that, in practice, adversaries extract the *full* secret key. As a result, a small-scale decrease of the first order success rate of an attack on one nibble leads to a significant reduction of the success probability of the attack on the full key. Therefore, designers of cryptographic primitives should consider optimisation methods minimising the success rate of physical attacks against S-boxes. As an example, let us take two 4×4 S-boxes

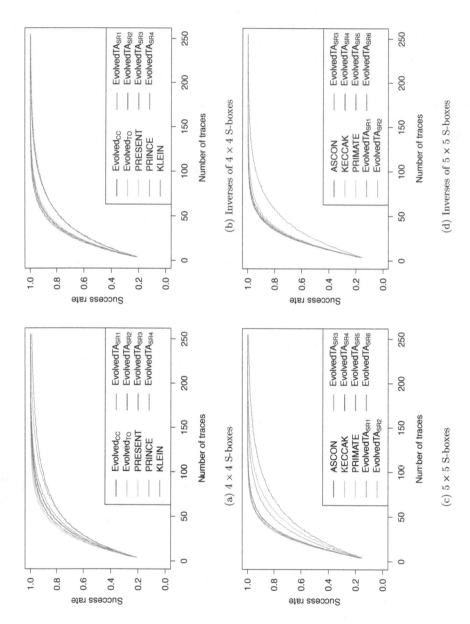

(a) 4×4 S-boxes

(b) Inverses of 4×4 S-boxes

(c) 5×5 S-boxes

(d) Inverses of 5×5 S-boxes

Fig. 2. Success rate of template attacks on 4×4 and 5×5 S-boxes.

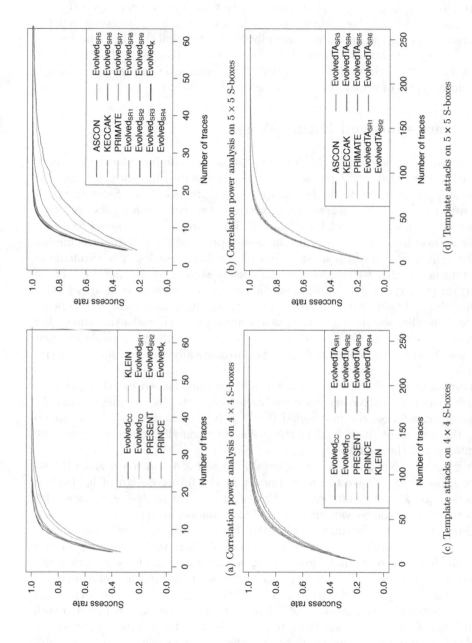

(a) Correlation power analysis on 4 × 4 S-boxes

(b) Correlation power analysis on 5 × 5 S-boxes

(c) Template attacks on 4 × 4 S-boxes

(d) Template attacks on 5 × 5 S-boxes

Fig. 3. Maximum success rate between attacks on the first round (S-box) and last round (inverse of the S-box) of an algorithm.

with similarly close success rates: Evolved$_K$ and the S-box of PRESENT. During a CPA using 15 attack traces, Evolved$_K$ results in success rate of 0.9820 while the S-box of PRESENT gives the success rate of 0.9605 (difference of about 0.02). However, it is important to note that this success rate corresponds to an attack on one 4-bit nibble. During an attack on a full cipher with 80-bit key, the adversary repeats the attack on each nibble (i.e., 20 times). Thus, the success rate of a complete attack results in the success rate of 0.4466 on the PRESENT S-box and 0.6954 in case of Evolved$_K$ which is a significant increase even though the success rates of attacks on one nibble are very close.

4   Conclusions and Future Work

In this paper, we investigate the design of S-boxes containing inherent resilience against various real-world physical attacks. The main difference between our work and the previous works lies in the design process of the S-boxes: previous works design S-boxes optimising metrics (e.g., confusion coefficient) that (according to the authors of these metrics) relate to the side-channel resilience while we take into account (during the design phase of the S-boxes) the quality of the generated S-boxes against actual physical adversaries. The rationale of our approach is that we remove the unnecessary step of connecting the value of a certain property (e.g., confusion coefficient) to a certain type of attack. Our results also highlight that such measures (e.g., confusion coefficient) can indicate the resilience but should not be used as a definitive guide in order to estimate the success probability of physical attacks [3, 24]. As a result, we provide the first S-boxes in which the countermeasure is automatically tailored for the device used by the implementers.

Our outcomes also generalise the results of previous works (that focus on the case where the adversary knows only the plaintexts) by considering that the adversary can target the first round (i.e., the S-box) as well as the last round (i.e., the inverse of the S-box) of the primitive when the adversary knows the plaintexts and the ciphertexts.

We conduct our analysis for S-boxes of sizes 4×4 and 5×5 since (1) we deem those sizes to have the most impact in the future design of lightweight ciphers, and (2) our results confirm that it is possible to design S-boxes with better resilience against various classes of side-channel attacks.

Several directions for future works exist. For example, an interesting perspective would be to work with involutive S-boxes. Ciphers with involutive S-boxes have smaller area cost than those having separate S-boxes for encryption and decryption. The main difficulty of this future work lies in the definition of a new search strategy in order to stay only in the involutive S-boxes search space. Other future works may explore additional criteria for the fitness function such as the size of an S-box in hardware (e.g., by counting the number of gates).

Another direction of future work considers other physical attacks like stochastic attacks [26], mutual information analysis [27] and machine learning attacks [28–30]. Going even one step further, a designer could find S-boxes that

(1) possess improved resilience against more than one type of attack, and (2) could be implemented in several devices (having different leakage functions).

Finally, the new proposed S-boxes can be combined with (more expensive) side-channel countermeasures such as masking. One of the easiest generic masking scheme (called *table re-computation*) computes a table look-up which associates to each masked input the output of the masked S-box [31]. Designers can easily combine this masking scheme with the new S-boxes. Other masking schemes tailored to the new S-boxes can also be applied, and constitute an interesting future work in order to investigate the resistance of physical cryptographic implementations generated by genetic algorithms against side-channel attacks.

Acknowledgments. L. Lerman is funded by the Brussels Institute for Research and Innovation (Innoviris) for the SCAUT project. S. Picek was supported in part by Croatian Science Foundation under the project IP-2014-09-4882.

References

1. Kocher, P.C.: Timing Attacks on Implementations of Diffie-Hellman, RSA, DSS, and Other Systems [32], pp. 104–113
2. Picek, S., Batina, L., Jakobovic, D.: Evolving DPA-resistant Boolean functions. In: Bartz-Beielstein, T., Branke, J., Filipič, B., Smith, J. (eds.) PPSN 2014. LNCS, vol. 8672, pp. 812–821. Springer, Cham (2014). doi:10.1007/978-3-319-10762-2_80
3. Picek, S., Papagiannopoulos, K., Ege, B., Batina, L., Jakobovic, D.: Confused by confusion: systematic evaluation of DPA resistance of various S-boxes. In: Meier, W., Mukhopadhyay, D. (eds.) INDOCRYPT 2014. LNCS, vol. 8885, pp. 374–390. Springer, Cham (2014). doi:10.1007/978-3-319-13039-2_22
4. Picek, S., Mazumdar, B., Mukhopadhyay, D., Batina, L.: Modified transparency order property: solution or just another attempt. In: Chakraborty, R.S., Schwabe, P., Solworth, J. (eds.) SPACE 2015. LNCS, vol. 9354, pp. 210–227. Springer, Cham (2015). doi:10.1007/978-3-319-24126-5_13
5. Young, A.L., Yung, M.: The Dark Side of "Black-Box" Cryptography, or: Should We Trust Capstone? [32], pp. 89–103
6. Carlet, C.: Boolean functions for cryptography and error correcting codes. In: Crama, Y., Hammer, P.L. (eds.) Boolean Models and Methods in Mathematics, Computer Science, and Engineering, 1st edn, pp. 257–397. Cambridge University Press, New York (2010)
7. Carlet, C.: Vectorial Boolean functions for cryptography. In: Crama, Y., Hammer, P.L. (eds.) Boolean Models and Methods in Mathematics, Computer Science, and Engineering, 1st edn, pp. 398–469. Cambridge University Press, New York (2010)
8. Leander, G., Poschmann, A.: On the classification of 4 bit S-boxes. In: Carlet, C., Sunar, B. (eds.) WAIFI 2007. LNCS, vol. 4547, pp. 159–176. Springer, Heidelberg (2007). doi:10.1007/978-3-540-73074-3_13
9. Borghoff, J., Canteaut, A., Güneysu, T., Kavun, E.B., Knezevic, M., Knudsen, L.R., Leander, G., Nikov, V., Paar, C., Rechberger, C., Rombouts, P., Thomsen, S.S., Yalçın, T.: PRINCE – a low-latency block cipher for pervasive computing applications. In: Wang, X., Sako, K. (eds.) ASIACRYPT 2012. LNCS, vol. 7658, pp. 208–225. Springer, Heidelberg (2012). doi:10.1007/978-3-642-34961-4_14

10. Bertoni, G., Daemen, J., Peeters, M., Assche, G.: Keccak. In: Johansson, T., Nguyen, P.Q. (eds.) EUROCRYPT 2013. LNCS, vol. 7881, pp. 313–314. Springer, Heidelberg (2013). doi:10.1007/978-3-642-38348-9_19

11. Dobraunig, C., Eichlseder, M., Mendel, F., Schläffer, M.: Ascon: CAESAR submission (2014). http://ascon.iaik.tugraz.at/

12. Andreeva, E., Bilgin, B., Bogdanov, A., Luykx, A., Mendel, F., Mennink, B., Mouha, N., Wang, Q., Yasuda, K.: PRIMATEs v1 Submission to the CAESAR Competition (2014). http://competitions.cr.yp.to/round1/primatesv1.pdf

13. Coron, J.-S., Kocher, P., Naccache, D.: Statistics and secret leakage. In: Frankel, Y. (ed.) FC 2000. LNCS, vol. 1962, pp. 157–173. Springer, Heidelberg (2001). doi:10.1007/3-540-45472-1_12

14. Chari, S., Rao, J.R., Rohatgi, P.: Template attacks. In: Kaliski, B.S., Koç, K., Paar, C. (eds.) CHES 2002. LNCS, vol. 2523, pp. 13–28. Springer, Heidelberg (2003). doi:10.1007/3-540-36400-5_3

15. Chakraborty, K., Sarkar, S., Maitra, S., Mazumdar, B., Mukhopadhyay, D., Prouff, E.: Redefining the transparency order. In: WCC2015-9th International Workshop on Coding and Cryptography 2015 (2015)

16. Fei, Y., Luo, Q., Ding, A.A.: A statistical model for DPA with novel algorithmic confusion analysis. In: Prouff, E., Schaumont, P. (eds.) CHES 2012. LNCS, vol. 7428, pp. 233–250. Springer, Heidelberg (2012). doi:10.1007/978-3-642-33027-8_14

17. Fei, Y., Ding, A.A., Lao, J., Zhang, L.: A statistics-based success rate model for DPA and CPA. J. Cryptographic Eng. 5(4), 227–243 (2015)

18. Standaert, F.-X., Malkin, T.G., Yung, M.: A unified framework for the analysis of side-channel key recovery attacks. In: Joux, A. (ed.) EUROCRYPT 2009. LNCS, vol. 5479, pp. 443–461. Springer, Heidelberg (2009). doi:10.1007/978-3-642-01001-9_26

19. Gong, Z., Nikova, S., Law, Y.W.: KLEIN: a new family of lightweight block ciphers. In: Juels, A., Paar, C. (eds.) RFIDSec 2011. LNCS, vol. 7055, pp. 1–18. Springer, Heidelberg (2012). doi:10.1007/978-3-642-25286-0_1

20. Bogdanov, A., Knudsen, L.R., Leander, G., Paar, C., Poschmann, A., Robshaw, M.J.B., Seurin, Y., Vikkelsoe, C.: PRESENT: an ultra-lightweight block cipher. In: Paillier, P., Verbauwhede, I. (eds.) CHES 2007. LNCS, vol. 4727, pp. 450–466. Springer, Heidelberg (2007). doi:10.1007/978-3-540-74735-2_31

21. Bertoni, G., Daemen, J., Peeters, M., Assche, G.V.: The Keccak reference Submission to NIST(Round 3) (2011)

22. Andreeva, E., Bilgin, B., Bogdanov, A., Luykx, A., Mendel, F., Mennink, B., Mouha, N., Yasuda, K., Wang, Q.: PRIMATEs v1.02: CAESAR submission, September 2014

23. Eiben, A.E., Smith, J.E.: Introduction to Evolutionary Computing. Springer, Berlin, Heidelberg, New York (2003)

24. Lerman, L., Markowitch, O., Veshchikov, N.: Comparing sboxes of ciphers from the perspective of side-channel attacks. In: 2016 IEEE Asian Hardware-Oriented Security and Trust (AsianHOST), 1–6 December 2016

25. Lerman, L., Poussier, R., Bontempi, G., Markowitch, O., Standaert, F.-X.: Template attacks vs. machine learning revisited (and the curse of dimensionality in side-channel analysis). In: Mangard, S., Poschmann, A.Y. (eds.) COSADE 2014. LNCS, vol. 9064, pp. 20–33. Springer, Cham (2015). doi:10.1007/978-3-319-21476-4_2

26. Schindler, W., Lemke, K., Paar, C.: A stochastic model for differential side channel cryptanalysis. In: Rao, J.R., Sunar, B. (eds.) CHES 2005. LNCS, vol. 3659, pp. 30–46. Springer, Heidelberg (2005). doi:10.1007/11545262_3

27. Gierlichs, B., Batina, L., Tuyls, P., Preneel, B.: Mutual information analysis. In: Oswald, E., Rohatgi, P. (eds.) CHES 2008. LNCS, vol. 5154, pp. 426–442. Springer, Heidelberg (2008). doi:10.1007/978-3-540-85053-3_27
28. Lerman, L., Bontempi, G., Markowitch, O.: Side channel attack: an approach based on machine learning. In: Second International Workshop on Constructive Side Channel Analysis and Secure Design, Center for Advanced Security Research Darmstadt, pp. 29–41 (2011)
29. Hospodar, G., Gierlichs, B., Mulder, E.D., Verbauwhede, I., Vandewalle, J.: Machine learning in side-channel analysis: a first study. J. Cryptograph. Eng. **1**(4), 293–302 (2011)
30. Lerman, L., Bontempi, G., Markowitch, O.: Power analysis attack: an approach based on machine learning. IJACT **3**(2), 97–115 (2014)
31. Messerges, T.S.: Securing the AES finalists against power analysis attacks. In: Goos, G., Hartmanis, J., Leeuwen, J., Schneier, B. (eds.) FSE 2000. LNCS, vol. 1978, pp. 150–164. Springer, Heidelberg (2001). doi:10.1007/3-540-44706-7_11
32. Koblitz, N. (ed.): Proceedings of 16th Annual International Cryptology Conference Advances in Cryptology - CRYPTO 1996, Santa Barbara, California, USA, 18–22 August 1996. LNCS, vol. 1109. Springer, Heidelberg (1996)

Efficient Conversion Method from Arithmetic to Boolean Masking in Constrained Devices

Yoo-Seung Won[1] and Dong-Guk Han[1,2(✉)]

[1] Department of Financial Information Security, Kookmin University, Seoul, Korea
{mathwys87,christa}@kookmin.ac.kr
[2] Department of Mathematics, Kookmin University, Seoul, Korea

Abstract. A common technique employed for preventing a side channel analysis is Boolean masking. However, the application of this scheme is not so straightforward when it comes to block ciphers based on Addition-Rotation-Xor structure. In order to address this issue, since 2000, scholars have investigated schemes for converting Arithmetic to Boolean (AtoB) masking and Boolean to Arithmetic (BtoA) masking schemes. However, these solutions have certain limitations. The time performance of the AtoB scheme is extremely unsatisfactory because of the high complexity of $\mathcal{O}(k)$ where k is the size of arithmetic operation. At the FSE 2015, an improved algorithm with time complexity $\mathcal{O}(\log k)$ based on the Kogge-Stone carry look-ahead adder was suggested. Despite its efficiency, this algorithm cannot consider for constrained environments. Although the original algorithm inherently extends to low-resource devices, there is no advantage in time performance; we call this variant as the generic variant. In this study, we suggest an enhanced variant algorithm to apply to constrained devices. Our solution is based on the principle of the Kogge-Stone carry look-ahead adder, and it uses a divide and conquer approach. In addition, we prove the security of our new algorithm against first-order attack.

By reducing the main loop complexity to $\lceil \log (l - 1) \rceil$ from $\lceil \log (k - 1) \rceil$ where l is the size of register bit, we can expect the reasonable time complexity for our variant algorithms. In implementation results based on this fact, when $k = 64$ and the register bit size of a chip is 8, 16 or 32, we obtain 58%, 72%, or 68% improvement, respectively, over the results obtained using the generic variant. When applying those algorithms to first-order SPECK, we also achieve roughly 40% improvement. Moreover, our proposal extends to higher-order countermeasures as previous study.

Keywords: Arithmetic to Boolean masking · Kogge-Stone carry look-ahead adder · ARX-based cryptographic algorithm

1 Introduction

Side channel analysis, which has been in the spotlight over the past decade, belongs to the genre of software and hardware implementation attacks.

© Springer International Publishing AG 2017
S. Guilley (Ed.): COSADE 2017, LNCS 10348, pp. 120–137, 2017.
DOI: 10.1007/978-3-319-64647-3_8

With respect to the properties of the cryptographic algorithm and physical information, the attack schemes of adversaries are diverse and they involve, for example, simple power analysis and differential power analysis attacks. To counteract these attacks, various countermeasures have been proposed in the literature. However, among those proposals, countermeasures such as Boolean masking, hiding, and threshold implementation have outlasted other methods and are usually recommended. Especially, in terms of software implementation, first-order attacks cannot destroy the (first-order) Boolean masking scheme regardless of the number of traces.

In another context, the advent of the Internet of Things (IoT) has encouraged cryptographic algorithms, in some instances, to play an important role in satisfying the needs of the IoT environment. In designing such algorithms, external conditions including time performance and gate size have become increasingly significant. Recently, cryptographic algorithms that satisfy these conditions have been published. In the case of block ciphers, most of their structures utilize S-boxes of nibble units or use addition to introduce a confusion factor. Moreover, Addition-Rotation-Xor(ARX)-based structure can enhance time performance in software implementation. For these reasons, several lightweight cryptographic algorithms such as SPECK [8] have emerged.

From the side channel countermeasure perspective, it is quite complicated to apply a Boolean masking scheme to a block cipher with an ARX-based structure. To overcome this challenge, other approaches are required. Initially, algorithms were developed for performing conversions between Boolean and arithmetic masking schemes. Goubin, in particular, described a very elegant algorithm for converting from Boolean to arithmetic (BtoA) masking [2], using only a constant number of operations, independent of the size of the arithmetic operation. This allows for the easy exploitation of the BtoA masking algorithm at low cost. On the other hand, there also have been several changes to the algorithm that convert from arithmetic to Boolean (AtoB) masking, in order to improve time performance and/or reduce memory consumption. First, Goubin reported an AtoB masking algorithm [2] which has $\mathcal{O}(k)$ complexity where k is the size of arithmetic operation. Later, at CHES 2003, an AtoB masking algorithm [3] with a precomputed table was introduced to reduce the time performance, rather than increasing the memory consumption. This algorithm can also be easily modified to suit the IoT environment due to the fact that the size of the precomputed table can be defined beforehand. Shortly thereafter, Neiße et al. in [4] suggested that their algorithm could result in lower memory consumption than the previous algorithm. Within the decade, Debraize proposed a new high performance algorithm as well as two modified algorithms with precomputed tables [6]. Despite these efforts, the problem of time complexity has endured at about $\mathcal{O}(k)$.

In recent years, the rising demand for higher-order masking countermeasures has led to the introduction of second-order masking countermeasures. At SPACE 2013, Vadnala et al. proposed two secure algorithms for converting between Boolean and arithmetic masking schemes against second-order attacks [7]. These algorithms apply the generic second-order secure countermeasure developed by Rivain et al. [5], but they were difficult to expand for a larger size of arithmetic operation because 2^k or $2^k/k$ bytes of RAM is required. To overcome this obstacle, handling of the carry bit was required and Vadnala et al. in [11] solved this problem without difficulty by utilizing a precomputed table to handle carry bit.

At about the same time, general approaches were achieved also for higher-order BtoA and AtoB schemes. The initial proposal [9] based on Goubin's conversion method [2] was described at the CHES 2014, although it has high time complexity of $\mathcal{O}(n^2 \cdot k)$ for $n = 2t + 1$ shares.

Another approach is to execute an arithmetic operation without converting the masking form. At the COSADE 2014, Karroumi et al. in [10] demonstrated that it was possible to use addition/subtraction for two Boolean masked inputs. As a result, an efficient algorithm was developed to satisfy the IoT environment, using lookup tables. Despite all these efforts, however, the arithmetic operation of algorithms without conversion continues to be characterized by a time complexity of about $\mathcal{O}(k)$.

Unlike the previous algorithms that were based on the classical ripple carry adder, algorithms based on the Kogge-Stone carry look-ahead adder with the logarithmic complexity $\mathcal{O}(\log k)$ [12] have recently been proposed. These algorithms not only easily expand to a higher-order masking scheme but can also be applied to algorithms for AtoB masking and arithmetic operation without conversion. To the best of our knowledge, these algorithms achieve an outstanding performance. Although the algorithms proposed by Coron et al. [12] require a low complexity of $\mathcal{O}(\log k)$, they become infeasible for direct implementation on low-resource devices. Practically, these algorithms cannot be applied to the constrained devices without any modifications if the register bit size of a chip is l bit is less than that of arithmetic operation k bit. However, by using an array concept, the original algorithm inherently extends to low-resource devices although there is no associated advantage in time performance. In this paper, we call this algorithm using an array concept as the generic variant. As a result, the generic variant algorithm might lose their merit when this algorithm is implemented on IoT devices.

Our Contributions. The arithmetic operation size of the most cryptographic algorithms with ARX-based structure does not correspond to the register size of IoT devices. Although the generic variant algorithm from [12] can directly be applied to a chip with the register l bit, we should endure a high cost. In response to this limitation, we suggest that the enhanced variant algorithms are significantly faster than the generic algorithm. Our solution follows the basic concept of the Kogge-Stone carry look-ahead adder, but uses a divide and conquer approach to prevent the time complexity from becoming too great. The currently proposed core concept involves a means of handling the carry occurrence of the Kogge-Stone carry look-ahead adder. Part of this procedure is to control the carries propagating from less to more significant words, which must also be protected by masking to prevent any first leakage. More precisely, after the carry value is generated from the previous block, it fills up the least significant bit in the next block. When the bitwidth is $k/2$, we also provide the advanced algorithm more than the enhanced variant algorithm. We demonstrate that this can be achieved in an efficient and secure fashion by using the principle of the Kogge-Stone carry look-ahead adder for the carries. We also verify the correctness and prove the security of our algorithms.

According to our analysis, the generic variant algorithm totally requires $36(k/l) \cdot \lceil \log (k-1) \rceil - 3(k/l) - 8\lceil \log (k-1) \rceil$ operations. However, our variant algorithm generally require $28(k/l) \cdot \lceil \log (l-1) \rceil + 11(k/l) + 28\lceil \log (l-1) \rceil - 10$. When $k/l = 2$ the computational complexity of our variant algorithm requires $28(k/l) \cdot \lceil \log (l-1) \rceil + 4(k/l) + 36$ operations. Therefore, compared with the generic variant algorithm, we can expect a better performance from our algorithms. Moreover, in our implementation results, the enhanced variant algorithms show a 58–72% improvement over the generic variant algorithms in terms of time performance. In addition, for $k/l = 2$, we acquired more than a 27% improvement as compared to our general solution algorithm is an enhanced variant algorithm. Finally, we applied the generic variant and enhanced variant algorithms to first-order masked SPECK. Consequently, we obtained an improvement of approximately 40% over the generic algorithm results.

2   Kogge-Stone Carry Look-Ahead Adder and Its Countermeasure

2.1   Notation

Before presenting a detailed description of our algorithms, we first describe the notations listed in Table 1. Note that the notations described in this section is maintained for consistency and used throughout the remainder of this paper.

Table 1. Notations used in this paper

Notation	Description
x	k-bit value
n_α	$\max(\lceil \log (\alpha - 1) \rceil, 1)$, $\alpha = k$ or l
l	Register bit size of CPU chip
m	Number of data blocks ($m = k/l$ ($0 < m \le l$))
$x^{(i)}$	Least significant i-th bit
$x_{(i)}$	i-th data block of x $\left(x_{(i)} = 2^{i \times l} \sum_{j=0}^{l-1} \left(2^j \times x^{(i \times l + j)} \right) \right)$
$X_{(i)}$	Modified i-th data block that includes the carry value generated from the previously modified data $2^{i(l-1)} \left(c^{(i(l-1))} + \sum_{j=1}^{l-1} 2^j x^{(i(l-1)+j)} \right) (1 \le i < m)$ $2^{i(l-1)} \left(c^{(i(l-1))} + \sum_{j=1}^{m-1} 2^j x^{(i(l-1)+j)} \right) (i = m)$ $x_{(0)}$ $(i = 0)$
$X'_{(i)}$	Modified i-th data block that doesn't include the carry value generated from the previously modified data $2^{i(l-1)} \sum_{j=1}^{l-1} 2^j x^{(i(l-1)+j)} (1 \le i < m)$ $2^{i(l-1)} \sum_{j=1}^{m-1} 2^j x^{(i(l-1)+j)} (i = m)$ $x_{(0)}$ $(i = 0)$

For example, if the register bit size is 32 and the size of the addition is 64, the number of data blocks is two. *i.e.*, $x = (x_{(1)}\|x_{(0)})$. Further, we denote x as $(X_{(2)}\|X_{(1)}\|X_{(0)})$, where $X_{(0)} = (x_{(31)} \cdots x_{(0)})_2$, $X_{(1)} = (x_{(62)} \cdots x_{(32)}c_{(31)})_2$, and $X_{(2)} = (x_{(63)}c_{(62)})_2$. Here, the carry value $c_{(i)}$ is generated from $x_{(i)} + y_{(i)}$.

2.2 Kogge-Stone Carry Look-Ahead Adder

In this section, we first recall the Kogge-Stone carry look-ahead adder [1,12] which is based on recurrence equations as follows:

$$\begin{cases} c^{(0)} = 0 \\ \forall i \geq 1, \ c^{(i)} = \{(x^{(i-1)} \oplus y^{i-1}) \wedge c^{(i-1)}\} \oplus (x^{(i-1)} \wedge y^{(i-1)}) \end{cases} \quad (1)$$

where $c^{(i)}$ is generated from the carry bit of $\sum_{j=0}^{i-1} 2^j x^{(j)} + \sum_{j=0}^{i-1} 2^j y^{(j)}$. We can therefore compute the carry bit. On the basis of Eq. (1), we can re-construct the k bit addition as

$$x + y = \sum_{i=0}^{k-1} 2^i x^{(i)} + \sum_{i=0}^{k-1} 2^i y^{(i)} = \sum_{i=0}^{k-1} (x^{(i)} \oplus y^{(i)} \oplus c^{(i)}) \quad (2)$$

According to Lemma 1 of [12], Eq. (1) can be converted to a Kogge-Stone recursive equation as follows:

$$\begin{cases} P_{i,j} = P_{i-1,j} \\ G_{i,j} = G_{i-1,j} \quad (0 \leq j < 2^{i-1}) \end{cases}$$

$$\begin{cases} P_{i,j} = P_{i-1,j} \wedge P_{i-1,j-2^{i-1}} \\ G_{i,j} = (P_{i-1,j} \wedge G_{i-1,j-2^{i-1}}) \oplus G_{i-1,j} \quad (2^{i-1} \leq j < k) \end{cases} \quad (3)$$

$$\Longrightarrow \begin{cases} c^{(0)} = 0, \quad c^{(1)} = G_{0,0} \\ c^{(j+1)} = G_{i,j} \quad (2^{i-1} \leq j < 2^i) \end{cases}$$

Here, when i corresponds to $\lceil \log (k-1) \rceil$, we obtain the most significant carry bit $c^{(k-1)}$.

Equation (3) inherently extends to the recurrence equation of the k bit addition; however, a time complexity of $\mathcal{O}(\log k)$ is still required. According to Theorem 3 in [12], the recurrence equation of the k bit addition is

$$\begin{cases} P_i = P_{i-1} \wedge (P_{i-1} \ll 2^{i-1}) \quad (1 \leq i < n) \\ G_i = (P_{i-1} \wedge (G_{i-1} \ll 2^{i-1})) \oplus G_{i-1} \end{cases} \Longrightarrow x + y = x \oplus y \oplus (2G_n) \quad (4)$$

where $P_i = \sum_{j=2^i-1}^{k-1} 2^j P_{i,j}$ with $P_0 = x \oplus y$, $G_i = \sum_{j=0}^{k-1} 2^j G_{i,j}$ with $G_0 = x \wedge y$, and n represents $\lceil \log (k-1) \rceil$. On the basis of Eq. (4), the sequence can be computed using Algorithm 1.

Algorithm 1. Kogge-Stone Adder

z **Input:** $x, y \in \{0,1\}^k$, $n_k = \max(\lceil \log(k-1) \rceil, 1)$
Output: $z = x + y \mod 2^k$

1: $P \leftarrow x \oplus y$
2: $G \leftarrow x \wedge y$
3: **for** $i := 1$ **to** $n_k - 1$ **do**
4: $\quad G \leftarrow (P \wedge (G \ll 2^{i-1})) \oplus G$
5: $\quad P \leftarrow P \wedge (P \ll 2^{i-1})$
6: **end for**
7: $G \leftarrow (P \wedge (G \ll 2^{n-1})) \oplus G$
8: **return** $x \oplus y \oplus (2G)$

2.3 Generic Variant for the Kogge-Stone Adder and AtoB Masking

In this section, we introduce the k bit addition when the size of register bit corresponds to the l bit, where l is less than k. Basically, a generic variant algorithm using the array concept is a direct application of the Kogge-Stone adder; further, we convert a generic variant algorithm into a first-order algorithm. Finally, we refer to the generic variant of the Kogge-Stone adder and AtoB masking in [12].

3 Enhanced Variant for AtoB Masking Based on Kogge-Stone Adder

3.1 Underlying Concept

In this section, we introduce the fundamental idea underlying our approach. Assuming k and l correspond to eight and four, respectively, we generate two arrays from the original data for the generic variant Kogge-Stone adder because the size of the register is smaller than that of addition; refer to Array1 and Array0 in Fig. 1. Therefore, the intermediate calculations of generic variant algorithm consume more cost than those of original algorithm because of the difference between the original Kogge-Stone adder and generic variant algorithm.

Our underlying concept is to split the original data into three arrays, as shown in Fig. 1, considering the bitwidth of a chip, and utilize the original Kogge-Stone adder without any modifications.

To acquire the most significant bit of each block, we review the correction of Theorem 3, as addressed in [12], for our new design. $2G_n$ is not exactly equal to $\sum_{j=0}^{k-1} 2^j c_j$, but rather $\sum_{j=0}^{k} 2^j c_j$. In other words, the $2G_n$ value that leaks the most significant k-th bit is the basis of our enhanced variant algorithm. That is, after the carry value is generated from the previous block, it fills up the least significant bit in the next block.

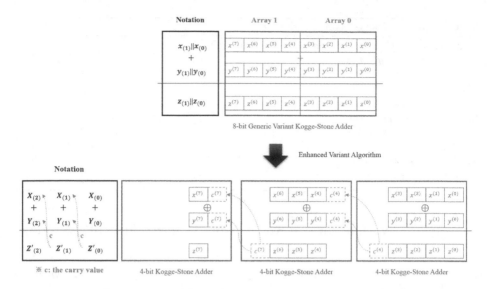

Fig. 1. Underlying concept for enhanced variant algorithm

3.2   Enhanced Variant for the Kogge-Stone Adder

Our proposed conversion algorithm is based on the principle of the Kogge-Stone adder. As mentioned earlier, owing to the missing value, we can build an enhanced variant of the Kogge-Stone adder and AtoB masking. The correctness of this technique is demonstrated in the following theorem.

Theorem 1. *Let* $x = (x_{(m-1)}|| \cdots ||x_{(0)}), y = (y_{(m-1)}|| \cdots ||y_{(0)}), z = x + y$ *be elements of* $\{0,1\}^k$. *Then* $\sum_{i=0}^{m-1} z_{(i)} = \sum_{i=0}^{m} Z'_{(i)} \left(= \sum_{i=0}^{m-1} \left(x_{(i)} + y_{(i)} \right) \right)$

Proof. From our notation, we rewrite the equation as:

$$\sum_{i=0}^{m-1} z_{(i)} \underset{\langle 1 \rangle}{=} \sum_{i=0}^{m} Z'_{(i)} \underset{\langle 2 \rangle}{=} \sum_{i=0}^{m} \left(X_{(i)} + Y_{(i)} \mod 2^{(i+1) \times l} \right)$$

For $\langle 1 \rangle$, we refer to Lemma 3 in Appendix A. Thus, we prove $\langle 2 \rangle$ in this theorem. Moreover, by proving Eq. (5), $\langle 2 \rangle$ can be inherently proven. Therefore, we only provide the proof of Eq. (5):

$$Z'_{(i)}/2^{i(l-1)} = \{X_{(i)} + Y_{(i)}\}/2^{i(l-1)} \mod 2^l \quad (0 \le i \le m) \tag{5}$$

$$= \left(c^{(i(l-1))} + \sum_{j=1}^{l-1} 2^j x^{(i(l-1)+(j-1))} + c^{(i(l-1))} + \sum_{j=1}^{l-1} 2^j y^{(i(l-1)+(j-1))} \right)$$

For simplicity, we re-index some of the variables,

$i.e., c^{(i(l-1))}$ as $a^{(0)}$ or $b^{(0)}, x^{(i(l-1)+j)} = a^{(j+1)}$, and $y^{(i(l-1)+j)} = b^{(j+1)}$

$$= \left(\sum_{j=0}^{l-1} 2^j a^{(j)} + \sum_{j=0}^{l-1} 2^j b^{(j)} \right) \mod 2^l$$

$$= \bigoplus_{j=0}^{l-1} 2^j \left(a^{(j)} \oplus b^{(j)} \oplus d^{(j)} \right)$$

$(d^{(j)}$ generated from $\displaystyle\sum_{p=0}^{j-1} a^{(p)} + \sum_{p=0}^{j-1} b^{(p)}$

is the carry value of the most significant bit)

$$= c^{(i(l-1))} \oplus c^{(i(l-1))} \oplus \bigoplus_{j=1}^{l-1} 2^j \left(x^{(i(l-1)+j-1)} \oplus y^{(i(l-1)+j-1)} \oplus c^{(i(l-1)+j-1)} \right)$$

(According to Lemma 4 in Appendix A)

$$= \bigoplus_{j=1}^{l-1} 2^j \left(z^{(i(l-1)+j-1)} \right) = Z'_{(i)}/2^{i(l-1)}$$

For $i = 0$ and m, we have excluded the proof as it is straightforward.   □

Based on Theorem 1, we build an enhanced variant algorithm and present it as Algorithm 2. In this algorithm, there is no need to calculate the carry value in Step 9 because $(X_{(j)} \oplus C) \oplus (Y_{(j)} \oplus C) = X_{(j)} \oplus Y_{(j)}$. Further, in Step 10, the result of the addition can obviously be acquired for only one operation using a carry value, $i.e., (X_{(j)} \oplus C) \wedge (Y_{(j)} \oplus C) = (X_{(j)} \wedge Y_{(j)}) \oplus C$. More importantly, the number of inner-loops is $\lceil \log (l-1) \rceil$ and the unit of operation is the l bit even though the number of outer-loops is $m + 1$. In other words, applying our underlying idea allows the number of inner-loops to be reduced from $\lceil \log (k-1) \rceil$ to $\lceil \log (l-1) \rceil$.

3.3   Enhanced Variant for AtoB Masking

Similarly, here we demonstrate how to secure the AtoB masking algorithm. We convert Algorithm 2 into a first-order secure algorithm by protecting all intermediate variables that utilize the AtoB masking algorithm in [12] and present it as Algorithm 3.

Some of the secure algorithms adopted in Algorithm 3, including $SecShift_l$, $SecAnd_l$, and $SecXor_l$ are based on the l bit unit using only one array, unlike the secure algorithms in [12].

In addition, there are some differences between Algorithms 2 and 3, because the δ value is required in order to prevent the leakage of the carry value. Thus, the carry value is masked by the δ value in Step 28. As such, in Step 11, we must eliminate the masked value δ because the G value had already been masked by

Algorithm 2. Enhanced Variant for Kogge-Stone Adder

Input: $x = \left(x_{(m-1)}||\cdots||x_{(0)}\right), y = \left(y_{(m-1)}||\cdots||y_{(0)}\right) \in \{0,1\}^k$
$\quad\quad n_l = \max(\lceil \log{(l-1)}\rceil, 1)$
Output: $z = \left(z_{(m-1)}||\cdots||z_{(0)}\right) = x + y \mod 2^k$

1: $X_{(0)} \leftarrow \left(x^{(l-1)}||\cdots||x^{(0)}\right)$
2: $Y_{(0)} \leftarrow \left(y^{(l-1)}||\cdots||y^{(0)}\right)$
3: $C \leftarrow 0$
4: **for** $i := 1$ **to** m **do**
5: $\quad X_{(i)} \leftarrow \left(x^{(i(l-1)+l-1)}||\cdots||x^{(i(l-1)+1)}||0\right)$
6: $\quad Y_{(i)} \leftarrow \left(y^{(i(l-1)+l-1)}||\cdots||y^{(i(l-1)+1)}||0\right)$
7: **end for**
8: **for** $j := 0$ **to** m **do**
9: $\quad P \leftarrow X_{(j)} \oplus Y_{(j)}$
10: $\quad G \leftarrow \left(X_{(j)} \wedge Y_{(j)}\right) \oplus C$
11: $\quad$ **for** $i := 1$ **to** $n_l - 1$ **do**
12: $\quad\quad G \leftarrow \left(P \wedge \left(G \ll 2^{i-1}\right)\right) \oplus G$
13: $\quad\quad P \leftarrow P \wedge \left(P \ll 2^{i-1}\right)$
14: $\quad$ **end for**
15: $\quad G \leftarrow \left(P \wedge \left(G \ll 2^{n-1}\right)\right) \oplus G$
16: $\quad Z'_{(j)} \leftarrow X_{(j)} \oplus Y_{(j)} \oplus (2G)$
17: $\quad$ **if** $j \neq 0$ **then**
18: $\quad\quad Z'_{(j)} \leftarrow Z'_{(j)} \gg 1$
19: $\quad$ **end if**
20: $\quad C \leftarrow G \gg (l-1)$
21: **end for**
22: $\left(z_{(m-1)}||\cdots||z_{(0)}\right) \leftarrow \left(Z'_{(m)}||\cdots||Z'_{(0)}\right) \mod 2^k$
23: **return** $z = \left(z_{(m-1)}||\cdots||z_{(0)}\right)$

s in Step 10. The following Lemma proves the security of our new algorithm against first-order attack.

Lemma 1. *When r is uniformly distributed in $\mathbb{F}_{2^k}$, any intermediate variable in Algorithm 3 has a distribution independent of $x = A + r \mod 2^k$.*

Proof. The proof is based on Lemma 5 of the previous paper [12], and also on the fact that all intermediate variables from Steps 10, 11, and 28 have a distribution independent of x.

Case 1: All variables from Step 28, have a distribution independent of x
$\quad G, G \gg (l-1), \{G \gg (l-1)\} \oplus \delta$, and $[\{G \gg (l-1)\} \oplus \delta] \oplus \{s \gg (l-1)\}$
$\quad \left(= c^{(j(l-1)+(i-2))} \oplus s\right)$ have a distribution independent of x, since $G \oplus s$ is
$\quad \left(c^{(j(l-1)+(i-2))}||\cdots||c^{(j(l-1)+0)}||0\right)$.
Case 2: All variables from Step 10, have a distribution independent of x
$\quad C$ has a distribution independent of x because of the masked value δ. Therefore, other intermediate variables have uniform distribution.

Algorithm 3. Enhanced Variant for AtoB Masking

Input: $a = (a_{(m-1)}||\cdots||a_{(0)})$, $r = (r_{(m-1)}||\cdots||r_{(0)}) \in \{0,1\}^k$
$\quad\quad n_l = \max(\lceil \log(l-1)\rceil, 1)$
Output: $x = (x_{(m-1)}||\cdots||x_{(0)})$ such that $x \oplus r = a + r \mod 2^k$

1: $s \leftarrow \{0,1\}^l,\ t \leftarrow \{0,1\}^l,\ u \leftarrow \{0,1\}^l,\ \delta \leftarrow \{0,1\}^l$
2: $a_{(0)} \leftarrow \left(a^{(l-1)}||\cdots||a^{(0)}\right),\ r_{(0)} \leftarrow \left(r^{(l-1)}||\cdots||r^{(0)}\right),\ C \leftarrow \delta$
3: **for** $i := 1$ **to** m **do**
4: $\quad a_{(i)} \leftarrow \left(a^{(i(l-1)+l-1)}||\cdots||a^{(i(l-1)+1)}||0\right)$
5: $\quad r_{(i)} \leftarrow \left(r^{(i(l-1)+l-1)}||\cdots||r^{(i(l-1)+1)}||0\right)$
6: **end for**
7: **for** $j := 0$ **to** m **do**
8: $\quad P \leftarrow a_{(j)} \oplus s$
9: $\quad P \leftarrow P \oplus r_{(j)}$
10: $\quad G \leftarrow s \oplus \left((a_{(j)} \oplus t) \wedge r_{(j)}\right) \oplus C$
11: $\quad G \leftarrow G \oplus \left(t \wedge r_{(j)}\right) \oplus \delta$
12: $\quad$ **for** $i := 1$ **to** $n_l - 1$ **do**
13: $\quad\quad H \leftarrow \mathsf{SecShift}_l[G, s, t, 2^{i-1}]$
14: $\quad\quad W \leftarrow \mathsf{SecAnd}_l[P, H, s, t, u]$
15: $\quad\quad G \leftarrow \mathsf{SecXor}_l[G, W, u]$
16: $\quad\quad H \leftarrow \mathsf{SecShift}_l[P, s, t, 2^{i-1}]$
17: $\quad\quad P \leftarrow \mathsf{SecAnd}_l[P, H, s, t, u]$
18: $\quad\quad P \leftarrow P \oplus s$
19: $\quad\quad P \leftarrow P \oplus u$
20: $\quad$ **end for**
21: $\quad H \leftarrow \mathsf{SecShift}_l[G, s, t, 2^{n-1}]$
22: $\quad W \leftarrow \mathsf{SecAnd}_l[P, H, s, t, u]$
23: $\quad G \leftarrow \mathsf{SecXor}_l[G, W, u]$
24: $\quad X'_{(j)} \leftarrow a_{(j)} \oplus (2G)$
25: $\quad X'_{(j)} \leftarrow X'_{(j)} \oplus (2s)$
26: $\quad$ **if** $j \neq 0$ **then** $X'_{(j)} \leftarrow X'_{(j)} \gg 1$
27: $\quad$ **end if**
28: $\quad C \leftarrow [\{G \gg (l-1)\} \oplus \delta] \oplus \{s \gg (l-1)\}$
29: **end for**
30: $(x_{(m-1)}||\cdots||x_{(0)}) \leftarrow (X'_{(m)}||\cdots||X'_{(0)}) \mod 2^k$
31: **return** $x = (x_{(m-1)}||\cdots||x_{(0)})$

Case 3: All variables from Step 11, have a distribution independent of x
$\quad$ Since G corresponds to $s \oplus \left((a_{(j)} \oplus t) \wedge r_{(j)}\right) \oplus C \oplus \left(t \wedge r_{(j)}\right) \oplus \delta$ $(= (a_{(j)} \wedge r_{(j)}) \oplus c^{(j(l-1)+(i-2))} \oplus s)$, G has a distribution independent of x, and other intermediate variables have uniform distribution. $\quad\quad\square$

Algorithm 3 provides a generalization of the l and k variables; however, when $m = 2$, we obtain more improvements in terms of time performance by modifying the operations on the most significant block. More specifically, the outer-loop can be reduced $m - 1$ times, because only a single bit remains in $X_{(m)}$ and $Y_{(m)}$,

Algorithm 4. Enhanced Variant for AtoB Masking ($m = 2$)

Input: $a = \left(a_{(m-1)}||\cdots||a_{(0)}\right), r = \left(r_{(m-1)}||\cdots||r_{(0)}\right) \in \{0,1\}^k$
$\quad n_l = \max(\lceil \log(l-1)\rceil, 1)$
Output: $x = \left(x_{(m-1)}||\cdots||x_{(0)}\right)$ such that $x \oplus r = a + r \mod 2^k$

1: $s \leftarrow \{0,1\}^l,\ t \leftarrow \{0,1\}^l,\ u \leftarrow \{0,1\}^l,\ \delta \leftarrow \{0,1\}^l$
2: $a_{(0)} \leftarrow \left(a^{(l-1)}||\cdots||a^{(0)}\right),\ r_{(0)} \leftarrow \left(r^{(l-1)}||\cdots||r^{(0)}\right),\ C \leftarrow \delta$
3: **for** $i := 1$ **to** m **do**
4: $\quad a_{(i)} \leftarrow \left(a^{(i(l-1)+l-1)}||\cdots||a^{(i(l-1)+1)}||0\right)$
5: $\quad r_{(i)} \leftarrow \left(r^{(i(l-1)+l-1)}||\cdots||r^{(i(l-1)+1)}||0\right)$
6: **end for**
7: **for** $j := 0$ **to** $m-1$ **do**
8: $\quad P \leftarrow a_{(j)} \oplus s$
9: $\quad P \leftarrow P \oplus r_{(j)}$
10: $\quad G \leftarrow s \oplus \left(\left(a_{(j)} \oplus t\right) \wedge r_{(j)}\right) \oplus C$
11: $\quad G \leftarrow G \oplus \left(t \wedge r_{(j)}\right) \oplus \delta$
12: $\quad$ **for** $i := 1$ **to** $n_l - 1$ **do**
13: $\quad\quad H \leftarrow \mathsf{SecShift}_l[G, s, t, 2^{i-1}]$
14: $\quad\quad W \leftarrow \mathsf{SecAnd}_l[P, H, s, t, u]$
15: $\quad\quad G \leftarrow \mathsf{SecXor}_l[G, W, u]$
16: $\quad\quad H \leftarrow \mathsf{SecShift}_l[P, s, t, 2^{i-1}]$
17: $\quad\quad P \leftarrow \mathsf{SecAnd}_l[P, H, s, t, u]$
18: $\quad\quad P \leftarrow P \oplus s$
19: $\quad\quad P \leftarrow P \oplus u$
20: $\quad$ **end for**
21: $\quad H \leftarrow \mathsf{SecShift}_l[G, s, t, 2^{n-1}]$
22: $\quad W \leftarrow \mathsf{SecAnd}_l[P, H, s, t, u]$
23: $\quad G \leftarrow \mathsf{SecXor}_l[G, W, u]$
24: $\quad X'_{(j)} \leftarrow a_{(j)} \oplus (2G)$
25: $\quad X'_{(j)} \leftarrow X'_{(j)} \oplus (2s)$
26: $\quad$ **if** $j \neq 0$ **then**
27: $\quad\quad X'_{(j)} \leftarrow X'_{(j)} \gg 1$
28: $\quad$ **end if**
29: $\quad C \leftarrow [\{G \gg (l-1)\} \oplus \delta] \oplus \{s \gg (l-1)\}$
30: **end for**
31: $x_{(m)} \leftarrow \left(a_{(m)} \oplus C\right) \oplus \delta$
32: $\left(x_{(m-1)}||\cdots||x_{(0)}\right) \leftarrow \left(X'_{(m)}||\cdots||X'_{(0)}\right) \mod 2^k$
33: **return** $x = \left(x_{(m-1)}||\cdots||x_{(0)}\right)$

---

excluding the carry value. The following Lemma provides more detail. Note also that Algorithm 4 could be used as an enhanced variant of AtoB masking for $m = 2$.

Lemma 2. *When $m = 2$, $X_{(m)} + Y_{(m)}$ is identical to $2^{m(l-1)+1} z^{(m(l-1)+1)}$. Moreover, the total operation requires only two XOR operations.*

Proof.

$$X_{(m)} + Y_{(m)}$$
$$= 2^{m(l-1)} \left(c^{(m(l-1))} + 2 \times x^{(m(l-1)+1)} \right) + 2^{m(l-1)} \left(c^{(m(l-1))} + 2 \times y^{(m(l-1)+1)} \right)$$
$$= 2^{m(l-1)+1} \left(c^{(m(l-1))} + x^{(m(l-1)+1)} + y^{(m(l-1)+1)} \right)$$
$$= 2^{m(l-1)+1} \left(c^{(m(l-1))} \oplus x^{(m(l-1)+1)} \oplus y^{(m(l-1)+1)} \right)$$
$$= 2^{m(l-1)+1} z^{(m(l-1)+1)} \qquad \qquad \square$$

4   Analysis and Implementations

We provide simulation results for the proposed algorithms and first-order masked lightweight block cipher SPECK [8] based on the ARX structure.

4.1   Computational Complexity for the Variant Algorithms

Table 2 compares computational complexity of Algorithms 3, 4 and the generic algorithm in [12]. We calculate the complexity of the subroutine functions; refer to Appendix B. Given these calculations, the total number of Xor operations from Step 2 to Step 19 in AtoB masking algorithm of [12] is then $5m + (n_k - 1)(18m - 4 + 2m) + (10m - 2) + 2m + 2(m - 1) = 20mn_k - m - 4n_k$. The final factor $2m + 2(m - 1)$ is derived from Step 18 and Step 19 in [12]. This is because Xor operation is required in order to apply Shift operation to array. Similarly, the total number of And and Shift operations can be calculated.

In addition, from Step 7 to Step 28 in Algorithm 3, the total number of Xor operations is $\{7 + (14 + 2)(n_l - 1) + (8 + 4)\}(m + 1) = 16mn_l + 3m + 16n_l + 3$. Moreover, our algorithms need to pre-processing and post-processing procedures. In Step 2–6 and Step 30–31, the total number of Xor operations is $(m - 3) + (m - 1) = 2m - 4$ if $m \neq 2$. Otherwise the total number is $0 + 1$. And and Shift operations can be applied in the same way.

Also, the number of random bits is only required $3l$ bits in our algorithms in comparison with the generic algorithm. Table 2 summarizes computational complexity for the proposed variant algorithms.

Table 2. Computational complexity for variant algorithms

Algorithm	Rand	Computational complexity		
		Xor	And	Shift
[12]	$3k$	$20mn_k - m - 4n_k$	$8mn_k - 2m$	$8mn_k - 4n_k$
Algorithm 3 ($m \neq 2$)	$4l$	$16mn_l + 5m + 16n_l - 1$	$8mn_l - m + 8n_l - 4$	$4mn_l + 7m + 4n_l - 5$
Algorithm 3 ($m = 2$)	$4l$	$16mn_l + 3m + 16n_l + 4$	$8mn_l - 2m + 8n_l - 1$	$4mn_l + 3m + 4n_l + 5$
Algorithm 4	$4l$	$16mn_l + 3m + 20$	$8mn_l - 2m + 7$	$4mn_l + 3m + 9$

As described in [12], identical weights are assigned to all operations. More specifically, all operations are assumed to incur the same computational cost. When considering these assumptions, we present the results for each k and l is Table 3. Thus, the proposed algorithms outperform the generic variant algorithm. However, for the case $l = 32$ and $k = 64$, the total complexity of Algorithm 3 could be higher than that of AtoB masking algorithm in [12]. Total complexity is the sum of each complexity of operations. More precisely, we definitely cannot decide the weight between operations. Thus, the proposed algorithms outperform the generic variant algorithm. However, for the case $l = 32$ and $k = 64$, the total complexity of Algorithm 3 could be higher than that of AtoB masking algorithm in [12]. Total complexity is the sum of each complexity of operations. More precisely, we definitely cannot decide the weight between operations.

Table 3. Computational complexity for each k and l for the variant algorithms

Algorithm	l	k	Computational complexity				
			Xor	And	Shift	Total	Penalty factor
[12]	8	64	928	368	360	1656	1.00
Algorithm 3	8	64	459	198	134	791	0.50
[12]	16	64	452	184	168	804	1.00
Algorithm 3	16	64	335	150	94	579	0.74
[12]	32	64	214	92	72	378	1.00
Algorithm 3	32	64	249	114	68	431	1.15
Algorithm 4	32	64	185	82	52	319	0.86

For example, all of our new algorithms outperform the generic variant algorithm if the ratio of weights between operations is (Xor : And : Shift) = (0.5 : 0.5 : 50). Especially, for $l = 32$ and $k = 64$, Algorithm 3 performs roughly 1% better than the generic variant algorithm. This is because the total number of Shift operations of Algorithm 3 is less than that of AtoB masking algorithm in [12]. Algorithm 3 is more efficient than the generic algorithm in realistic device, although computational complexity of Algorithm 3 is higher than that of generic algorithm; refer to Tables 2, 3 and 4. As mentioned earlier, there may be some of the reasons the difference between theoretical complexity and simulated results. For example, the difference of compiler language or the ratio of weights between operations. In conclusion, in any case, our algorithms outperform the generic algorithm.

4.2   Simulation Results for the Variant Algorithms

We implemented the proposed variant algorithms incorporating the Kogge-Stone adder. Considering low-resource devices such as the smart card, we implemented three possible partitionings for 64 bit additions, with register length $l = 8, 16$, and 32. To compare time performance, we used the AVR Studio 6.2 simulation for $l = 8$, the IAR Embedded Workbench Evaluation simulation for $l = 16$, and CodeWarrior for ARM Developer Suite v1.2 for $l = 32$. To ensure fairness, we did not consider the time performance for the generation of random numbers.

We obtained 58%, 72%, and 68% improvement over the generic variant algorithm; for $l = 8, 16$, and 32, respectively. We also showed 27% improvement over a general enhanced variant algorithm compared to between Algorithms 3 and 4.

Table 4. Simulation results for the variant algorithms

Algorithm	l	k	Clock cycle	Penalty factor
[12]	8	64	2864	1.00
Algorithm 3	8	64	1217	0.42
[12]	16	64	2705	1.00
Algorithm 3	16	64	765	0.28
[12]	32	64	1196	1.00
Algorithm 3	32	64	526	0.44
Algorithm 4	32	64	384	0.32

4.3   Simulation Results for First-Order SPECK

We apply our countermeasures to realize the first-order secure implementation of SPECK-128/128. Since the structure of SPECK has ARX operation, we must convert between Boolean and arithmetic masking scheme. For BtoA masking scheme, we used Goubin's algorithm in [2]. Considering the SPECK structure, we must perform two BtoA and one AtoB maskings per round, totaling 64 BtoA and 32 AtoB maskings overall.

Table 5 summarizes the results for computing the SPECK-128/128 for a single message block. Similar to previous simulation results, we achieved approximately 36%, 43%, and 37% improvements over the masked SPECK for $l = 8, 16$ and 32, respectively.

Table 5. Simulation results for non-masked and masked SPECK

Algorithm	l	k	Clock cycle	Penalty factor
Non-masked SPECK	8	64	24,360	1.00
Masked SPECK with [12]	8	64	177,303	7.27
Masked SPECK with Algorithm 3	8	64	112,951	4.64
Non-masked SPECK	16	64	21,446	1.00
Masked SPECK with [12]	16	64	143,642	6.70
Masked SPECK with Algorithm 3	16	64	81,562	3.80
Non-masked SPECK	32	64	10,279	1.00
Masked SPECK with [12]	32	64	71,006	6.91
Masked SPECK with Algorithm 3	32	64	49,470	4.81
Masked SPECK with Algorithm 4	32	64	44,926	4.37

5   Conclusion

For low-resource devices, we proposed two enhanced variant algorithms based on the principle of the Kogge-Stone carry look-ahead adder. One is a generalized algorithm for the bitwidth l and the other, in the case $l = k/2$, provides greater improvement over a general enhanced variant algorithm in time performance. All variant algorithms not only keep logarithmic complexity, i.e., $\mathcal{O}(\log k)$ but also outperform the generic variant algorithm of the Kogge-Stone adder. We proved the proposed algorithms were secure against first-order attacks, and that they inherently extend to higher-order AtoB masking scheme and arithmetic operation without conversion. In both analysis and implementation, we achieved remarkable improvements. In particular, implementation performance increases approximately 58–72% over the generic variant algorithm results. When applied to lightweight block cipher SPECK, we obtained improvements of approximately 36–43% compared to the generic variant algorithm from [12].

Acknowledgements. The authors would like to thank the anonymous reviewers for their useful comments that improved the quality of the paper. This work was partially supported by the Institute for Information and Communications Technology Promotion (IITP) grant funded by the Korea government (MSIP) (No. B0126-15-1008) and by the Basic Science Research Program through the National Research Foundation of Korea (NRF) funded by the Ministry of Education (NRF-2013R1A1A2A10062137).

A   Remaining Proof of Theorem 1

Here, we provide the remaining proof of Theorem 1, and thereby prove the following Lemma.

Lemma 3. $\sum_{i=0}^{m-1} x_{(i)}$ is equal to $\sum_{i=0}^{m} X'_{(i)}$.

Proof.

$$\sum_{i=0}^{m-1} x_{(i)}$$

$$= x_{(0)} + x_{(1)} + \cdots + x_{(m-2)} + x_{(m-1)}$$

$$= 2^0 \left(2^0 x^{(0)} + \cdots + 2^{l-1} x^{(l-1)} \right)$$

$$+ 2^l \left(2^0 x^{(l+0)} + \cdots + 2^{l-1} x^{(l+l-1)} \right) + \cdots$$

$$+ 2^{l(m-2)} \left(2^0 x^{(l(m-2)+0)} + \cdots + 2^{l-1} x^{(l(m-2)+l-1)} \right)$$

$$+ 2^{l(m-1)} \left(2^0 x^{(l(m-1)+0)} + \cdots + 2^{l-1} x^{(l(m-1)+l-1)} \right)$$

$$= 2^0 \left(2^0 x^{(0)} + \cdots + 2^{l-1} x^{(l-1)} \right)$$

$$+ 2^{l-1} \left(2^1 x^{((l-1)+1)} + \cdots + 2^{l-1} x^{((l-1)+l-1)} \right) + \cdots$$

$$+ 2^{(l-1)(m-1)} \left(2^1 x^{((l-1)(m-1)+1)} + \cdots + 2^{l-1} x^{((l-1)(m-1)+l-1)} \right)$$

$$+ 2^{(l-1)m} \left(2^1 x^{((l-1)m+1)} + \cdots + 2^{l-1} x^{((l-1)m+(m-1))} \right)$$

$$= X'_{(0)} + \cdots + X'_{(m-1)} + X'_{(m)} = \sum_{i=0}^{m} X'_{(i)} \qquad \square$$

Lemma 4. $d^{(j)}$ *is equal to* $c^{(i(l-1)+j-1)}$.

Proof. We proceed by mathematical induction.
  Lemma 4 holds for $j = 0$,

$$d^{(1)} = \{(a^{(0)} \oplus b^{(0)}) \wedge d^{(0)}\} \oplus (a^{(0)} \wedge b^{(0)})$$

$$= \{(c^{(i(l-1))} \oplus c^{(i(l-1))}) \wedge d^{(0)}\} \oplus (c^{(i(l-1))} \wedge c^{(i(l-1))})$$

$$= (0 \wedge d^{(0)}) \oplus (c^{(i(l-1))} \wedge c^{(i(l-1))}) = c^{(i(l-1))}$$

  If Lemma 4 holds for $j = k$, then it holds for $j = k + 1$, *i.e.,* $d^{(k)} = c^{(i(l-1)+k-1)}$.

$$d^{(k+1)}$$

$$= \{(a^{(k)} \oplus b^{(k)}) \wedge d^{(k)}\} \oplus (x^{(i(l-1)+k-1)} \wedge y^{(i(l-1)+k-1)})$$

$$= \{(x^{(i(l-1)+k-1)} \wedge y^{(i(l-1)+k-1)}) \wedge c^{(i(l-1)+k-1)}\} \oplus (x^{(i(l-1)+k-1)} \wedge y^{(i(l-1)+k-1)})$$

$$= c^{(i(l-1)+k)} \qquad \square$$

B   Secure Computation of Basic Operation and Its Computational Complexity

We provide the basic operations used in Algorithms 3, 4, and the AtoB masking algorithm of [12]. For AtoB masking countermeasure, the basic operation required for sensitive information is secure computation. We show how to secure the Shift operations against first-order attacks, SecXor and SecAnd operations are a direct application of secure computations in [12].

Algorithm 5. SecShift

Input: $x' = \left(x'_{(m-1)} || \cdots || x'_{(0)} \right), s = \left(s_{(m-1)} || \cdots || s_{(0)} \right), t = \left(t_{(m-1)} || \cdots || t_{(0)} \right), j, l$
    such that $x' = x \oplus s$, $j > 0$, and l is the size of register bit
Output: $y' = \left(y'_{(m-1)} || \cdots || y'_{(0)} \right)$ such that $y' = (x \ll j) \oplus t$
1: $st \leftarrow j \bmod l$
2: $bk \leftarrow \lfloor j/l \rfloor$
3: **for** $i := 0$ **to** $bk - 1$ **do**
4:     $y'_{(i)} \leftarrow t_{(i)}$
5: **end for**
6: $y'_{(bk)} \leftarrow t_{(bk)} \oplus \left(x'_{(0)} \ll st \right)$
7: $y'_{(bk)} \leftarrow y'_{(bk)} \oplus \left(s_{(0)} \ll st \right)$
8: **for** $i := bk + 1$ **to** $m - 1$ **do**
9:     $y'_{(i)} \leftarrow t_{(i)} \oplus \left(x'_{(i-bk)} \ll st \right) \oplus \left(x'_{(i-bk-1)} \gg (l - st) \right)$
10:     $y'_{(i)} \leftarrow y'_{(i)} \oplus \left(s_{(i-bk)} \ll st \right) \oplus \left(s_{(i-bk-1)} \gg (l - st) \right)$
11: **end for**
12: **return** $y' = \left(y'_{(m-1)} || \cdots || y'_{(0)} \right)$

SecXor and SecAnd algorithms are very straightforward. However, in contrast to [12] the SecShift algorithm should be some modified.

Table 6. Computational complexity of basic operations

Algorithm	Computational complexity		
	Xor	And	Shift
SecXor in [12]	$2m$	-	-
SecAnd in [12]	$4m$	4m	-
Algorithm 5	$4m - 2$	-	$4m - 2$

Without loss of generality, we assume that bk in Algorithm 5 is identical to 0. Then, the SecShift algorithm requires $(4m - 2)$ Xor operations and $(4m - 2)$ Shift operations in Table 6. More precisely, by the loop of m times, $4(m - 1)$ is required. However, only two Xor and Shift are required in the initial loop.

References

1. Kogge, P.M., Stone, H.S.: A parallel algorithm for the efficient solution of a general class of recurrence equations. IEEE Trans. Comput. **100**(8), 786–793 (1973)
2. Goubin, L.: A sound method for switching between Boolean and arithmetic masking. In: Koç, Ç.K., Naccache, D., Paar, C. (eds.) CHES 2001. LNCS, vol. 2162, pp. 3–15. Springer, Heidelberg (2001). doi:10.1007/3-540-44709-1_2
3. Coron, J.-S., Tchulkine, A.: A new algorithm for switching from arithmetic to Boolean masking. In: Walter, C.D., Koç, Ç.K., Paar, C. (eds.) CHES 2003. LNCS, vol. 2779, pp. 89–97. Springer, Heidelberg (2003). doi:10.1007/978-3-540-45238-6_8
4. Neiße, O., Pulkus, J.: Switching blindings with a view towards IDEA. In: Joye, M., Quisquater, J.-J. (eds.) CHES 2004. LNCS, vol. 3156, pp. 230–239. Springer, Heidelberg (2004). doi:10.1007/978-3-540-28632-5_17
5. Rivain, M., Dottax, E., Prouff, E.: Block ciphers implementations provably secure against second order side channel analysis. In: Nyberg, K. (ed.) FSE 2008. LNCS, vol. 5086, pp. 127–143. Springer, Heidelberg (2008). doi:10.1007/978-3-540-71039-4_8
6. Debraize, B.: Efficient and provably secure methods for switching from arithmetic to Boolean masking. In: Prouff, E., Schaumont, P. (eds.) CHES 2012. LNCS, vol. 7428, pp. 107–121. Springer, Heidelberg (2012). doi:10.1007/978-3-642-33027-8_7
7. Vadnala, P.K., Großschädl, J.: Algorithms for switching between Boolean and arithmetic masking of second order. In: Gierlichs, B., Guilley, S., Mukhopadhyay, D. (eds.) SPACE 2013. LNCS, vol. 8204, pp. 95–110. Springer, Heidelberg (2013). doi:10.1007/978-3-642-41224-0_8
8. Beaulieu, R., Shors, D., Smith, J., Treatman-Clark, S., Weeks, B., Wingers, L.: The SIMON and SPECK families of lightweight block ciphers. Cryptology ePrint Archive, Report 2013/404 (2013)
9. Coron, J.-S., Großschädl, J., Vadnala, P.K.: Secure conversion between Boolean and arithmetic masking of any order. In: Batina, L., Robshaw, M. (eds.) CHES 2014. LNCS, vol. 8731, pp. 188–205. Springer, Heidelberg (2014). doi:10.1007/978-3-662-44709-3_11
10. Karroumi, M., Richard, B., Joye, M.: Addition with blinded operands. In: Prouff, E. (ed.) COSADE 2014. LNCS, vol. 8622, pp. 41–55. Springer, Cham (2014). doi:10.1007/978-3-319-10175-0_4
11. Vadnala, P.K., Großschädl, J.: Faster mask conversion with lookup tables. In: Mangard, S., Poschmann, A.Y. (eds.) COSADE 2014. LNCS, vol. 9064, pp. 207–221. Springer, Cham (2015). doi:10.1007/978-3-319-21476-4_14
12. Coron, J.-S., Großschädl, J., Tibouchi, M., Vadnala, P.K.: Conversion from arithmetic to Boolean masking with logarithmic complexity. In: Leander, G. (ed.) FSE 2015. LNCS, vol. 9054, pp. 130–149. Springer, Heidelberg (2015). doi:10.1007/978-3-662-48116-5_7

Side-Channel Analysis of Keymill

Christoph Dobraunig$^{(\boxtimes)}$, Maria Eichlseder, Thomas Korak,
and Florian Mendel

Graz University of Technology, Graz, Austria
`christoph.dobraunig@iaik.tugraz.at`

Abstract. One prominent countermeasure against side-channel attacks,
especially differential power analysis (DPA), is fresh re-keying. In such
schemes, the so-called re-keying function takes the burden of protecting
a cryptographic primitive against DPA. To ensure the security of the
scheme against side-channel analysis, the re-keying function has to withstand both simple power analysis (SPA) and differential power analysis
(DPA). Recently, at SAC 2016, Taha et al. proposed Keymill, a side-
channel resilient key generator (or re-keying function), which is claimed
to be inherently secure against side-channel attacks. In this work, however, we present a DPA attack on Keymill, which is based on the dynamic
power consumption of a digital circuit that is tied to the $0 \rightarrow 1$ and $1 \rightarrow 0$
switches of its logical gates. Hence, the power consumption of the shift-
registers used in Keymill depends on the $0 \rightarrow 1$ and $1 \rightarrow 0$ switches
of its internal state. This information is sufficient to obtain the internal differential pattern (up to a small number of bits, which have to be
brute-forced) of the 4 shift-registers of Keymill after the nonce has been
absorbed. This leads to a practical key-recovery attack on Keymill.

Keywords: Side-channel analysis · Fresh re-keying · Differential power
analysis

1 Introduction

Side-channel attacks like differential power analysis (DPA) pose a serious threat
to devices operating in a hostile environment. Such scenarios quite naturally
appear in our current information infrastructure whenever an entity has physical
access to a device which uses a cryptographic key that must be kept secret from
this entity. Hence, it is necessary to protect such devices against the extraction
of the secret key by means of side-channel analysis like SPA and DPA [7]. In
particular, for resource-constrained or low-cost devices that are used for the
Internet of Things or in RFID applications, the use of protection mechanisms
is not straightforward, since applied protection mechanisms have to be cheap
and efficient. One protection mechanism that suits such applications very well
is fresh re-keying.

Fresh re-keying [9] is an approach for precluding DPA on cryptographic primitives. The resistance against DPA is achieved by a separation-of-duties principle,

© Springer International Publishing AG 2017
S. Guilley (Ed.): COSADE 2017, LNCS 10348, pp. 138–152, 2017.
DOI: 10.1007/978-3-319-64647-3_9

where a re-keying function takes the burden of protection against DPA away from the cryptographic primitive. In this construction, the re-keying function processes a nonce and master key to compute a fresh session key. This session key is then used by the cryptographic primitive. The nonce, or initial value (IV), is generated uniquely for each encryption, and must never be reused for another encryption. The nonce is considered public information and has to be transmitted to (or synchronized with) the decrypting recipient together with the ciphertext. Since the cryptographic primitive is only called once per session key, DPA attacks are naturally prevented, and only dedicated countermeasures against SPA are needed. However, the re-keying function has to provide resistance against SPA and DPA attacks, either by its design, or by application of countermeasures like threshold implementations [10], masking [12], hiding [2], shuffling [6], etc. The intention behind re-keying schemes is that the re-keying function itself can be protected more easily against DPA than the cryptographic scheme, or that it can even be designed to provide inherent security against DPA. Both options profit from the fact that the re-keying function itself does not need to fulfill strong cryptographic requirements [9].

Re-keying Functions. Medwed et al. [9] proposed polynomial multiplication as re-keying function, which has further been extended to the multi-user setting [8]. While such a polynomial multiplication lacks inherent protection against DPA, it is easy to mask and additionally allows easy-to-implement countermeasures against SPA, such as shuffling [9]. However, Pessl and Mangard [11] showed at CT-RSA 2016 that this multiplication is vulnerable to side-channel analysis, in particular at the point where its masks have to be combined and the session key is used in the cryptographic scheme. Additionally, the original scheme by Medwed et al. is susceptible to time-memory trade-off attacks [3]. Recently at Crypto 2016, Dziembowski et al. [4] presented a more formal treatment of re-keying functions and proposed two schemes. The first is based on learning parity with leakage, the second on learning with rounding, and both are efficient and easy to mask.

Keymill. In contrast to designs relying on side-channel countermeasures like masking for side-channel protection, Keymill [14] claims to be secure against side-channel analysis inherently by design without requiring any redundant circuit. Having a re-keying function which provides inherent security against side-channel analysis is beneficial with respect to implementation metrics. Since such schemes do not require masking to withstand DPA, no randomness is needed to create and update masks, and masks do not have to be stored and processed in the first place. A comparison of a modular multiplication and Keymill by Taha et al. [14] shows that a hardware implementation of Keymill requires 775 gate equivalents (GE), while an implementation of a modular multiplication with first-order masking requires 7300 GE [9].

To achieve such low implementation costs, Keymill only uses 4 nonlinear feedback shift-registers taken from the stream cipher Achterbahn [5]. The shift-registers are connected via a rotating cross-connect, which shifts the output of each shift-register's nonlinear feedback function into another shift-register. This cross-connect joins the function outputs with shift-register inputs cyclically per clock. For this construction and also for a toy example consisting of two 8-bit registers involving a similar rotating cross-connect, the authors claim that no DPA attacks are feasible without constructing a hypothesis for the whole key, or equivalently for the whole internal state of the four shift-registers, and thus render DPA attacks infeasible.

Our Contribution. In this work, we present a DPA attack on Keymill. Our attack shows that the claim of Keymill to be inherently secure against side-channel attacks without the need of additional circuits does not hold. The basic idea of the attack is as follows. Instead of making a hypothesis about the exact values of the internal state bits or the secret key, we target the internal difference between neighboring bits of the shift-registers. As observed by Burman et al. [1], and Zadeh and Heys [15], the dynamic power consumption of shift-registers depends on the number of internal differences of neighboring bits. The more internal differences we have, the more power the shift-register consumes. We recover those internal differences bit by bit by comparing the power consumption of a reference nonce (e.g., 0), with power traces of a modified nonce where a single bit has been flipped. Knowing these internal differences allows to recover the full state and consequently the master key by guessing a few additional bits.

Our attack requires the attacker to obtain traces for related (partially chosen) pairs of nonce values, but without violating the single-use requirement for nonces. This scenario is explicitly covered by the security claim of Keymill, although similar to chosen-plaintext attacks, it might not be easy to collect such data in a practical application. We verified the validity and robustness of the attack both for simulated data and for measurements from an FPGA implementation of Keymill.

Outline. In Sect. 2, we give a brief background on fresh re-keying and restate the specification of Keymill. Then, we describe the side-channel attack on Keymill and on a variant of Toy Model II given in the Keymill specification in Sect. 3. Section 4 gives experiments for our attack and discusses the influence of different levels of noise. Finally, we conclude in Sect. 5.

2   Background

In this section, we first give a brief introduction to the concept of fresh re-keying, where we restate the requirements on re-keying functions. Then, we briefly summarize the specification of Keymill and finally, discuss time-memory trade-off attacks on such re-keying schemes.

2.1 Fresh Re-keying

Fresh re-keying has been proposed by Medwed et al. [9] as a countermeasure against side-channel and fault attacks for low-cost devices. A typical scenario where fresh re-keying can be applied is the communication of an RFID tag with an RFID reader. Typically, RFID tags are low-cost devices that additionally have strict requirements regarding power consumption, not allowing costly protection mechanisms against side-channel and fault attacks of the implemented cryptographic primitives. This stands in contrast to the more expensive RFID readers, where costly protection mechanisms like masking are usually affordable.

Figure 1 shows the working principle of fresh re-keying in a communication scenario between an RFID reader and an RFID tag. For sending a message, the tag generates a nonce and derives a session key k^* by using a re-keying function g. This session key is then used by the block cipher E to encrypt the message m. The ciphertext c together with the nonce is sent to the reader, where it can be decrypted.

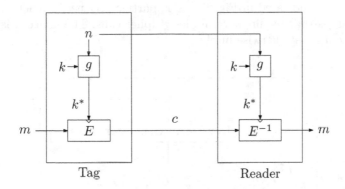

Fig. 1. Fresh re-keying scheme of Medwed et al. [9].

Since the nonce is generated by the tag, the tag can ensure that the block cipher E is always used with a new session key k^*, which will preclude DPA on the block cipher. However, in the case of the reader, having a unique nonce cannot be ensured, because the nonce is received over the communication channel and thus, might be chosen by an attacker. Therefore, the implementation of the block cipher E of the reader has to be protected against DPA by other means. Apart from that, the implementation of g for both entities has to withstand DPA, because here, the master key k is processed with a different nonce. On the designer's side, the challenge is to find a suitable re-keying function g which fulfills the following six properties given by Medwed et al. [9]:

1. Good diffusion of the master key k.
2. No synchronization between parties. Hence, g should be stateless.
3. No need for additional key material.
4. Little hardware overhead. Total costs lower than protecting E alone.
5. Easy protection against side-channel attacks.
6. Regularity.

One option for a re-keying function is the polynomial multiplication in $\mathbb{F}_{2^8}[y]$ modulo $p(y)$ proposed by Medwed et al. [9]:

$$g \; : \; (\mathbb{F}_{2^8}[y]/p(y))^2 \rightarrow \mathbb{F}_{2^8}[y]/p(y), \quad (k, n) \mapsto k \cdot n.$$

2.2 Brief Description of Keymill

Keymill [14] is a new keystream generator recently proposed by Taha et al. at SAC 2016. In contrast to the fresh re-keying scheme by Medwed et al. discussed in Sect. 2.1, Keymill does not only provide one session key k^*, instead it provides a keystream. As indicated in Fig. 2, this is particularly useful when encrypting longer messages that require several block cipher calls. The nonce n is required to be unique, but is otherwise public.

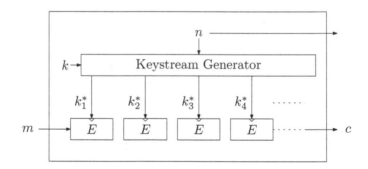

Fig. 2. Re-keying using a keystream generator as shown in [14].

Keymill operates on an internal state of 128 bits, composed of 4 NLFSRs as shown in Fig. 3. Shift-register R_0 has 31 bits, shift-registers R_1 and R_2 have 32 bits, and shift-register R_3 has 33 bits. The feedback functions F_0, F_1, F_2 and F_3 are selected from the set of feedback functions used for the stream cipher Achterbahn [5]:

$$F_0(S) = s_0 + s_2 + s_5 + s_6 + s_{15} + s_{17} + s_{18} + s_{20} + s_{25} + s_8 s_{18} + s_8 s_{20}$$
$$+ s_{12} s_{21} + s_{14} s_{19} + s_{17} s_{21} + s_{20} s_{22} + s_4 s_{12} s_{22} + s_4 s_{19} s_{22}$$
$$+ s_7 s_{20} s_{21} + s_8 s_{18} s_{22} + s_8 s_{20} s_{22} + s_{12} s_{19} s_{22} + s_{20} s_{21} s_{22}$$
$$+ s_4 s_7 s_{12} s_{21} + s_4 s_7 s_{19} s_{21} + s_4 s_{12} s_{21} s_{22} + s_4 s_{19} s_{21} s_{22}$$
$$+ s_7 s_8 s_{18} s_{21} + s_7 s_8 s_{20} s_{21} + s_7 s_{12} s_{19} s_{21} + s_8 s_{18} s_{21} s_{22}$$
$$+ s_8 s_{20} s_{21} s_{22} + s_{12} s_{19} s_{21} s_{22}$$

$$F_1(S) = F_2(S) = s_0 + s_3 + s_{17} + s_{22} + s_{28} + s_2 s_{13} + s_5 s_{19} + s_7 s_{19}$$
$$+ s_8 s_{12} + s_8 s_{13} + s_{13} s_{15} + s_2 s_{12} s_{13} + s_7 s_8 s_{12} + s_7 s_8 s_{14}$$
$$+ s_8 s_{12} s_{13} + s_2 s_7 s_{12} s_{13} + s_2 s_7 s_{13} s_{14} + s_4 s_{11} s_{12} s_{24}$$
$$+ s_7 s_8 s_{12} s_{13} + s_7 s_8 s_{13} s_{14} + s_4 s_7 s_{11} s_{12} s_{24} + s_4 s_7 s_{11} s_{14} s_{24}$$

$$F_3(S) = s_0 + s_2 + s_7 + s_9 + s_{10} + s_{15} + s_{23} + s_{25} + s_{30} + s_8 s_{15} + s_{12} s_{16}$$
$$+ s_{13} s_{15} + s_{13} s_{25} + s_1 s_8 s_{14} + s_1 s_8 s_{18} + s_8 s_{12} s_{16} + s_8 s_{14} s_{18}$$
$$+ s_8 s_{15} s_{16} + s_8 s_{15} s_{17} + s_{15} s_{17} s_{24} + s_1 s_8 s_{14} s_{17} + s_1 s_8 s_{17} s_{18}$$
$$+ s_1 s_{14} s_{17} s_{24} + s_1 s_{17} s_{18} s_{24} + s_8 s_{12} s_{16} s_{17} + s_8 s_{14} s_{17} s_{18}$$
$$+ s_8 s_{15} s_{16} s_{17} + s_{12} s_{16} s_{17} s_{24} + s_{14} s_{17} s_{18} s_{24} + s_{15} s_{16} s_{17} s_{24}$$

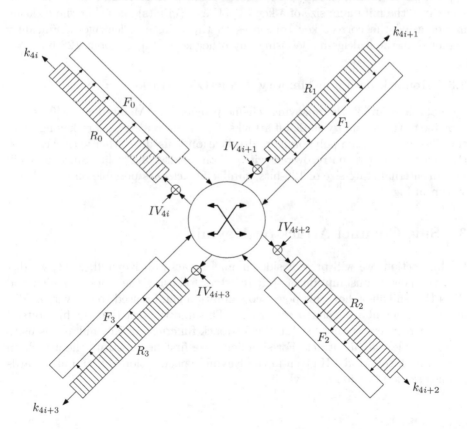

Fig. 3. Structure of Keymill

Note that all feedback functions are nonsingular and additionally do not depend on the first bit $s_{\ell-1}$ of each ℓ-bit register, that is, they are of the form

$$F_j(S) = F_j(s_0, \ldots, s_{\ell-1}) = s_0 + F_j'(s_1, \ldots, s_{\ell-2}).$$

The outputs of the feedback functions are then mixed via a rotating cross-connect, depending on the current clock cycle index i:

$$F_j \rightarrow R_{j+i} \pmod 4 \quad \text{for } j = 0, 1, 2, 3.$$

After loading the 128-bit secret key into the internal state, 4 bits of the 128-bit nonce that can be monitored (or controlled) by the attacker are added to the feedback functions of the shift-registers in each clock cycle. After absorbing the nonce in 32 clock cycles, the internal state is clocked 33 more times before producing any output. Afterwards 4 bits of output are generated (one from each shift-register) in each clock cycle. We refer to the specification of Keymill [14] for a more detailed description.

The designers claim that this construction "expands the size of any useful key hypothesis to the full entropy" [14]. More specifically, they claim that the SCA-security ("the minimum size of a key hypothesis (in bits) such that the leakage-model using the correct key correlates to the measured leakage significantly higher than the leakage-model using any other key" [14]) is about 128 bits.

2.3 Remark on Time-Memory Trade-Off Attacks

As elaborated in [3], the re-keying scheme proposed by Medwed et al. [9] is susceptible to time-memory trade-off attacks dependent on the used re-keying function. For instance, if a polynomial multiplication is used together with AES-128, the master key can be recovered with a complexity of 2^{65} [3]. Since Keymill has an internal state-size of 128-bits, similar attacks are possible on the scheme shown in Fig. 2.

3 Side-Channel Attack on Keymill

In this section, we will present side-channel attacks on Keymill. First, we discuss the power consumption of shift-registers following the work of Zadeh and Heys [15] and show how this power consumption can be used to recover the differences of neighboring shift-register bits. This and the fact that the first bits of the shift-registers are not used in the feedback functions of Keymill allows us to mount a side-channel attack. For simplicity, we first demonstrate the attack on a variant of Toy Model II given in the Keymill specification [14] and afterwards discuss the application to Keymill.

3.1 Power Consumption of a Shift-Register

In all our attacks, we exploit the dynamic power consumption of the shift-registers at the triggering edge of the clock (i.e., positive edge). More specifically, we observe the dynamic power consumption of the building blocks of the shift-registers, the D-flip-flops. As shown by Zadeh and Heys [15], the dynamic power consumption of a D-flip-flop at the triggering edge depends on whether its state changes or not. If the state of the D-flip-flop changes, more power is consumed than if it remains the same. As an example, Zadeh and Heys [15] analyze a D-flip-flop constructed out of 6 NAND gates. For such a flip-flop, 3 gates change if the flip-flop changes its state, whereas only one gate changes if not.

Next, we have a look at the power consumption of a shift-register. For simplicity, consider a 4-bit shift-register consisting of 4 flip-flops D_0, D_1, D_2, and D_3. In the following, we assume that D_4 is the input of our shift-register, which is shifted towards D_0. For instance, let us consider the power consumption of the change from state $S_0 = 0110_2$ to state $S_1 = 1101_2$. For this transition, D_0 changes its state, D_1 keeps its state, D_2 changes its state, and D_3 changes its state. Since the power consumption of the flip-flops is higher if they change their state, the power consumption of the shift-register is correlated with the Hamming weight of $S_0 \oplus S_1 (= 1011_2)$. In this example, 3 flip-flops change their state.

Now, we want to consider a state change from S_0 to S_1', where we shift in a 0 instead of a 1 as before. So we observe the power consumption for the change from state $S_0 = 0110_2$ to state $S_1' = 1100_2$. If this transition happens, only two flip-flops change their state. Thus, we observe for the transition $S_0 \rightarrow S_1'$ a smaller power consumption than for $S_0 \rightarrow S_1$. This allows us to derive information about the difference of the bits stored in D_4 and D_3 of S_1' and S_1, respectively. In more detail, we know that they are equal for S_1' and different for S_1. We will use this observation in our side-channel attack on a variant of Toy Model II and Keymill itself in the following sections.

3.2 Attack on Toy Model II

For the sake of simplicity, we first describe the working principle of our attack on a slightly modified variant of Toy Model II given in the Keymill specification [14], which has only two 8-bit shift-registers. In the attack, we assume that similar to Keymill, the output of the first flip-flop of each shift-register is not connected to the feedback function, as shown in Fig. 4. Besides nonsingularity, this is the only assumption on the feedback function that is necessary to mount our attack. We do not rely on any other specific properties of the feedback functions. The shift-register is preinitialized with the secret key. After that, the 16-bit nonce is absorbed, 2 bits per clock cycle. Our goal is to recover all *internal differences* of both shift-registers after the nonce (e.g., $n = 0000_{16}$) has been absorbed.

First, we collect two power traces, one for a nonce starting with 00_2 and one for a nonce starting with 10_2. We look at the power consumption when the first two bits of the nonce are absorbed in the first cycle. Here, we have a difference

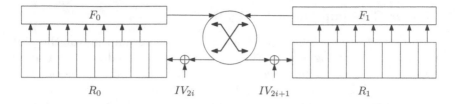

Fig. 4. Structure of modified Toy Model II

in n_0 for R_0, but equal values in n_1 for R_1. Since the first flip-flop of each shift-register is not connected to the feedback function, the circuit processes the same information for both initial values, except for the first flip-flop of the left shift-register R_0. As already discussed in Sect. 3.1, this gives us information about the difference of the first two bits of R_0 after absorbing the first two bits of the nonce. If the power consumption when absorbing 00_2 is higher than in the 10_2 case, we know that the first two bits of R_0 are different after 00_2 is absorbed. If the power consumption is lower, then they are equal.

Next, we use two initial values starting with 00_2 and 01_2. This allows us to learn the internal difference of the first two bits of the shift-register R_1 after 00_2 is absorbed. Then, we use 0000_2 and 0010_2 to learn information of the difference of the first two bits after 0000_2 has been absorbed, still preserving the information of the difference of the now second and third bits of both shift-registers learned in the steps before. By continuing in this way, we can learn the differences of all neighboring bits of R_0 and R_1 after the nonce 0000_{16} has been absorbed.

Now, guessing one bit in each shift-register determines the other 7 bits in each shift-register. Hence, we are left with only 4 possible internal states. From this states on, we can invert Toy Model II step by step until we get 4 key candidates in total. Note that inversion of a fully known state is trivial due to the nonsingularity of the feedback functions, which allows to recover the previous last bit s_0 from the known feedback output and the known values of the other taps. Overall, if we are able to obtain noiseless measurements for about 16 chosen nonces (one per bit of the state), we can recover the entire key k.

3.3   Attack on Keymill

Compared to Toy Model II, Keymill is essentially the same, except everything is larger. As described in Sect. 2.2, we have 4 shift-registers: one 31-bit shift-register, two 32-bit shift-registers, and one 33-bit shift-register. The 128-bit nonce is absorbed in 32 cycles, each cycle taking 4 bits. Furthermore, the 4 feedback functions of Keymill do not consider the outputs of the first flip-flop of each shift-register. As mentioned before, this fact is exploited in our side-channel attack. Again, we want to recover the internal differential pattern of the used shift-registers after a certain nonce, e.g., $n = 0 \cdots 0$ has been absorbed. Please note that the all 0 nonce is just an example taken for simplicity. The attack works for every other choice of the nonce.

The attack proceeds in a similar way as described in Sect. 3.2. First, we record a power trace for a nonce starting with 0000_2 and a second trace for a nonce starting with 1000_2. We compare the power consumption for the two traces at the time the first nibble of the nonce is absorbed. At this time, for both traces, the processed values are equal except for the inputs of shift-register R_0. Since the output of the first flip-flop of R_0 is not fed back into the feedback function, the power consumption differs only because of the state changes of this flip-flop. As discussed in Sect. 3.1, this is sufficient to recover the difference of the first two bits of shift-register R_0. The power traces of nonces starting with 0100_2, 0010_2, and 0001_2 can be used to learn the difference of shift-registers R_1, R_2 and R_3, respectively.

When the second nibble of the nonce is absorbed, those differences are shifted by one position, but are still known, if the first nibble of the nonce starts with 0000_2. Hence, we can use nonces starting with $0000\,0000_2$, $0000\,1000_2$, $0000\,0100_2$, $0000\,0010_2$, and $0000\,0001_2$ and learn the differences of the first two bits of each shift-register, while retaining the knowledge of the differences between the second and the third bits. Proceeding this way, we can learn at most 32 differences of neighboring bits per shift-register.

This means that we can learn all internal differences of all 4 shift-registers, since one shift-register has 31 bits, two have 32 bits and one has 33 bits. So, at most 30, two times 31, and 32 differences have to be learned. Since we know all internal differences of each shift-register, a guess of one state bit in each shift-register determines all others. Thus, guessing 4 bits in total leads to 16 different states we recover. From these states, we can invert Keymill, resulting in 16 possible key candidates in total.

Summarizing, if we can obtain noiseless measurements for about 128 chosen nonces, then we can recover the full internal state and consequently the secret key k. In particular, we recover the internal state bit by bit by making a hypothesis on 1 bit of "equivalent key information", instead of an actual key bit value: The xor difference of two neighboring state bits.

3.4   A Note on Filtering the Noise

The success of our attacks crucially depends on the ability to distinguish power consumption changes for a change of the input values. This means that the noise level has to be small enough to reliably identify these changes. If the attacker is allowed to repeat nonces, averaging the traces and filtering the noise is no problem. Even if the nonce is required to be unique (as usually the case), this can easily be done, since the state of the shift-registers only depends on bits of the nonce that have already been absorbed. Hence, we can use all the remaining nonce bits after the relation we want to recover to average the power consumption for this cycle. For Keymill, we can average over up to 16 power traces even if we recover bit relations in the penultimate nonce-absorbing cycle. Dependent on the noise level, it might happen that the last few internal differences of the state cannot be recovered anymore, since there are too few traces to filter the noise. So these bits might have to be guessed additionally at the end of the attack.

4   Practical Evaluation

In order to show the practicability of the attacks discussed in Sect. 3, we present two experiments. First, we run the attack based on simulated leakage traces to analyze the impact of noise on the success of the attack. For the second evaluation, we use power measurements from an FPGA implementation of Keymill to evaluate the practicability of the attack targeting real hardware.

First, we simulate the described attack targeting the proposed Keymill design as shown in Fig. 3. Therefore, the four registers $R_0 \ldots R_3$ and the corresponding feedback functions $F_0 \ldots F_3$, which compose the four NLFSRs, have been modelled in software. At the start of the simulation, the registers are initialized with the secret key. Then, for every clock cycle, the simulation returns the Hamming distance produced by the shift registers. The current Hamming distance depends on the values in the shift register, the results of the feedback functions $F_0 \ldots F_3$ and the nonce.

Gaussian noise with zero mean ($\mu_{\text{noise}} = 0$) and varying standard deviation σ_{noise} can be added to the noise-free Hamming-distance measurements ($\text{HD}_{\text{noisefree}}$) in order to simulate measurements captured from real hardware, i.e. HD_{meas} (see Eq. 1). In order to minimize the influence of the noise it is possible to repeat the simulation with a similar nonce t times for calculating the mean of the measurements.

$$\text{HD}_{\text{meas}} = \text{HD}_{\text{noisefree}} + \text{noise}, \qquad \text{where noise} \leftarrow \mathcal{N}(0, \sigma_{\text{noise}}). \qquad (1)$$

For every setting (specific σ_{noise} and specific t), we performed $N_{\text{full}} = 500$ experiments with randomly chosen initial states of the four shift-registers $R_0 \ldots R_3$ to calculate the success rate SR of the attack,

$$\text{SR} = \frac{N_{\text{success}}}{N_{\text{full}}},$$

where N_{success} is the number of successful state recoveries. Figure 5 depicts the results of this simulation. It is clearly visible that SR decreases with increasing noise. This effect can be compensated by repeating the attack with the same nonce t times and calculate the mean of the measurements. For $t = 1$, the success rate starts to decrease for noise levels above $\sigma_{\text{noise}} = 0.1$. For $t = 50$, the success rate remains 1 up to a noise level of $\sigma_{\text{noise}} = 1.3$.

Figure 6 shows the influence of σ_{noise} on the Hamming-distance measurements (HD_{meas}). For this specific plot, $\text{HD}_{\text{noisefree}} = 64$ has been selected. The '+' markers represent single HD measurements. In the noise-free scenario, i.e. $\sigma_{\text{noise}} = 0$, all HD measurements have the value 64. For a high noise level, i.e. $\sigma_{\text{noise}} = 2$, the HD measurements are in the range between 58 and 70.

In a final experiment, Keymill is evaluated on real hardware. We chose the Sakura G board [13], which is the reference platform for side-channel evaluations of cryptographic hardware designs on FPGAs. The main FPGA (Xilinx Spartan-6 LX75) has been configured with the Keymill design and the power consumption during the initialization (i.e. the first 33 clock cycles where the bits

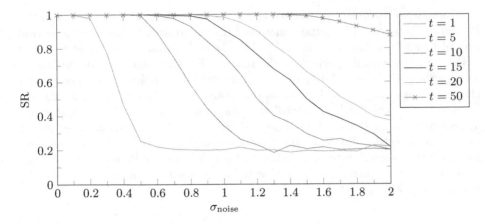

Fig. 5. Success rate (SR) for increasing noise levels (σ_{noise}). For the graphs different numbers (1–50) of Hamming-distance measurements have been used for calculating the mean Hamming distance.

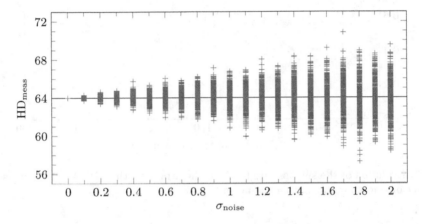

Fig. 6. Hamming distance measurements for increasing σ_{noise}, $HD_{noisefree} = 64$.

of the nonce are shifted into the shift registers, four bits per clock cycle) has been measured with an oscilloscope. For every bit position of the nonce, two trace sets have been recorded, one with the corresponding bit set to '0' and one with the corresponding bit set to '1'. In order to evaluate the number of traces required for reaching a specific success rate, 10 000 traces have been recorded for every nonce. The results of the evaluations are depicted in Fig. 7. It shows that for the given FPGA implementation, at least 220 measurements for every nonce are required for reaching a success rate of 1. In scenarios where repeated measurements of the same nonce are prohibited, iterating over the last 8 bits of the nonce can be done to average the measurements. This leads to 256 traces per fixed 120 bits that can be used to filter the noise.

Comparing the means of the two trace sets allows to distinguish between the Hamming distances. The higher amount of traces required for reaching a success rate of 1 indicates that the noise on real hardware is significantly larger than the noise during previously performed simulations. For the sake of completeness we have performed the simulations for $t = 220$ and larger noise levels. The results show that for $\sigma_{\text{noise}} \geq 3.6$ the success rate starts to decrease for $t = 220$. Experiments on the real hardware reveal that for recovering the whole initial state, approximately $220 \cdot 128 = 28\,160$ measurements are required in total. The applied measurement setup allows us to collect the required amount of measurements for reaching a success rate of 1 within an hour. With some improvements of the setup the measurement time could be reduced to a few minutes, but this was not the goal of this work.

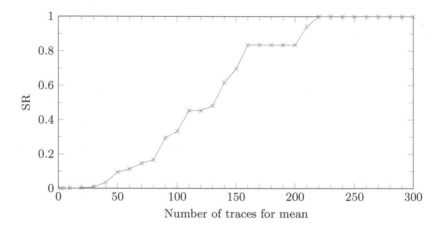

Fig. 7. Evolution of the success rate (SR) for the attack on the FPGA with increasing number of traces for calculating the mean.

5   Conclusion

In this work, we showed that a DPA attack on Keymill is feasible. In contrast to the DPA attacks that are claimed to be thwarted by the specification of Keymill, we do not make hypotheses on the actual values of Keymill's key or internal state. Instead, we first recover the internal differences of neighboring bits step by step from side-channel measurements, and then take advantage of the resulting entropy reduction to recover the actual values. Our attack violates the claim by the designers that Keymill is inherently secure against side-channel attacks by design. Indeed, we show that Keymill needs dedicated countermeasures against DPA attacks exploiting internal differences.

Our attack requires the ability of an attacker to choose the nonces. Therefore, guaranteeing that only random nonces can be used seems to be an efficient countermeasure. Although this prevents a straightforward application of our attack to recover all differences between state-bits, the recovery of just a fraction of the differences of the first few bits still remains possible. Hence, it is part of future work to evaluate if extensions of the presented attack concept are applicable for random nonces.

Acknowledgments. This work has been supported in part by the Austrian Science Fund (project P26494-N15) and by the Austrian Research Promotion Agency (FFG) under grant number 845589 (SCALAS).

References

1. Burman, S., Mukhopadhyay, D., Veezhinathan, K.: LFSR based stream ciphers are vulnerable to power attacks. In: Srinathan, K., Rangan, C.P., Yung, M. (eds.) INDOCRYPT 2007. LNCS, vol. 4859, pp. 384–392. Springer, Heidelberg (2007). doi:10.1007/978-3-540-77026-8_30
2. Clavier, C., Coron, J.-S., Dabbous, N.: Differential power analysis in the presence of hardware countermeasures. In: Koç, Ç.K., Paar, C. (eds.) CHES 2000. LNCS, vol. 1965, pp. 252–263. Springer, Heidelberg (2000). doi:10.1007/3-540-44499-8_20
3. Dobraunig, C., Eichlseder, M., Mangard, S., Mendel, F.: On the security of fresh re-keying to counteract side-channel and fault attacks. In: Joye, M., Moradi, A. (eds.) CARDIS 2014. LNCS, vol. 8968, pp. 233–244. Springer, Cham (2015). doi:10.1007/978-3-319-16763-3_14
4. Dziembowski, S., Faust, S., Herold, G., Journault, A., Masny, D., Standaert, F.-X.: Towards sound fresh re-keying with hard (physical) learning problems. In: Robshaw, M., Katz, J. (eds.) CRYPTO 2016. LNCS, vol. 9815, pp. 272–301. Springer, Heidelberg (2016). doi:10.1007/978-3-662-53008-5_10
5. Gammel, B.M., Göttfert, R., Kniffler, O.: Achterbahn-128/80. eSTREAM, ECRYPT Stream Cipher Project (2006)
6. Herbst, C., Oswald, E., Mangard, S.: An AES smart card implementation resistant to power analysis attacks. In: Zhou, J., Yung, M., Bao, F. (eds.) ACNS 2006. LNCS, vol. 3989, pp. 239–252. Springer, Heidelberg (2006). doi:10.1007/11767480_16
7. Kocher, P., Jaffe, J., Jun, B.: Differential power analysis. In: Wiener, M. (ed.) CRYPTO 1999. LNCS, vol. 1666, pp. 388–397. Springer, Heidelberg (1999). doi:10.1007/3-540-48405-1_25
8. Medwed, M., Petit, C., Regazzoni, F., Renauld, M., Standaert, F.-X.: Fresh re-keying II: securing multiple parties against side-channel and fault attacks. In: Prouff, E. (ed.) CARDIS 2011. LNCS, vol. 7079, pp. 115–132. Springer, Heidelberg (2011). doi:10.1007/978-3-642-27257-8_8
9. Medwed, M., Standaert, F.-X., Großschädl, J., Regazzoni, F.: Fresh re-keying: security against side-channel and fault attacks for low-cost devices. In: Bernstein, D.J., Lange, T. (eds.) AFRICACRYPT 2010. LNCS, vol. 6055, pp. 279–296. Springer, Heidelberg (2010). doi:10.1007/978-3-642-12678-9_17
10. Nikova, S., Rijmen, V., Schläffer, M.: Secure hardware implementation of non-linear functions in the presence of glitches. In: Lee, P.J., Cheon, J.H. (eds.) ICISC 2008. LNCS, vol. 5461, pp. 218–234. Springer, Heidelberg (2009). doi:10.1007/978-3-642-00730-9_14
11. Pessl, P., Mangard, S.: Enhancing side-channel analysis of binary-field multiplication with bit reliability. In: Sako, K. (ed.) CT-RSA 2016. LNCS, vol. 9610, pp. 255–270. Springer, Cham (2016). doi:10.1007/978-3-319-29485-8_15

12. Prouff, E., Rivain, M.: Masking against side-channel attacks: a formal security proof. In: Johansson, T., Nguyen, P.Q. (eds.) EUROCRYPT 2013. LNCS, vol. 7881, pp. 142–159. Springer, Heidelberg (2013). doi:10.1007/978-3-642-38348-9_9
13. Sakura-G - Side-Channel Evaluation Board. http://satoh.cs.uec.ac.jp/SAKURA/hardware/SAKURA-G.html, Accessed 28 Nov 2016
14. Taha, M., Reyhani-Masoleh, A., Schaumont, P.: Keymill: side-channel resilient key generator. In: Avanzi, R., Heys, H. (eds.) SAC 2016. LNCS, Springer (2016), (to appear). eprint version. http://eprint.iacr.org/2016/710
15. Zadeh, A.A., Heys, H.M.: Simple power analysis applied to nonlinear feedback shift registers. IET Inf. Secur. 8(3), 188–198 (2014)

On the Easiness of Turning Higher-Order Leakages into First-Order

Thorben Moos[✉] and Amir Moradi

Horst Görtz Institute for IT Security, Ruhr-Universität Bochum, Bochum, Germany
{thorben.moos,amir.moradi}@rub.de

Abstract. Applying random and uniform masks to the processed intermediate values of cryptographic algorithms is arguably the most common countermeasure to thwart side-channel analysis attacks. So-called
masking schemes exist in various shapes but are mostly used to prevent
side-channel leakages up to a certain statistical order. Thus, to learn any
information about the key-involving computations a side-channel adversary has to estimate the higher-order statistical moments of the leakage
distributions. However, the complexity of this approach increases exponentially with the statistical order to be estimated and the precision of
the estimation suffers from an enormous sensitivity to the noise level.
In this work we present an alternative procedure to exploit higher-order
leakages which captivates by its simplicity and effectiveness. Our approach, which focuses on (but is not limited to) univariate leakages of
hardware masking schemes, is based on categorizing the power traces
according to the distribution of leakage points. In particular, at each
sample point an individual subset of traces is considered to mount ordinary first-order attacks. We present the theoretical concept of our approach based on simulation traces and examine its efficiency on noisy
real-world measurements taken from a first-order secure threshold implementation of the block cipher PRESENT-80, implemented on a 150 nm
CMOS ASIC prototype chip. Our analyses verify that the proposed technique is indeed a worthy alternative to conventional higher-order attacks
and suggest that it might be able to relax the sensitivity of higher-order
evaluations to the noise level.

1 Introduction

It has become a general knowledge that implementations of cryptographic algorithms are in danger of being attacked by means of side-channel analysis (SCA)
key-recovery attacks, if dedicated countermeasures have not (or incorrectly) been
integrated. Amongst the known and common SCA countermeasures, *masking*
is by far the most-widely studied scheme and has interested both academia
and industry. Its underlying sound proofs and theoretical foundation should be
named among the reasons for such a popularity. Except particular constructions
(e.g., [7,12]), the security of masking schemes is based on the uniformity of the
masks. More precisely, in an $(s+1)$-sharing construction, which is called s-order

© Springer International Publishing AG 2017
S. Guilley (Ed.): COSADE 2017, LNCS 10348, pp. 153–170, 2017.
DOI: 10.1007/978-3-319-64647-3_10

masking, for a particular x each $(x_1, \ldots, x_{s+1})$ with $x = \bigoplus_{i=1}^{s+1} x_i$ should occur equally likely[1]. Otherwise, it can be pretended that the randomness source is biased, which potentially leads to exploitable leakage.

With respect to the adversary model, security of masking schemes is evaluated based on two different models: (i) probing model [10], and (ii) bounded moment model [2]. The former one is primarily used for security proofs and more conservative than the later one, which is usually applied in practical evaluations. Our focus is mainly on the *bounded moment* model, and we call a device *without first-order leakage* if the leakages associated to two different given sets of operands x and y (of the same operation[2]) are not distinguishable[3] from each other through average, i.e., first-order statistical moment. Similarly the leakages should **further** not be distinguishable through variance, i.e., second-order centered moment, for second-order security, and likewise for higher orders. Optionally, the described setting can be incorporated by a pre-processing step, which combines different leakage points. Compared to *univariate* settings, where the combination of leakage points is not required, in a *multi-variate* scenario two (or more) different leakage points are combined prior to evaluation/attack (see [14] for more details).

In short, in order to attack an s-order masked implementation, multi-variate $(s+1)$-order statistical moments should be observed if the operations are serially performed on the shares (i.e., a typical software implementation with sequential nature). On the other hand, in case of a hardware implementation usually univariate $(s + 1)$-order statistical moments are observed due to the inherent parallel processing fashion. It is noteworthy that the complexity of higher-order evaluations increases exponentially with s. Further, estimation of higher-order statistical moments becomes extremely hard in practice when the leakages are sufficiently noisy [21].

Instead of a conventional higher-order attack, we present in this work a trick that converts higher-order leakages to the first order and exploits them for key recovery. The focus of our scheme is **univariate** higher-order leakages, i.e., mainly targeting masked hardware implementations. It is essentially based on the principle of pruning the traces according to the distribution of leakage points. Its detailed expression is given in Sect. 3. Indeed, a similar approach has initially been considered in [23], to exploit the leakage of a masked dual-rail logic style (MDPL) [19]. We review the relevant state of the art in Sect. 2. Compared to a classical higher-order attack (e.g., mean-free square as an optimal second-order univariate attack) our scheme can be more efficient in particular cases. More precisely, it can exploit the leakage and recover the key while the classical higher-order attacks fail. As a case study, given in Sect. 4, we present practical results based on an ASIC prototype chip of a provably first-order secure threshold implementation (TI) [17] of the block cipher PRESENT [4].

---

[1] In case of Boolean masking.
[2] For example, two different plaintexts of an AES encryption with a fixed key.
[3] t-test can be used to detect the distinguishability [24].

2   State of the Art

For the majority of masking schemes it is a mandatory requirement that the masks are drawn from a uniform distribution. If this distribution is not uniform, but rather stems from a biased randomness source, vital security claims are not met and exploitable first-order leakage can emerge. Thus, an adversary might be interested in compromising the security of masked implementations deliberately by forcing a bias into the masks that conceal key-dependent intermediate values. One way of achieving this goal is to attack the randomness source directly by means of fault attacks. Of course, the feasibility of this approach depends highly on the particular implementation that is investigated. Another, more generic strategy, which has mainly been applied to compromise software-based masking schemes on microcontrollers, is to categorize the traces that are recorded in a power analysis attack into groups that only contain a biased subset of all possible masks. Intuitively, such an attack can be performed on a software-based masking scheme by determining a point in the power traces where the mask value is processed and then discarding all traces with a measured power consumption above (or below) a certain threshold at that sample point. Assuming now that the investigated device leaks information about the processed intermediate values by means of the Hamming weight (HW) model (which is a reasonable assumption for microcontrollers, see [13]), one has selected a subset of traces with a probability different from $\frac{1}{2}$ for each mask bit to be 1 (or 0). This allows a better-than-random guess what the mask value would be, e.g. all-one (or all-zero), which enables successful first-order attacks on the reduced set of traces. Hence, without preprocessing the power values in the traces, but only by ignoring a subset of the acquired measurements, one has moved the higher-order leakages to a setting where they can be exploited in the first order. Technically, due to the prior selection of power traces, this is still a higher-order attack, but in fact does not require the estimation of higher-order statistical moments. This kind of attack, which we extend and generalize for a different setting in the following course of this work, is referred to as biased mask attack, e.g. in [13,25]. Regardless of the surprisingly simple attack procedure, biased mask attacks have not gained much popularity since multi-variate higher-order attacks, utilizing the higher-order statistical moments of the *full* set of traces, are considered more powerful in the general case. Indeed, the loss of information due to disregarding a subset of the measurements is undeniable. Additionally, some kind of initial profiling has to be performed to find a sample point in the power traces where the mask value is leaked.

The described procedure can not be mapped directly to hardware implementations, because in parallel designs the mask is not processed discretely but usually together with the masked data and a number of further intermediate values at the same time. Consequently, only the cumulative leakage of mask and masked data can be observed in a univariate fashion and is not only buried in electronic noise, like for software implementations, but also in the switching noise originating from the remaining parts of the circuit (see [13]). On the one hand, due to the univariate nature of the leakages, the necessity for a profiling phase is removed, but on the other hand the categorization of the traces based on the leakage of the mask value is much less precise. Nevertheless several attempts have been made to perform biased mask attacks on hardware implementations of gate- and algorithmic-level masking schemes. In [26], such an approach is considered for the first time. It is shown by toggle count simulations of a small test circuit (S-box+key XOR) that categorizing power traces with a simple threshold filter is sufficient to remove the one bit of entropy that is introduced by the use of the logic style Random Switching Logic (RSL). The affiliated work in [23] utilizes gate-level simulations of an AES chip design to show that routing imbalances in the DPA-resistant logic style MDPL [19] can be exploited to estimate the mask bit. Again, this can be used to remove the effect of the masking scheme by performing conventional first-order DPA attacks exclusively on the subset of traces that is obtained through a simple filtering operation. In [8] the authors extend their approach to an algorithmic-level hardware masking scheme for the first time. In accordance to the biased mask attacks on software-based implementations the authors are able to verify that a secure hardware masking scheme can equally be compromised by means of simple first-order distinguishers, when only a subset of the traces is considered. Unfortunately, the article fails to investigate how to select a suitable subset of traces that is most informative for an attack. Even more importantly it is not examined at all whether a first-order attack on their specific (or any other choice of) subset can outperform a univariate second-order attack using the mean-free square on the full set of traces. Finally, none of the listed works on hardware masking schemes verified the described attack procedures with practical measurements, taken from a physical hardware device. To the best of our knowledge, no subsequent work explores any of these data points either.

The last branch of research that can be considered related to our approach uses a subset of power traces to enhance the correlation in CPA [6] attacks in general, without concentrating on protected implementations or circumventing specific countermeasures in particular. These works, presented e.g., in [11,18], focus on selecting power traces with a high Signal-to-Noise ratio (SNR). They come to the conclusion that, considering the distribution of power values at the point of interest, especially those traces with a small probability density function value, have the highest SNR. In a simplified phrasing this means that concentrating on the power traces whose value at the point of interest is extraordinarily low or high (leftmost or rightmost slices of the leakage distribution) leads to the best correlation for the correct key candidate.

3   Underlying Approach

In this section we introduce and define our novel approach to exploit higher-order leakages. For the sake of simplicity, let us focus on a single sample point of side-channel leakages. The main idea is to observe the distribution of the **univariate** leakages, categorize them into e.g., two non-overlapping parts, and then perform the attack(s) on each part independently. This indeed is the same concept which has been applied in [11] on *unprotected* implementations with the goal of improving the attacks with respect to the required number of traces (see Sect. 2). However, we employ more-or-less the same technique to exploit higher-order leakages. Let us express the underlying concept with simulation results. Suppose that the leakage of a device under test (DUT) can be represented by a noisy Hamming weight (HW) model as

$$l(x) = HW(x) + \mathcal{N}(\mu, \delta^2),$$

with mean $\mu = 0$ and standard deviation δ. Further, suppose that the intermediate values of the DUT are masked following the concept of first-order Boolean masking. Hence, every value x is represented by (x_m, m) with $x_m = x \oplus m$ and m being a random mask with uniform distribution. In a univariate setting, the leakage of the DUT associated to x is represented by

$$l(x_m) + l(m) = HW(x_m) + HW(m) + \mathcal{N}(0, \delta^2).$$

If we simulate 1,000,000 times the leakage for two different $x \in \{0, 1\}^8$ values and a particular $\delta = 2$, two different distributions are observed, that are depicted in Fig. 1(a). These two distributions are not distinguishable from each other through their means, i.e., a first-order distinguisher would not be able to differentiate them. Along the same lines, t statistics of a Welch's t-test would give a low-confidence result as well, i.e., t being smaller than 4.5.

However, if we consider only those leakages which are less than a threshold, see Fig. 1(b), the leakages are distinguishable from each other through their means. For example, in this case the t statistics yields the value 133, i.e., high confidence of a first-order distinguisher. The threshold in this example has been defined in such a way that 20% of the leakages are below the threshold and the remaining 80% above. As shown in Fig. 1(c), (d) and (e), considering the upper 80%, lower 80% or upper 20% leakages would lead to distinguishability through means as well. However, in case of Fig. 1(f) and (g) when the middle part or the side parts of the distributions are considered, the mean does not reveal any distinguishability. This is indeed due to the symmetric form of the original distributions shown in Fig. 1(a).

We should highlight that these observations are not limited to first-order masking. As an example, we repeated the same simulation under second-order Boolean masking with univariate leakage

$$l(x_m) + l(m_1) + l(m_2) = HW(x_m) + HW(m_1) + HW(m_2) + \mathcal{N}(0, \delta^2),$$

where $x_m = x \oplus m_1 \oplus m_2$ and the uniform distribution for m_1 and m_2. The distributions and the t statistics as distinguishability measure after classifying the leakages based on a particular threshold are shown in Fig. 2. Following the concept of second-order masking, the distributions are distinguishable only through their skewness (see Fig. 2(a)). However, by categorizing them based on a 20%

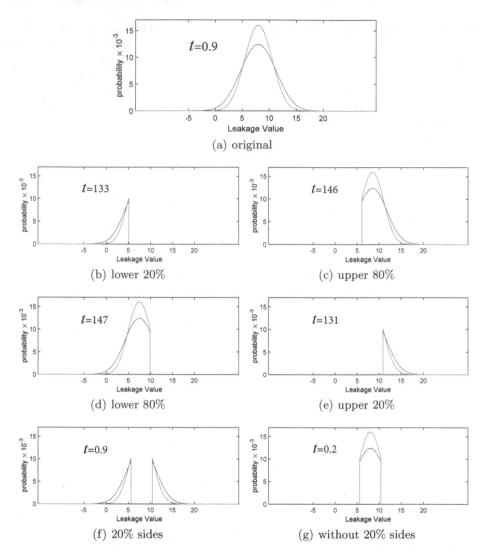

Fig. 1. Simulated leakage distributions of two different values represented by first-order masking, t represents the statistics of the t-test.

threshold (either above or below the threshold) the means reveal the difference between the distributions. Interestingly, the symmetric forms, i.e., middle part or the sides (Fig. 2(f) and (g)), also lead to high-evidence first-order distinguishability.

When evaluating the effectiveness of this approach it is important to know for which threshold value the attack performs best. To identify the optimal threshold, we conducted another simulation based on first-order masking. We have randomly selected a vector of n elements as $X : (x^1, \ldots, x^n)$,

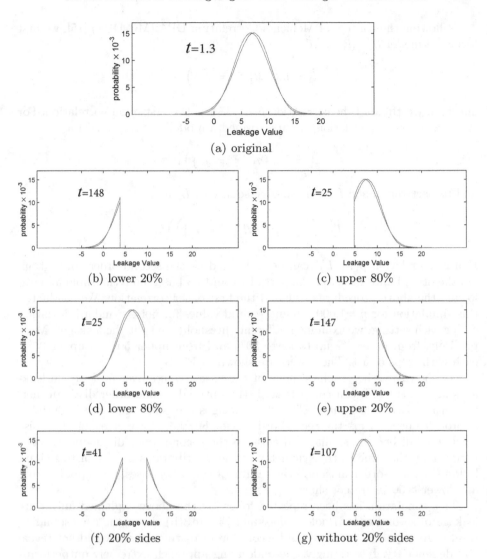

Fig. 2. Simulated leakage distributions of two different values represented by second-order masking, t represents the statistics of the t-test.

where $x^i \in \{0,1\}^8$. Then, by two separate uniformly-distributed n−element mask vectors M_1 and M_2 we formed $X_{M_1} = (x_{m_1}^1, \ldots, x_{m_1}^n)$, where $x_{m_1}^i = x^i \oplus m_1^i$ (resp. for X_{M_2}). Following the univariate noisy Hamming weight leakage model, we formed two leakage vectors $L_1 : (l_1^1, \ldots, l_1^n)$ and $L_2 : (l_2^1, \ldots, l_2^n)$ in such a way that for example

$$l_1^i = HW(x_{m_1}^i) + HW(m_1^i) + \mathcal{N}(0, \delta^2).$$

Following the concept of Moments-Correlating DPA (MC-DPA) [15], we first formed a model $\dot{L}_1 : (\dot{l}_1^1, \ldots, \dot{l}_1^n)$ as

$$\dot{l}_1^i = \mu\left(\{\forall l_1^j | x^j = x^i\}\right),$$

and finally estimated the correlation $\rho(\dot{L}_1, L_2)$ as the first-order correlation. For the second-order correlation, we first formed a model $\ddot{L}_1 : (\ddot{l}_1^1, \ldots, \ddot{l}_1^n)$ as

$$\ddot{l}_1^i = \delta^2\left(\{\forall l_1^j | x^j = x^i\}\right),$$

and respectively made L_2' as mean-free square of L_2 as

$$l_2'^i = \left(l_2^i - \mu\left(\{\forall l_2^j | x^j = x^i\}\right)\right)^2.$$

Hence, correlation $\rho(\ddot{L}_1, L_2')$ can be estimated as the second-order correlation. On the other hand, we selected a part of L_1 and L_2 based on a threshold and following the above procedure estimated the first-order correlation. We conducted this simulation for $n = 1{,}000{,}000$ and several values for noise standard deviation δ. For each setting, we examined different thresholds to split the leakages. More precisely, from lower 5% up to lower 50% and from upper 50% to upper 95%, each with steps of 5%. The results are shown in Fig. 3(a).

As shown by the graphics, none of the cases, where over 50% of the leakages are considered, can compete with the optimal second-order distinguisher. In contrast, when less than 50% of the leakages are considered, the underlying approach outperforms the second-order one. Further, by increasing the noise level they all become similar and close to the second-order distinguisher. It is noteworthy that due to the symmetry of the distributions in case of this simulation (i.e., first-order masking) the results of the other cases, i.e., upper $< 50\%$ and lower $> 50\%$, are not shown.

This simulation has been repeated following the above-explained univariate leakage of second-order Boolean masking. Figure 3(b) shows the corresponding results. As expected, the first- and second-order distinguishers would not reveal any dependency. Interestingly, the underlying approach extremely outperforms the optimal third-order distinguisher, and even by increasing the noise standard deviation it still performs better.

We should note that any other distinguisher, where instead of any particular statistical moment the distribution of the leakages are considered, would also differentiate the univariate higher-order leakages. But, these distinguishers (e.g., MIA [9]) would need to predict the probability distributions, e.g., by histogram where the number of bins and the size of each bin play an important role for the efficiency of the distinguisher, alternatively by Kernel where the important issues include the type of the Kernel function and the associated parameters. The diversity of their results based on the selected parameters can make such distinguishers more complicated or less efficient compared to higher-order attacks. However, in the approach presented here we just consider the

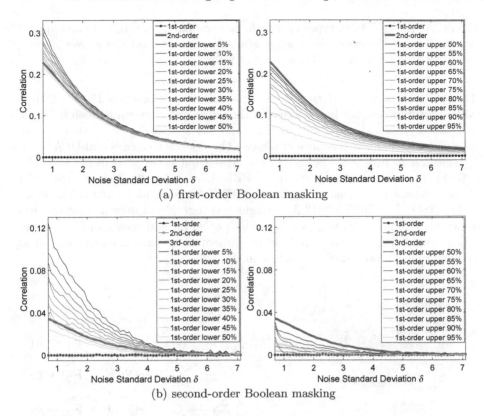

(a) first-order Boolean masking

(b) second-order Boolean masking

Fig. 3. Correlation (based on MC-DPA), simulated univariate (a) second-order and (b) third-order leakages, comparison between different distinguishers for different threshold values over noise standard deviation.

distribution obtained based on pure histogram. More precisely, the histogram made by the nature of the SCA measurements (i.e., 256 bins as the result of the 8-bit ADC[4] of the acquisition equipment digital oscilloscope) would suffice to find the threshold for a given percentage, e.g., lower 20%.

4   Practical Results

Now that we have presented the theoretical concept of our approach, it is time to evaluate the soundness of the technique based on real-world measurements taken from the physical implementation of a hardware masking scheme. After a description of the target device and the measurement setup we analyze the side-channel leakage of the test chip by means of conventional higher-order attacks, which are based on the estimation of higher-order statistical moments. As a second step we present the results of our novel approach for different threshold

---

[4] Analog to Digital Converter.

values. At the end, both types of attacks are compared in terms of the required number of measurements for a successful key recovery and the convenience of the procedure from an attacker's point of view.

Target. The target platform for our practical evaluations is a 150 nm CMOS ASIC prototype chip. A layered view of the fabricated chip can be seen in Fig. 4. The prototype contains 6 different cores and was specifically developed to evaluate the side-channel resistance of state-of-the-art block ciphers and DPA countermeasures in practice. The core of the ASIC that is targeted in the following experiments realizes the block cipher PRESENT-80 under 3-share first-order threshold implementation concept. PRESENT-80 is an ultra-lightweight block cipher (ISO/IEC 29192-2:2012 lightweight cryptography standard) that features a block size of 64 bit as well as a key length of 80 bit and consists of 31 computation rounds [4], whereas threshold implementations have been introduced as an efficient hardware masking scheme in [17].

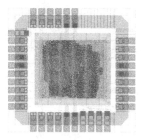

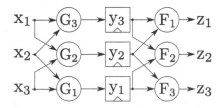

Fig. 4. ASIC prototype with 6 cores in 150 nm CMOS.

Fig. 5. Threshold implementation of the 4-bit PRESENT S-box with 3 shares.

Concerning hardware implementations of masking schemes, it has historically been a challenging task to ensure that glitches in the combinatorial parts of the circuit do not recombine the shares and thus lead to exploitable leakage. Threshold implementations prevent this issue by adding the so-called non-completeness property to the masked computations [2]. Non-completeness means here that each fully combinatorial circuit must be independent of at least one of the shares. This is achieved by splitting the non-linear parts of a circuit into several shared functions that do not operate on all shares at once, but rather perform only one part of the overall computation that refers to its respective inputs. Accordingly, glitches can never recombine all shares at once, meaning that an adversary is not able to learn any information about the secret from the side-channel leakage of only one of these circuits. Indeed, multiple leakages of multiple combinatorial (sub-) circuits need to be combined to perform a successful (higher-order) attack. Following this concept, which is based on Boolean secret sharing and multi-party computation, the threshold implementation technique can be used to implement non-linear functions of symmetric block ciphers in such a way that provable security against first-order power analysis attacks can be guaranteed, even in the presence of glitches. Higher-order threshold implementations can furthermore be used to conceal the leakages at higher-order statistical moments [3]. A second property that has to be fulfilled when sharing a non-linear function is the uniformity of the outputs. For each unshared input to the non-linear function, each shared output should occur equally likely. In this way the output of the shared functions is still uniformly distributed and a remasking is not required. More precisely during the full execution of a block cipher that is implemented in this masking scheme no fresh masks needs to be fed. The plaintext is split up into the required number of shares at the beginning of the algorithm (see [17]), which implies the generation of two or more plaintext-sized masks, and all further computations are performed on those shares. Compared to conventional masking, the drawback of this method is a higher number of required shares. In particular at least three shares (two masks) are required to realize each non-linear part of a circuit[5]. Additionally the number of shares increases with the degree of the function that needs to be implemented [17]. Hence, larger S-boxes, e.g. 8-bit, are difficult to implement efficiently in this scheme [5]. Nevertheless, for ciphers with small S-boxes, e.g., PRESENT-80, threshold implementation has become the de facto standard for hardware masking [2].

The realization of the PRESENT-80 block cipher as a threshold implementation was introduced in [20]. The authors proposed several implementation profiles with different levels of security. Our targeted ASIC core implements profile 2, which refers to a nibble-serial implementation of the block cipher with a shared data path (with 3 shares) but an unshared key schedule. Hence, one instance of the shared S-box is implemented and the 4-bit nibbles of the cipher state are processed in a pipelined manner. A schematic view of the shared S-box, based on a decomposition to quadratic functions F and G with $S(x) = F(G(x))$, can be seen in Fig. 5. Due to the register stage between the G- and the F functions one full cipher-round takes 18 clock cycles[6]. It is noteworthy that although first-order threshold implementation corresponds to Boolean masking with 3 shares it provides only first-order security due to its underlying quadratic functions (i.e., G and F in Fig. 5). In other words, this implementation is supposed to exhibit second- and third-order leakages.

Measurement Setup. We performed our measurements on a Side-channel Attack Standard Evaluation Board (SASEBO-R) [1] that was specifically developed to evaluate the side-channel resistance of cryptographic hardware. For this purpose it provides a socket for an ASIC prototype, which is connected by a 16-bit bidirectional data bus as well as a 16-bit address signal to a Xilinx Virtex-II Pro control FPGA, clocked by a 24-MHz oscillator. For the side-channel measurements a Teledyne LeCroy HRO 66zi oscilloscope was used. We collected 5 million measurements for random plaintexts and a fixed key by measuring the voltage drop over a $1\,\Omega$ resistor in the Vdd path, while the ASIC was operated

---

[5] Lower number of shares can be achieved at the price of additional fresh masks [22].
[6] The permutation layer in one separate clock cycle.

at a frequency of 3 MHz and a supply voltage of 1.8 V. Each of the power traces contains 100,000 sample points recorded at a sampling rate of 500 MS/s with a resolution of 8 bits. Due to a very low amplitude of the signal two ×10 AC amplifiers in series have been employed, resulting in a ×100 gain. Figure 6(a) depicts a sample trace over the two clock cycles that we are referring to in the following course of this analysis. The two random and uniform 64-bit masks that are needed for the initial sharing of the plaintext are generated and delivered by a PRNG (AES-128 in counter mode) on the control FPGA of the SASEBO-R, which in turn is seeded by the PC via UART.

Results of Conventional Attacks. To evaluate the effectiveness of the presented approach on noisy real-world measurements it is necessary to assess the vulnerability of the underlying hardware masking scheme by means of conventional DPA attacks in a first step. To this end, we performed first-, second- and third-order Correlation Power Analysis (CPA) attacks [6] using the Hamming weight (HW) of the S-box output (which is the same as the output of the F function in Fig. 5). This did not lead to a successful recovery of any key nibble. Hence, we performed the same attack using the HW of the output of the G function (i.e., the value of the intermediate register) and obtained the results which are depicted in Fig. 6. All results are plotted over the two clock cycles that leak the targeted intermediate value. This is on the one hand the clock cycle in which the G-boxes are evaluated in parallel and on the other hand the succeeding clock cycle where the outputs of the F-boxes are computed based on the G-box outputs. As expected the first-order attack is not successful. The second-order CPA, on the other hand, reveals the correct key nibble, but only by a slight margin. The third-order attack does not succeed since the correct key candidate does not lead to the overall highest correlation during the targeted two clock cycles. In particular several ghost peaks with a higher correlation can be identified. For both, the second- and the third-order CPA, we have plotted the evolution of the correlation for the most leaking time sample (marked by a cross in Fig. 6(c) and (e)). In this way we obtain a quantitative measure to express how many traces are required to reveal the higher-order leakages. For the second-order attack at least 200,000 traces are required, whereas for the third-order attack even with the entire 5,000,000 measurements the correct candidate might not be detectable. We observed the same results targeting several other key nibbles. Indeed, it can be concluded that our measurements are sufficiently noisy to serve as a suitable data source for our further analysis.

The efficiency of CPA attacks relies on the linear dependency between the hypothetical power model (here HW of the G-box output) and the actual leakage of the device. Alternatively, Moments-Correlating DPA (MCDPA) [15] can relax such a necessity at the price of (usually) requiring more traces compared to a corresponding CPA with a suitable power model. To examine whether a collision setting can improve the number of required measurements here, which would indicate an imperfect choice of the leakage model in the CPA evaluations, we performed an MCDPA on the same traces. Hereby, the leakage of one S-box is

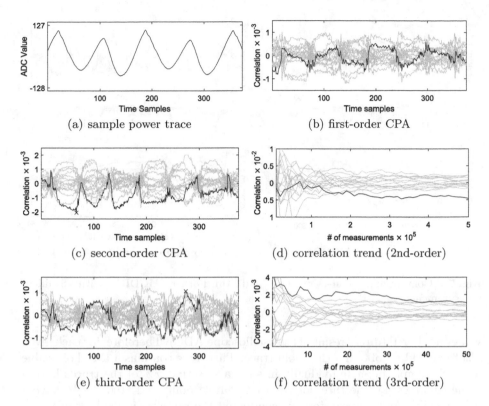

Fig. 6. Sample power trace and conventional first-, second- and third-order CPA with 5 million measurements using the HW of the G-box output.

used to build a model which is then used in an attack on another S-box, leading to a recovery of the linear difference between the corresponding key nibbles. In our case the same hardware instance of the S-box is used for both steps, which ensures a similar leakage model. Figure 7 shows the results indicating that only the third-order MCDPA is able to reveal the correct key difference with 5 million measurements[7]. And even this is only true when exclusively the second leaking clock cycle is considered. Otherwise, there are again ghost peaks with a higher correlation. Nevertheless, 1.5 million measurements are required to exploit the third-order leakage. This result enhances our confidence that the Hamming weight of the output of the G-box is a suitable leakage model for our target.

Results of Our Novel Approach. Hereafter, we concentrate on applying our novel approach (expressed in Sect. 3) on the same traces. In this regard we first obtained a histogram for each sample point using all 5,000,000 traces. The histograms – as given before – have been made by 256 bins, i.e., the full range of signed 8-bit integers -128 to 127 which reflect the sampled power consumption

---

[7] Only positive correlation values indicate a collision in an MCDPA attack.

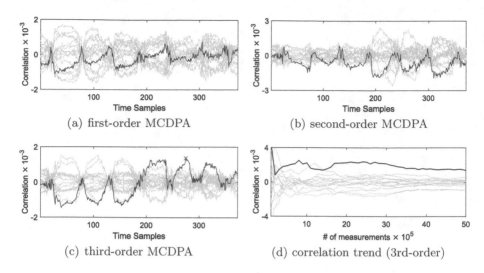

(a) first-order MCDPA

(b) second-order MCDPA

(c) third-order MCDPA

(d) correlation trend (3rd-order)

Fig. 7. Conventional first-, second- and third-order MCDPA with 5 million measurements.

values unaltered (direct result of the oscilloscope ADC). Therefore, for each given $x\%$ threshold we obtain a threshold trace. This trace contains a threshold value for each sample point individually in such a way that $x\%$ of the traces have a value smaller than the threshold at that sample point and $(100 - x)\%$ have a higher value. As the next step, we conducted the attacks on a subset of traces either as "lower $x\%$" or "upper $(100 - x)\%$". It should be noted that such a separation of traces as well as the attack is performed on each sample point separately. In other words, for each sample point it is individually decided which traces to be considered in the attack.

We have examined the threshold values between 5% and 95% with intervals of 5%. In Fig. 8 we represent the result of the attacks (CPA with HW of the G-box output) for the most successful settings, i.e., 20% and 30% thresholds. Interestingly it can be noted that attacks on subsets with a power consumption below the threshold, i.e., lower 80% and lower 70%, lead to a positive correlation for the correct key candidate, and vise versa for the corresponding upper 20% and upper 30%. This is in fact due to the different biases that are introduced into the three shares by selecting measurements with a power consumption either above or below a certain threshold.

Comparison. When comparing our approach to the corresponding conventional second-order CPA, the value of the highest correlation for the correct key candidate is not very meaningful. Due to the fact that a much smaller number of measurements contributes to the results of our approach the correlation values are usually significantly higher compared to the conventional attacks. Hence we have to rely on the required number of measurements as well as a visual inspection of

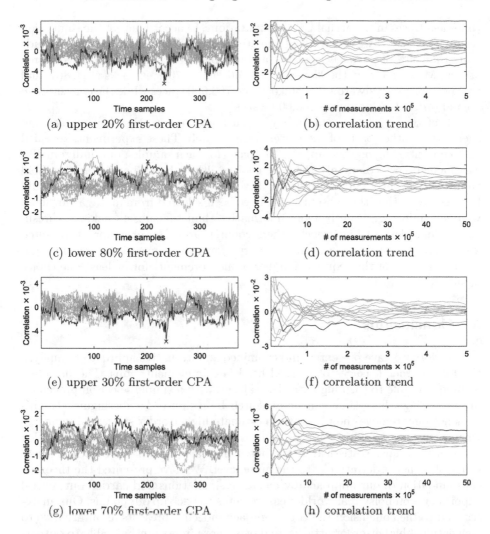

Fig. 8. First-order CPA on different slices of the 5 million measurements using the Hamming weight of the G-box output.

the results as the only available metrics for a comparison. Regarding the required number of measurements we can refer to Fig. 8(b) and (f) that only 50,000 and respectively 70,000 measurements are required to reveal the leakage with our approach. It should be noted that these numbers as well as Fig. 8(b) and (f) reflect the number of traces used to both, find the threshold and perform the attack on. In other words, when it is shown that 50,000 traces are required for a "upper 20%" attack, all 50,000 traces are used to find the threshold. Amongst them, around $50{,}000 \times 20\% = 10{,}000$ traces are used in the attack. Hence, compared to the conventional second-order attack, the attack with "upper 20%" required 4 times less traces altogether and, due to the fact that only a subset is considered, includes 20 times less traces in the actual CPA computations. In accordance to the simulation results (in Sect. 3) we can see that the attacks on subsets of traces, that include more than 50% of the measurements, are not able to outperform the conventional attack. More precisely, the "lower 80%" and "lower 70%" attacks (Fig. 8(d) and (h)) need respectively around 2,500,000 and 700,000 traces while the conventional second-order attack requires 200,000 traces.

All of the presented attacks have been repeated for other key nibbles and therefore on other parts of the power traces as well. These experiments revealed that concentrating on the "upper 30%" part (for each sample point individually) was indeed most commonly the best choice, although the particular threshold values vary slightly between different key nibbles. Another tendency that could be observed is that the subsets which have been selected from above a threshold were generally significantly more informative than the subsets below a threshold (independent of being each others counterpart). However, for all targeted key nibbles our approach was able to outperform the conventional second-order attack in terms of the required number of measurements for at least one choice of subset.

5   Conclusions

In this work we have presented and examined an alternative approach to analyze the higher-order leakages of masked hardware implementations. The proposed technique is able to turn higher-order leakages with a simple selection procedure into a setting where they can be exploited by a first-order distinguisher. This does not only remove the necessity to estimate higher-order statistical moments when attacking masking schemes, which becomes exponentially more complex with an increasing statistical order, but it may also be able to relax the sensitivity of higher-order attacks to the noise level. We have presented the theoretical foundation of our approach by means of simulations and carried out several experiments on noisy real-world measurements to back up our claims. Our analyses lead to the conclusion that our approach indeed represents an alternative to conventional higher-order attacks, and even more importantly is able to outperform them in specific settings. In our setup for example a standard first-order CPA on the subset of traces, that contains only the 20% highest power consumption values (individuality at each sample point), is able to exploit the leakage with 4 times less traces than the conventional second-order CPA attack (i.e., by mean-free square). Hence, a significant improvement could be achieved by simply ignoring a specific part of the traces (at each sample point).

It has been given in literature that masking and hiding countermeasures should be combined to achieve a high level of security. In works like [16] hardware masking is implemented by power-equalization schemes to practically complicate higher-order attacks. As a future work, we will investigate the feasibility of the approach introduced here on such implementations. Another interesting approach to explore is whether it is worthwhile to combine the result of the attacks after splitting the traces. More precisely, we have shown the result of the attacks for "upper 20%" and "lower 80%". The question is whether combining these results would lead to a more effective attack.

Acknowledgements. The authors would like to acknowledge Axel Poschmann for the hardware designs and Stefan Heyse for his help on taping out the prototype chip. This work is partly supported by the German Research Foundation (DFG) through the project "NaSCA: Nano-Scale Side-Channel Analysis".

References

1. Side-channel Attack Standard Evaluation Board SASEBO-R Specification - Version 1.0. Research Center for Information Security, National Institute of Advanced Industrial Science and Technology, Japan. http://www.risec.aist.go.jp/project/sasebo/download/SASEBO-R_Spec_Ver1.0_English.pdf
2. Barthe, G., Dupressoir, F., Faust, S., Grégoire, B., Standaert, F.-X., Strub, P.-Y.: Parallel implementations of masking schemes and the bounded moment leakage model. In: Coron, J.-S., Nielsen, J.B. (eds.) EUROCRYPT 2017. LNCS, vol. 10210, pp. 535–566. Springer, Cham (2017). doi:10.1007/978-3-319-56620-7_19
3. Bilgin, B., Gierlichs, B., Nikova, S., Nikov, V., Rijmen, V.: Higher-order threshold implementations. In: Sarkar, P., Iwata, T. (eds.) ASIACRYPT 2014. LNCS, vol. 8874, pp. 326–343. Springer, Heidelberg (2014). doi:10.1007/978-3-662-45608-8_18
4. Bogdanov, A., Knudsen, L.R., Leander, G., Paar, C., Poschmann, A., Robshaw, M.J.B., Seurin, Y., Vikkelsoe, C.: PRESENT: an ultra-lightweight block cipher. In: Paillier, P., Verbauwhede, I. (eds.) CHES 2007. LNCS, vol. 4727, pp. 450–466. Springer, Heidelberg (2007). doi:10.1007/978-3-540-74735-2_31
5. Boss, E., Grosso, V., Güneysu, T., Leander, G., Moradi, A., Schneider, T.: Strong 8-bit Sboxes with efficient masking in hardware. In: Gierlichs, B., Poschmann, A.Y. (eds.) CHES 2016. LNCS, vol. 9813, pp. 171–193. Springer, Heidelberg (2016). doi:10.1007/978-3-662-53140-2_9
6. Brier, E., Clavier, C., Olivier, F.: Correlation power analysis with a leakage model. In: Joye, M., Quisquater, J.-J. (eds.) CHES 2004. LNCS, vol. 3156, pp. 16–29. Springer, Heidelberg (2004). doi:10.1007/978-3-540-28632-5_2
7. Carlet, C., Danger, J.-L., Guilley, S., Maghrebi, H.: Leakage squeezing of order two. In: Galbraith, S., Nandi, M. (eds.) INDOCRYPT 2012. LNCS, vol. 7668, pp. 120–139. Springer, Heidelberg (2012). doi:10.1007/978-3-642-34931-7_8
8. Chen, Z., Schaumont, P.: Slicing up a perfect hardware masking scheme. In: HOST 2008, pp. 21–25. IEEE (2008)
9. Gierlichs, B., Batina, L., Tuyls, P., Preneel, B.: Mutual information analysis. In: Oswald, E., Rohatgi, P. (eds.) CHES 2008. LNCS, vol. 5154, pp. 426–442. Springer, Heidelberg (2008). doi:10.1007/978-3-540-85053-3_27
10. Ishai, Y., Sahai, A., Wagner, D.: Private circuits: securing hardware against probing attacks. In: Boneh, D. (ed.) CRYPTO 2003. LNCS, vol. 2729, pp. 463–481. Springer, Heidelberg (2003). doi:10.1007/978-3-540-45146-4_27
11. Kim, Y., Sugawara, T., Homma, N., Aoki, T., Satoh, A.: Biasing power traces to improve correlation in power analysis attacks. COSADE **2010**, 77–80 (2010)
12. Maghrebi, H., Guilley, S., Danger, J.-L.: Leakage squeezing countermeasure against high-order attacks. In: Ardagna, C.A., Zhou, J. (eds.) WISTP 2011. LNCS, vol. 6633, pp. 208–223. Springer, Heidelberg (2011). doi:10.1007/978-3-642-21040-2_14

13. Mangard, S., Oswald, E., Popp, T.: Power Analysis Attacks: Revealing the Secrets of Smart Cards. Springer, New york (2007)
14. Moradi, A., Mischke, O.: On the simplicity of converting leakages from multivariate to univariate. In: Bertoni, G., Coron, J.-S. (eds.) CHES 2013. LNCS, vol. 8086, pp. 1–20. Springer, Heidelberg (2013). doi:10.1007/978-3-642-40349-1_1
15. Moradi, A., Standaert, F.-X.: Moments-correlating DPA. In: Workshop on Theory of Implementation Security, TIS 2016, pp. 5–15. ACM (2016)
16. Moradi, A., Wild, A.: Assessment of hiding the higher-order leakages in hardware. In: Güneysu, T., Handschuh, H. (eds.) CHES 2015. LNCS, vol. 9293, pp. 453–474. Springer, Heidelberg (2015). doi:10.1007/978-3-662-48324-4_23
17. Nikova, S., Rijmen, V., Schläffer, M.: Secure hardware implementation of nonlinear functions in the presence of Glitches. J. Cryptology 24(2), 292–321 (2011)
18. Ou, C., Wang, Z., Sun, D., Zhou, X., Ai, J., Pang, N.: Enhanced correlation power analysis by biasing power traces. In: Bishop, M., Nascimento, A.C.A. (eds.) ISC 2016. LNCS, vol. 9866, pp. 59–72. Springer, Cham (2016). doi:10.1007/978-3-319-45871-7_5
19. Popp, T., Mangard, S.: Masked dual-rail pre-charge logic: DPA-resistance without routing constraints. In: Rao, J.R., Sunar, B. (eds.) CHES 2005. LNCS, vol. 3659, pp. 172–186. Springer, Heidelberg (2005). doi:10.1007/11545262_13
20. Poschmann, A., Moradi, A., Khoo, K., Lim, C.-W., Wang, H., Ling, S.: Side-channel resistant crypto for less than 2,300 GE. J. Cryptology 24(2), 322–345 (2011)
21. Prouff, E., Rivain, M., Bevan, R.: Statistical analysis of second order differential power analysis. IEEE Trans. Comput. 58(6), 799–811 (2009)
22. Reparaz, O., Bilgin, B., Nikova, S., Gierlichs, B., Verbauwhede, I.: Consolidating masking schemes. In: Gennaro, R., Robshaw, M. (eds.) CRYPTO 2015. LNCS, vol. 9215, pp. 764–783. Springer, Heidelberg (2015). doi:10.1007/978-3-662-47989-6_37
23. Schaumont, P., Tiri, K.: Masking and dual-rail logic don't add up. In: Paillier, P., Verbauwhede, I. (eds.) CHES 2007. LNCS, vol. 4727, pp. 95–106. Springer, Heidelberg (2007). doi:10.1007/978-3-540-74735-2_7
24. Schneider, T., Moradi, A.: Leakage assessment methodology. In: Güneysu, T., Handschuh, H. (eds.) CHES 2015. LNCS, vol. 9293, pp. 495–513. Springer, Heidelberg (2015). doi:10.1007/978-3-662-48324-4_25
25. Tillich, S., Herbst, C., Mangard, S.: Protecting AES software implementations on 32-bit processors against power analysis. In: Katz, J., Yung, M. (eds.) ACNS 2007. LNCS, vol. 4521, pp. 141–157. Springer, Heidelberg (2007). doi:10.1007/978-3-540-72738-5_10
26. Tiri, K., Schaumont, P.: Changing the odds against masked logic. In: Biham, E., Youssef, A.M. (eds.) SAC 2006. LNCS, vol. 4356, pp. 134–146. Springer, Heidelberg (2007). doi:10.1007/978-3-540-74462-7_10

Side-Channel Attacks Against the Human Brain: The PIN Code Case Study

Joseph Lange, Clément Massart$^{(\boxtimes)}$, André Mouraux,
and Francois-Xavier Standaert

Université catholique de Louvain, 1348 Louvain-la-Neuve, Belgium
{clement.massart,andre.mouraux,fstandae}@uclouvain.be

Abstract. We revisit the side-channel attacks with Brain-Computer Interfaces (BCIs) first put forward by Martinovic et al. at the USENIX 2012 Security Symposium. For this purpose, we propose a comprehensive investigation of concrete adversaries trying to extract a PIN code from electroencephalogram signals. Overall, our results confirm the possibility of partial PIN recovery with high probability of success in a more quantified manner (i.e., entropy reductions), and put forward the challenges of full PIN recovery. They also highlight that the attack complexities can significantly vary in function of the adversarial capabilities (e.g., supervised/profiled vs. unsupervised/non-profiled), hence leading to an interesting tradeoff between their efficiency and practical relevance. We then show that similar attack techniques can be used to threat the privacy of BCI users. We finally use our experiments to discuss the impact of such attacks for the security and privacy of BCI applications at large, and the important emerging societal challenges they raise.

1 Introduction

State-of-the-Art. The increasing deployment of Brain Computer Interfaces (BCIs) allowing to control devices based on cerebral activity has been a permanent trend over the last decade. While originally specialized to the medical domain (e.g., [13,22]), such interfaces can now be found in a variety of applications. Notorious examples include drowsiness estimation for safety driving [19] and gaming [9]. Quite naturally, these new capabilities come with new security and privacy issues, since the signals BCIs exploit can generally be used to extract various types of sensitive information [7,15]. For example, at the USENIX 2012 Security Symposium, Martinovic et al. showed empirical evidence that electroencephalogram (EEG) signals can be exploited in simple, yet effective attacks to (partially) extract private information such as credit card numbers, PIN codes, dates of birth and locations of residence from users [21]. These impressive results leveraged a broad literature in neuroscience, which established the possibility to extract such private information (e.g., see [14] for lie detection and [16] for neural markers of religious convictions). Or less invasively, they can be connected to linguistic research on the reactions of the brain to semantic associations and

© Springer International Publishing AG 2017
S. Guilley (Ed.): COSADE 2017, LNCS 10348, pp. 171–189, 2017.
DOI: 10.1007/978-3-319-64647-3_11

incongruities (e.g., [6,17,18]). All these threats gain concrete relevance with the availability of EEG-based gaming devices to the general public [1,2].

Motivation and Goals. Based on this state-of-the-art, the next step is to push the evaluation of the side-channel threat model in the context of BCI-based applications further. In this respect, the seminal work of Martinovic et al. clearly puts forward the existence of an exploitable bias for various types of private information extraction. But quantifying the impact of this bias in advanced adversarial contexts was left as an important challenge. Typical questions include:

- Can we exactly extract private information with high success rate by increasing the number of observations in side-channel attacks exploiting BCIs?
- How does the effectiveness of unsupervised (aka non-profiled) side-channel attacks exploiting BCIs compare to supervised (aka profiled) ones?
- How efficiently can an adversary build a sufficiently accurate model for supervised (aka profiled) side-channel attacks exploiting BCIs?

Interestingly, these are typically questions that have been intensively studied in the context of side-channel attacks against cryptographic devices (see [20] for an engineering survey and the proceedings of the CHES conference for regular advances in the field [3]). In particular, a recurring problem in the analysis of such implementations is to determine their worst-case security level, in order to bound the probability of success of any adversary in the most accurate manner [27]. This implies very different challenges than in the standard cryptographic setting, since the efficiency of such physical attacks highly depends on the adversary's understanding and knowledge of his target device. Hence, a variety of tools have been developed in order to ensure that side-channel security evaluations are "good enough" (as described next). Our goal in this paper is to investigate the applicability of such tools in order to answer the previous questions regarding the efficiency and impact of side-channel attacks against the human brain.

Contributions. For this purpose, we propose an in-depth study of (a variation of) one of the case studies in [21], namely side-channel PIN code recovery attacks, that share some similarities with key recovery attacks against embedded devices. In this respect, our contributions are threefold. After a description of our experimental settings (Sect. 2), we first describe a methodology allowing us to analyze the informativeness of EEG signals and their impact on security with confidence (Sect. 3). While this methodology indeed borrows tools from the field of side-channel attacks against cryptographic implementations, it also deals with new constraints (e.g., the limited amount of observations available for the evaluations, and the less regular distribution of these observations, for which a very systematic and principled approach is particularly important). Second, we provide a comprehensive experimental evaluation of our side-channel attacks against the human brain using this methodology (Sect. 4). We combine information theoretic and security analyzes in the supervised/profiled and unsupervised/non-profiled contexts, provide quantified estimates for the complexity of the attacks, and pay a particular attention to the stability of and confidence in our results. We conclude by discussing consequences the consequences of our work for the security and privacy of BCI-based applications Sect. 5).

Admittedly, and as will be discussed in detail next, our results can be seen as positive or negative. That is, we show in the same time that partial information about PINs can be extracted with confidence, and that full PIN extractions are challenging because of the high cardinality of the target and risks of false positive. So they should mostly be viewed as a warning flag that such partial information is possible and may become critical when the cardinality of the target decreases and/or large amounts of data are available to the adversary.[1]

2   Experimental Setting, Threat Model and Limitations

In our experiments, eight people (next denoted as users) agreed to provide the 4-digit PIN code that they consider the most significant to them, meaning the one they use the most frequently in their daily life. This PIN code was given by the users before the experiment started, stored during the experiment, and deleted afterward for confidentiality reasons. Five other random 4-digit codes were generated for each user (meaning a total of six 4-digit codes per user).

Each (real or random) PIN was then shown on a computer exactly 150 times to each user (in a random order), meaning a total of 900 events for which we recorded the EEG signal in sets of 300, together with a tag T ranging from 1 to 6 (with $T = 1$ the correct PIN and $T = 2$ to 6 the incorrect ones). We used 32 Ag-AgCl electrodes for the EEG signals collection. These were placed on the scalp using a Waveguard cap from Cephalon, using the international 10-10 system. The Stimulus Onset Asynchrony (SOA) was set to 1,009 s (i.e., slightly more than one second, to reduce the environmental noise). The time each PIN was shown was set to 0,5 s. When no PIN was displayed on the screen, a + sign was maintained in order to keep the focus of the user on the center of the screen. We additionally ensured that two identical 4-digit codes were always separated by at least two other 4-digit codes. The split of our experiments in sub-experiments of 300 events was motivated by a maximum duration of 5 min, during which we assumed the users to remain focused on the screen. The signals were amplified and sampled at a 1000 Hz rate with a 32-channel ASA-LAB EEG system from Advanced Neuro Technologies. Eventually, and in order to identify eye-blinks which potentially perturb the EEG signal, we added two bipolar surface electrodes on the upper left and lower right sides of the right eye, and rejected the records for which such an artifact was observed. This slightly reduced the total number of events stored for each user (precisely, this number was reduced to 900, 818, 853, 870, 892, 887, 878, 884, for users 1 to 8).

This simplified setting naturally comes with limitations. First and concretely, the number of possible PIN codes for a typical smart card would of course be much larger than the 6 ones we investigate (e.g., 10,000 for a 4-digit PIN). In this respect,

---

[1] The experiments described next were approved by the local Research Ethics Committee and performed in compliance with the Code of Ethics of the World Medical Association (Declaration of Helsinki). All participants gave written informed consent.

we first insist that the primary goal of the following experiments is to investigate the information leakages in EEG signals thoroughly, and this limited number of PIN codes allowed us to draw conclusions with good statistical confidence. Yet, we also note that this setting could be extended to a reasonable threat model. For example, one could target ≈1000 different users by repeatedly showing them ≈10 PIN codes among the 10,000 possible ones, and recover one PIN with good confidence. Second, and since the attacks we carry out essentially test familiar vs. unfamiliar information, there is also a risk of false positives (e.g., an all zero code or a close to correct code). This is in fact something we observed in our experiments. In this respect, our mitigation plan is to exploit statistical tools minimizing the number of false negatives, therefore potentially allowing enumeration among the most likely candidates [28].

3   Methodology

In this section, we describe the methodology we used in order to assess and better quantify the feasibility of side-channel attacks against the human brain. Concretely, and contrary to the case of embedded devices where the leakage distributions are supposed to be stable and the number of observations made by the adversary can be large, we deal with a very different challenge. Namely, we need to cope with irregular distributions possibly affected by outliers, and can only assume a limited number of observations.

As a result, the following sections mainly aim to convince the reader that our treatment of the EEG signals is not biased by dataset-specific overfitting. For this purpose, our strategy is twofold. First, we apply the same (pre)processing methods to the measurements of all the users. This means the same selection of electrodes, the same dimensionality reduction and Probability Density Function (PDF) estimation tools (with identical parameters), and the same outliers definition. Second, we systematically verified that our results were in the same time consistent with neurophysiological expectations, and stable across a sufficient range of (pre)processing parameters. As a result, our primary focus is on the confidence in and stability of the results, more than on their optimality (which is an interesting scope for further research). In other words, we want to guarantee that EEG signals provide exploitable side-channel information for PIN code recovery, and to evaluate a sufficient number of observations for which such an attack can be performed with good success probability.

3.1   Notations

We denote the (multivariate) EEG signals of our experiments with a random variable O, a sample EEG signal as o, and the set of all the observations available for evaluation as $\mathcal{O}$. These observations depend on (at least) three parameters: the user under investigation, next denoted with a random variable U such that $u \in \{1, 2, \ldots, 8\}$; the nature of the 4-digit code observed (i.e., whether it is correct or a random PIN), next denoted with a random variable P such that

$p \in \{0, 1\}$; and a noise random variable N. Each observation is initially made of 32 vectors of 1,000 samples, corresponding to 32 electrodes and $\approx 1\,$s per event.

3.2   Supervised (aka Profiled) Evaluation

In order to best evaluate the actual informativeness of the EEG signals regarding the PIN displayed in our experiments, and inspired by the worst-case side-channel security evaluations of cryptographic devices, our work first investigates so-called profiled attacks, which correspond to a supervised machine learning context. For this purpose, a part of the observations in $\mathcal{O}$ are used to estimate a (probabilistic) model $\hat{\Pr}_{\mathrm{model}}[P = p|O = o]$. The adversary/evaluator then uses this model in order to try extracting the PIN from the remaining observations. Note that our profiling is based on the binary random variable p, where $p = 0$ if the PIN is random and $p = 1$ if the PIN is real, and not based on the value of the PIN tag itself. This is motivated by the following practical and neurophysiological reasons:

- From a practical point-of-view, building a model for all the PINs and users seems impractical in real-world settings: this would require being able to collect multiple observations for each of the 10,000 possible values of a 4-digit code. Furthermore, and as discussed in Sect. 3.3, our real vs. random profiling allowed us to lean towards realistic (non-profiled) attacks.
- From a neurophysiological point-of-view, the information we aim to extract is based on Event-Related Potentials (ERPs) that have been shown to reflect semantic associations and incongruities [6,17,18]. In this respect, while we can expect a user to react differently to real and random 4-digit codes, there is no reason for him to treat the random codes differently.

A. Evaluation Metrics. Following the general principles put forward in [27], our evaluations will be based on a combination of information theoretic and security analyzes. The first ones aim at evaluating whether exploitable information is available in the EEG signals; the second ones at evaluating how efficiently this information can be exploited to mount a side-channel attack. Note that since we do not assume the users to behave identically, these metrics will always be evaluated and discussed for each user independently.

Perceived Information. The Perceived Information (PI) was introduced in the context of side-channel attacks against cryptographic devices, of which the goal is to recover some secret data (aka key) given some physical leakage [23]. The PI aims at quantifying the amount of information about the secret key, independent of the adversary who will exploit this information. Informally, we will use this metric in a similar way, by just considering P as a bit to recover, and the observations as leakages. Using the previous notations, we define the PI between the PIN random variable P and the observation random variable O:

$$\mathrm{PI}(P; O) = \mathrm{H}[P] + \sum_{p} \Pr[p] \cdot \int_{o} \mathrm{f}(o|p) \cdot \log_2 \Pr_{\mathrm{model}}[p|o] \; do,$$

where we use the notation $\Pr[X = x] =: \Pr[x]$ for conciseness, and $f(o|p)$ is the (continuous) PDF of the observations given the value of p. In the ideal case where the model is perfect, the PI is identical to Shannon's mutual information. In the practical cases where the model differs from the observation's true distribution, the PI captures the amount of information that is extracted from these observations, biased by the model (assumption & estimation) errors [11].

Of course, concretely the true distribution $f(o|p)$ is unknown to the adversary/evaluator and can only be sampled. Therefore, the approach in side-channel analysis, that we repeat here, is to split the set of observations $\mathcal{O}$ in k non-overlapping sets $\mathcal{O}^{(i)}$. We then define the profiling sets $\mathcal{O}_{\mathsf{p}}^{(j)} = \bigcup_{i \neq j} \mathcal{O}^{(i)}$ and the test sets $\mathcal{O}_{\mathsf{t}}^{(j)} = \mathcal{O} \setminus \mathcal{O}_{\mathsf{p}}^{(j)}$. The PI is computed in two phases:

1. The observations' conditional distribution is estimated from a profiling set. We denote this phase with $\hat{f}_{\mathrm{model}}^{(j)}(o|p) \leftarrow \mathcal{O}_{\mathsf{p}}^{(j)}$. Note that the $\Pr_{\mathrm{model}}[p|o]$ factor involved in the PI definition is directly derived via Bayes' theorem as:

$$\hat{\Pr}_{\mathrm{model}}[p|o] = \frac{\hat{f}_{\mathrm{model}}^{(j)}(o|p) \cdot \Pr[p]}{\sum_{p^*} \hat{f}_{\mathrm{model}}^{(j)}(o|p^*) \cdot \Pr[p^*]}.$$

2. The model is then tested by computing the PI estimate:

$$\hat{\mathrm{PI}}^{(j)}(P;O) = \mathrm{H}[P] + \sum_{p=0}^{1} \Pr[p] \cdot \sum_{o \in \mathcal{O}_{\mathsf{t}}^{(j)}|p} \frac{1}{n_p^j} \cdot \log_2 \hat{\Pr}_{\mathrm{model}}[p|o],$$

where n_p^j is the number of observations in the test set $\mathcal{O}_{\mathsf{t}}^{(j)}|p$.

Eventually, the k outputs $\hat{\mathrm{PI}}^{(j)}(P;O)$ are averaged to get an unbiased estimate, and their spread characterizes the accuracy of the result (see Paragraph G). Note that concretely, the maximum size for the profiling set in our experiments equals ≈ 899, leading to a cross-validation parameter $k \approx 900$ and a test set of size 1. In this case, the model building phase is repeated ≈ 900 times, and each model is tested once against an independent sample. (We use the $\approx$ symbol to reflect the fact that these values are approximated, due to the rejection of eye blinks mentioned in Sect. 2). This "leave one our" strategy has a large cross-validation parameter compared to current practice (e.g., in side-channel attacks against cryptographic implementations a value of $k = 10$ was selected [11]), leading to computationally intensive evaluations. Yet, it is justified in our study because of the limited number of samples available in our experiments.

Success Rate and Average Rank. In order to confirm that the estimated PI indeed leads to concrete attacks, we consider two simple security metrics. Here, the main challenge is that we only have models for the real and random PIN codes, while the actual observations in the test set naturally come from six different events. As a result, we first considered the success rate event per event. For this purpose, the ≈ 900 observations are split in 6 sets of ≈ 150 observations that correspond to the six different tag values. Based on these 6 sets, we can compute the probability that the observations are correctly classified as real or random in function of the number of observations exploited in the attack, next denoted as q. This is done by averaging a success function S that is computed as follows. If $q = 1$: $S(o_1) = 1$ if $\hat{Pr}_{model}[p|o_1] > \hat{Pr}_{model}[\bar{p}|o_1]$ and $S(o_1) = 0$ otherwise (where $\bar{p}$ denotes the incorrect event); if $q = 2$: $S(o_1, o_2) = 1$ if $\hat{Pr}_{model}[p|o_1] \times \hat{Pr}_{model}[p|o_2] > \hat{Pr}_{model}[\bar{p}|o_1] \times \hat{Pr}_{model}[\bar{p}|o_2]$; ... Concretely, this success rate is an interesting metric to check whether the observations generated by different incorrect PIN values indeed behave similarly.

Of course, an adversary eventually wants to compare the likelihoods of different PIN values. For this purpose, we also considered the average rank of the correct PIN in an experiment where we gradually increase the number of observations per tag q, but this time consider sets of 6 observations at once, that we classify only according to the model for the real PIN. This leads to vectors $(\hat{Pr}_{model}[p|o_1^1], \hat{Pr}_{model}[p|o_1^2], \hat{Pr}_{model}[p|o_1^3], \ldots, \hat{Pr}_{model}[p|o_1^6])$ if $q = 1$, $(\hat{Pr}_{model}[p|o_1^1] \times \hat{Pr}_{model}[p|o_2^1], \ldots, \hat{Pr}_{model}[p|o_1^6] \times \hat{Pr}_{model}[p|o_2^6])$ if $q = 2$, ..., where the superscripts denote the tag from which the observations originate. The average rank is then obtained by sorting this vector and estimating the sample mean of the position of the tag 1 in the sorted vector.

Connecting the Metrics (Sanity Check). Note that as discussed in [10], information theoretic and security metrics can be connected (i.e., a model that leads to a positive PI should lead to successful attacks).[2] We consider both types of metrics in our experiments because the first ones allow a better assessment of the confidence in the evaluations (see Paragraph G) while the second ones lead to simpler intuitions regarding the concrete impact of the attacks.

B. Preprocessing. As a first step, all the observations were preprocessed using a bandpass filter. We set the low-frequency cut-off to 0.5 Hz to remove the slow drifts in the EEG signals, and the high-frequency cut-off to 30 Hz to remove muscle artifacts and 50 Hz environmental noise.

C. Selection of Electrodes. As mentioned in introduction, each original observation is made of 32 vectors of 1,000 samples, leading to a large amount of data to process. To simplify our treatments, we started by analyzing the different electrodes independently. Among the 32 ones of our cap, electrodes P7, P8, Pz, O1 and O2 gave rise to non-negligible signal (see Fig. 1), which is consistent with the existing literature where ERPs related to semantic associations and incongruities were exhibited in the central/parietal zones [6,17,18]. Our following analyzes are based on the exploitation of the electrodes P7 and P8 which provided the most regular information across the different users.

---

[2] More precisely, the PI is an average metric, so what is needed is that each line of the PI matrix defined in [27] (corresponding to 6 different events in our study) are positive, which we observed and confirmed with the success rate analysis.

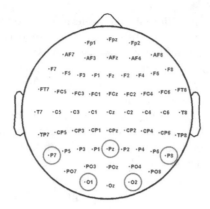

Fig. 1. Repartition of the electrodes on the scalp.

For illustration, Figs. 2 and 3 represent the mean and standard deviation traces corresponding to two different users. From these examples, a couple of relevant observations can already be extracted (and will be useful for the design and interpretation of our following evaluations). First, we see (on the left parts of Fig. 2) that the EEG signals may be more or less informative depending on the users and electrodes. More precisely, we generally noticed informative ERP components after 300 to 600 ms (known as the P300) for most users and electrodes, which is again consistent with the existing literature [6,17,18]. Yet, our measurements also put forward user-specific differences in the shape of the mean traces corresponding to the correct PIN value. (Note that the figure only shows

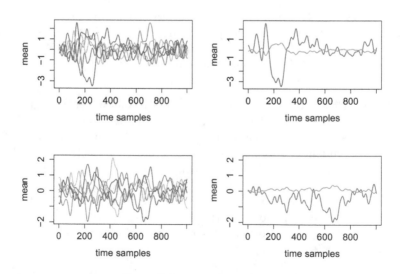

Fig. 2. Exemplary mean traces for different tag (left) and PIN (right) values. Top: User 8, Electrode P7. Bottom: User 6, Electrode P7.

examples of informative EEG signals, but for some other users and electrodes, no such clear patterns appear). Second, and quite importantly, the difference between the left and right parts of the figures illustrates the significant gain when moving from an unsupervised/unprofiled evaluation context to a supervised/profiled one. That is, while in the first case, we need the traces corresponding to the correct PIN value to stand out, in the second case, we only need it to behave differently than the others.

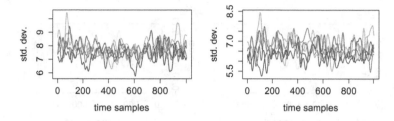

Fig. 3. Exemplary standard deviation traces for different tag values corresponding to User 8, Electrode P7 (left) and User 6, Electrode P7 (right).

Eventually, a look at the standard deviation curves in Fig. 3 suggests that the measurements are quite noisy, hence non-trivial to exploit with a limited amount of observations. This will be confirmed in our following PDF estimation phase, and therefore motivates the dimensionality reduction in the next section (intuitively because using more dimensions can possibly lead to better signal extraction, which can mitigate the effect of a large noise level).

D. Dimensionality Reduction. The evaluation of our metrics requires to build a probabilistic model, which may become data intensive as the number of dimensions in the observations increases. For example, directly estimating a 2000-dimensional PDF corresponding to our selected electrodes is not possible. In order to deal with this problem, we follow the standard approach of reducing dimensionality. More precisely, we use the Principal Component Analysis (PCA) that was shown to provide excellent results in the context of side-channel attacks against cryptographic devices [4]. We investigate two options in this direction.

First, and looking at the observations in Fig. 2, it appears that the mean traces corresponding to the different tags are quite discriminant regarding the value of p. Hence, and as in [4], a natural option is to compute the projection vectors of the PCA based on these mean traces. This implies computing average vectors $\bar{o}^j = \mathsf{E}_{i\approx1}^{150} o_i^j$, and then to derive the PCA eigenvectors based on the $\bar{o}^j$'s, which we denote as $\boldsymbol{R}_{1:N_d} \leftarrow \mathsf{PCA}(\{\bar{o}^j\}_{j=1:6})$, where N_d is the number of dimensions to extract. Due to the limited number of mean traces (i.e., 6), we can only compute $N_d = 5$ eigenvectors, and therefore are limited to 5-dimensional attacks in this case.[3] However, it turned out that in our experiments, this version

---

[3] Because we used the small sample size variant of PCA in [4].

of the PCA extracts most of the relevant samples in the first dimension. This is intuitively witnessed by Fig. 4 which represents the first and fifth eigenvectors corresponding to User 8 and Electrode P7 (i.e., $\boldsymbol{R}_1$ and $\boldsymbol{R}_5$): we indeed observe that the first dimension corresponds to the points of interest in Fig. 2, while the fifth one seems to be dominated by noise. In the following, we will denote this solution as the "average PCA". Note that such a dimensionality reduction does not take advantage of any secret information (i.e., it is not a supervised/profiled one) since it builds the mean traces based on public tags.

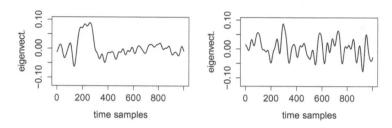

Fig. 4. Examplary eigenvectors for the average PCA, corresponding to User 8, Electrode P7. Left: first dimension. Right: fifth dimension.

Yet, one possible drawback of the previous method is that estimating the average traces $\bar{\boldsymbol{o}}^j$ becomes expensive when the number of PIN codes increases. In order to deal with and quantify the impact of this limitation, we also considered a "raw PCA", where we directly reduce the dimensionality based on raw traces, next denoted as $\boldsymbol{R}_{1:N_d} \leftarrow \mathsf{PCA}(\{\boldsymbol{o}_i\}_{i\approx1:900})$. While this approach is not expected to extract the information as effectively, it allows deriving a much larger number of dimensions than in the previous (average) case. Concretely though, exploiting dimensions 1 to 5 only was a good tradeoff between the informativeness of the dimensionality reduction, the risk of ovefitting (useless) dataset-dependent patterns and the risk of outliers in our experiments (see Paragraph F).

As a result of this dimensionality reduction phase, the observation vectors $\boldsymbol{o}(1:2000)$ (which correspond to the concatenation of the measurements for our two selected electrodes) are reduced to smaller vectors $\boldsymbol{R}_{1:N_d} \times \boldsymbol{o}$ (i.e., each dimension $o(d)$ corresponds to the scalar product between the original observations $\boldsymbol{o}$ and a 2000-element vector $\boldsymbol{R}_d$). We recall that PCA is not claimed to be an optimal dimensionality reduction, since it optimizes a criteria (i.e., the variance between the raw or mean traces) which does not capture all the information in our measurements. However, it is a natural first step in our investigations, and we could verify that our following conclusions are not affected by slight variations of the number of extracted dimensions (i.e., adding one or two dimensions), which therefore fits our (primary) confidence and stability goal.

E. PDF Estimation. We now describe the main ingredient of our supervised/profiled evaluation, namely the PDF estimation for which we exploit the knowledge of the p values for the observations in the profiling sets.

In order to build a model $\hat{f}_{model}(\boldsymbol{o}_{1:N_d}|p)$, we first take advantage of the fact that the dimensions of the $\boldsymbol{o}_{1:N_d}$ vectors after PCA are orthogonal. By additionally considering them as independent, this allows us to reduce the PDF estimation problem from one N_d-variate one to N_d univariate ones. Based on this simplification, the standard approach in side-channel analysis is to assume the observations to be normally distributed, and to build Gaussian templates [8]. Yet, in our experiments no such obvious assumption on the distributions in hand was a priori available. As a result, we first considered a (non-parametric) kernel density estimation as used in [5], which has slower convergence but avoids any risk of biased evaluations [11]. Kernel density estimation is a generalization of histograms. Instead of bundling samples together in bins, it adds (for each observation) a small kernel centered on the value of the observation to the estimated PDF. The resulting estimation that is a sum of kernels is smoother than histograms and usually converges faster. Concretely, kernel density estimation requires selecting a kernel function (we used a Gaussian one) and to set the bandwidth parameter (which can be seen as a counterpart to the bin size in histograms). The optimal choice of the bandwidth depends on the distribution of the observations, which is unknown in our case. So we need to rely on a heuristic, and used Silverman's rule-of-thumb for this purpose [24].

F. Outliers. As mentioned in Paragraph D, the main drawback of the raw PCA is that it extracts the useful EEG information less efficiently, which we mitigate by using more dimensions. Unfortunately, this comes with an additional caveat. Namely, the less informative information extraction combined with the addition of more dimensions increases the risk of outliers (i.e., observations that would classify the correct PIN value very badly for some dimensions, possibly leading to a negative PI). In this particular case, we considered an additional post-processing (after the dimensionality reduction and model building phases). Namely, given the ≈ 900 probabilities $\hat{\Pr}[p|\boldsymbol{R}_{1:N_d} \times \boldsymbol{o}_i]$, we rejected the ones below 0.001 and beyond 0.999. This choice is admittedly heuristic, yet did consistently lead to positive results for all the users. It is motivated by limiting the weight of the log probabilities for the outliers in the PI estimation. We insist that this treatment of outliers is only needed for the raw PCA. For the average PCA, we did not reject any observation (other than the ones in Sect. 2).

G. Confidence. By using ≈ 900-fold cross-validation, we can guarantee that our PI estimates will be based on 900 observations, leading to 900 values for the log probabilities $\log_2(\hat{\Pr}[p|\boldsymbol{R}_{1:N_d} \times \boldsymbol{o}_i])$. Since this remains a limited amount of data compared to the case of side-channel attacks against cryptographic implementations, and the extracted PI values are small, we completed our information theoretic evaluations by computing a confidence interval for the PI estimates. To avoid any distribution-specific assumption, we computed a 10% bootstrap confidence interval [12], by resampling 100 bootstrap samples out of our 900 log probabilities, computing 100 mean bootstrap samples, sorting them, and using the 95th and 5th percentiles as the endpoints of the intervals. For simplicity, this was only done for the PI metric and not for the success rate and average rank since (i) successful Bayesian attacks are implied by the information theoretic analysis [10], (ii) these metrics are more expensive to sample (e.g., we have only one evaluation of the success function with $q \approx 150$ per user), and (iii) they are only exhibited to provide intuitions regarding the exploitability of the observations (i.e., the attack complexities).

3.3 Unsupervised (aka Non-profiled) Analysis

While supervised (aka profiled) analyzes are the method of choice to gain understanding about the information available in a side-channel, their practical applicability is of course questionable. Indeed, building a model for a target user may not always be feasible, and this is particularly true in the context of attacks against the human brain since (as discussed the long version of this paper), models built for one user are not always (directly) exploitable against another user. In this section, we therefore propose an unsupervised/non-profiled extension of the information theoretic evaluation outlined in Sect. 3.2. To the best of our knowledge, this variation was never described as such in the open literature (although it shares some similarities with the non-profiled attacks surveyed in [5]). For this purpose, our starting point is the observation from Fig. 2, that in an unsupervised/non-profiled context, one can take advantage of the fact that the (e.g., mean) traces of the EEG signals corresponding to the correct PIN value may stand out. As a result, a natural idea is to compute the PI metric 6 times independently, each time assuming a different (possibly random) tag to be correct during an "on-the-fly" modeling phase. If the traces corresponding to the (truly) correct PIN are more singular (comparatively to the others), we can expect the PI estimated with this PIN to be larger, leading to a successful attack.

Of course, such an attack implies an additional neurophysiological assumption (while in the supervised/profiled setting, we just exploit any information available). Yet, it nicely fits the intuitions discussed in the rest of this section, which makes it a good candidate for concrete evaluation. Furthermore, we mention that directly recovering the correct PIN value may not always be necessary: as in the case of side-channel analysis, reducing the rank of the correct PIN value down to an enumerable one may be sufficient [28].

4 Experimental Results

4.1 Supervised (aka Profiled) Evaluation

As in the previous section, we start with the results of our supervised/profiled evaluations, which will be in two (information theoretic and security) parts. Beforehand, there is one last choice regarding the computation of $\hat{\Pr}[p|\boldsymbol{R}_{1:N_d} \times \boldsymbol{o}_i]$ via Bayes' theorem described in Sect. 3.2, Paragraph A. Namely, should we consider maximum likelihood or maximum a posteriori attacks (i.e., should we take advantage of the a priori knowledge of $\Pr[p]$ or consider a uniform a priori).

Interestingly, in our context ignoring this a priori and performing maximum likelihood attacks is more relevant, since we mostly want to avoid false negatives (i.e., correct PINs that would be classified as random ones), which prevent efficient enumeration. Since the a priori on P increases the amount of such errors (due to the a priori bias of $5/6$ towards random PIN values), the rest of this section reports on the results of maximum likelihood attacks.

A. Perceived Information. As a first step in our evaluations, we estimated the PI using the methodology described in the previous section. We started by looking at the evolution of the PI estimation in function of the number of observations in the profiling set used to build the model. The results of this analysis are in Fig. 5 from which two quantities must be observed:

- The value of the PI estimate using the maximum profiling set (i.e., the extreme right values in the graphs). It reflects the informativeness of the model built in the profiling phases, and is correlated with the success rate of the online (maximum likelihood) attack using this model [10]. Positive PI values indicate that the model is sound (up to Footnote 2) and should lead to successful online attacks if the number of observations (i.e., the q parameter in our notations of Sect. 3.2) used by the adversary is sufficient.
- The number of traces in the profiling set required to reach a positive PI. It reflects the (offline) complexity of the model estimation (profiling) phase [26].

In this respect, the results in Fig. 5 show a positive convergence for the two illustrated users, yet towards different PI values which indicates that the informativeness of the EEG signals differs between them. Next, and quite interestingly, we

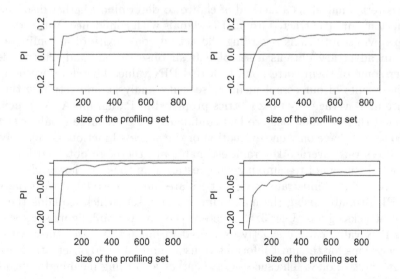

Fig. 5. Evolution of the PI in function of the size of the profiling set for Users 3 (top) and 6 (bottom), using average PCA (left) and raw PCA (right).

Table 1. Estimated PI values with maximum profiling set.

User	$\hat{PI}(P; O)$ with avg. PCA	$\hat{PI}(P; O)$ with raw PCA
1	0.0739	0.0618
2	0.1643	0.1315
3	0.1494	0.1398
4	0.0920	0.0228
5	∅	∅
6	0.0521	0.0214
7	0.0759	0.0568
8	0.1697	0.0458

also see that the difference between average PCA (in the left part of the figure) and raw PCA (in the right side) confirms the expected intuitions. Namely, the fact that raw PCA reduces dimensionality based on a less meaningful criteria and requires more dimensions implies a slower model convergence. Typically, model convergence was observed in the 100 observations' range with average PCA and required up to 400 traces with raw PCA. For completeness, Table 1 contains the estimated PI values with maximum profiling set, for the different users and types of PCA. Excepted for one user (User 5) for which we could never reach a positive PI value,[4] this analysis suggests that all the users lead to exploitable information and confirms the advantage of average PCA.

B. Success Rate and Average Rank. As discussed in Sect. 3.2, our information theoretic analysis is a method of choice to determine whether discriminant information can be extracted from EEG signals with confidence. Yet, it does not lead to obvious intuitions regarding the actual complexity of an online attack where an adversary obtains a set of q fresh observations and tries to detect whether some of them correspond to a real PIN value. Therefore, we now provide the results of our complementary security analysis, and estimate the success rate and average key rank metrics proposed in Paragraph A. As previously mentioned these evaluations are less confident, since for large q values such as $q = 150$ we can have only one evaluation of the success function. Concretely, the best success rate/average key rank estimates are therefore obtained for $q = 1$. We took advantage of re-sampling when estimating them for larger q's.

Figures 6 and 7 illustrate these metrics are indeed correlated with the value of the PI estimates using the maximum profiling set, which explains the more efficient attacks against User 3. Concretely, the average rank figure suggests that correct PIN value can be exactly extracted in our 6-PIN case study with 5 to 10 observations for the most informative users and 30 to 40 observations for the least informative ones. The success rate curves also bring meaningful intuitions

---

[4] As mentioned in Sect. 2, this is due to the presence of another familiar event for this user, which he mentioned to us after the experiments were performed.

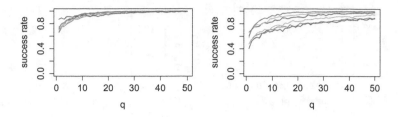

Fig. 6. Success rates per tag value for User 3 (left) and User 6 (right).

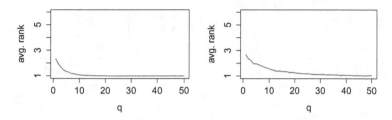

Fig. 7. Avg. rank of the correct PIN for User 3 (left) and User 6 (right).

since they highlight that all (correct and random) PIN values can be correctly classified with our profiled models (in slightly more traces). This confirms our neurophysiological assumption from the previous section that the users react similarly to all random values.[5] Besides, Fig. 6 is interesting since it shows how confidently the correct PIN value is classified independent of the others. Hence, its results would essentially scale with larger number of PIN values.

4.2 Unsupervised (aka Non-profiled) Analysis

We now move to the more challenging problem of unsupervised/non-profiled attacks. For this purpose, we first applied the attack sketched in Sect. 3.3 with the maximum number of traces in the profiling set. That is, we repeated our evaluation of the PI metric six times, assuming each of the tag values to be the real one. Furthermore, we computed the confidence intervals for each of the PI estimates according to Sect. 3.2, Paragraph G. The results of this experiment are in Fig. 8 for two users and lead to three observations.

First, looking at the first line of the figure, which corresponds to the correct PIN value, we can now confirm that the PI estimates of Sect. 4.1 are sufficiently accurate (e.g., the confidence intervals clearly guarantee a positive PI). Second, the confidence intervals for the random PIN values (i.e., tags 2 to 6) confirm the observation from our success rate curves (Fig. 6) that the users react similarly to all random values. Third, the middle and bottom parts of the figure show

---

[5] We may expect more singularities (such as the one of User 5) to appear and launch false alarms in case studies with more PIN values. Yet, this would not contradict the trend of a significantly reduced average rank for the correct PIN value.

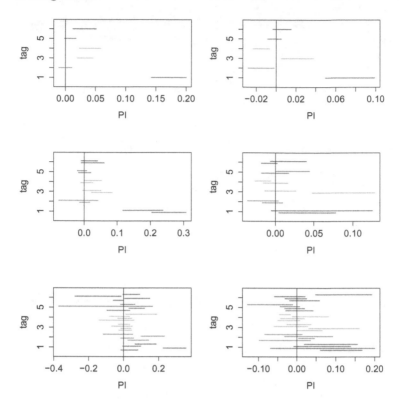

Fig. 8. Confidence intervals for the (non-profiled) PI evaluation of Sect. 3.3 with ≈900 observations (top), ≈450 observations (middle) and ≈225 observations (bottom), for Users 8 (left) and 6 (right).

the results of two (resp. 4) non-profiled attacks where the profiling set was split in 2 (resp. 4) independent parts (without re-sampling), therefore leading to the evaluation of 2 (resp. 4) confidence intervals for each tag value. As expected, it indicates that the information extraction is significantly more challenging in this unsupervised/non-profiled context. Concretely, the PI estimate for the correct PIN value consistently started to overlap with the ones of random PINs for all users, as soon as the number of attack traces q was below 200, and no clear gain for the correct PIN could be noticed below $q = 100$. This confirms the intuition that unsupervised/non-profiled side-channel attacks are generally more challenging than supervised/profiled ones (here, by an approximate factor 5 to 10 depending on the users).

This conclusion also nicely matches the one in Sect. 4.1, Fig. 5, where we already observed that the (offline) estimation of an informative model is more expensive than its (online) exploitation for PIN code recovery as measured by the success rate and average rank (by similar factors). Indeed, in the unsupervised/non-profiled context such an estimation has to be performed "on-the-fly".

5   Consequences and Conclusions

The results in this paper lead to two important conclusions.

First, and from the security point-of-view, our experiments show that extracting concrete PIN codes from EEG signals, while theoretically feasible, may not be a very critical threat. PIN extraction attacks using BCIs indeed require several observations to succeed with high probability. Furthermore, the difference between the complexity of successful supervised/profiled attacks (around 10 correct PIN observations) and unsupervised/non-profiled attacks (more in the hundreds range) is noticeable. Yet, our results generally confirm the existence of exploitable information in EEG signals, which may become more worrying in case of targets with smaller cardinalities (e.g., extracting the knowledge of one relative among a set of unknown people displayed on a screen).

Second, and given the importance of profiling for efficient information extraction from EEG signals, our experiments underline that privacy issues may be even more worrying than security ones in BCI-based applications. Indeed, when it comes to privacy, the adversary trying to identify a user is less limited in his profiling abilities. In fact, any correlation between his target user and some feature found in a dataset is potentially exploitable. In this context, the data minimization principle does not seem to be a sufficient answer: it may be that the EEG signals collected for one (e.g., gaming) activity can be used to reveal various other types of (e.g., medical, political, ...) correlations. Anonymity is probably not the right answer either (since correlations with groups of users may be as discriminant as personal ones). And such issues are naturally amplified in case of malicious applications (e.g., it seem possible to design a BCI-based game where situations lead the users to incidentally reveal preferences). So overall, it appears as an important challenge to design tools that provide evidence of "fair treatment" when manipulating EEG signals, which can be connected to emerging challenges related to computations on encrypted data [25].

References

1. http://emotiv.com/. Accessed July 2016
2. http://neurosky.com/. Accessed July 2016
3. http://www.chesworkshop.org/. Accessed July 2016
4. Archambeau, C., Peeters, E., Standaert, F.-X., Quisquater, J.-J.: Template attacks in principal subspaces. In: Goubin, L., Matsui, M. (eds.) CHES 2006. LNCS, vol. 4249, pp. 1–14. Springer, Heidelberg (2006). doi:10.1007/11894063_1
5. Batina, L., Gierlichs, B., Prouff, E., Rivain, M., Standaert, F., Veyrat-Charvillon, N.: Mutual information analysis: a comprehensive study. J. Cryptol. 24(2), 269–291 (2011)
6. Berlad, I., Pratt, H.: P300 in response to the subject's own name. Electroencephalogr. Clin. Neurophysiol. 96(5), 472–474 (1995)
7. Bonaci, T., Calo, R., Chizeck, H.J.: App stores for the brain: privacy and security in brain-computer interfaces. IEEE Technol. Soc. Mag. 34(2), 32–39 (2015)

8. Chari, S., Rao, J.R., Rohatgi, P.: Template attacks. In: Kaliski, B.S., Koç, K., Paar, C. (eds.) CHES 2002. LNCS, vol. 2523, pp. 13–28. Springer, Heidelberg (2003). doi:10.1007/3-540-36400-5_3

9. Coyle, D., Príncipe, J.C., Lotte, F., Nijholt, A.: Guest editorial: brain/neuronal - computer game interfaces and interaction. IEEE Trans. Comput. Intell. AI Games 5(2), 77–81 (2013)

10. Duc, A., Faust, S., Standaert, F.-X.: Making masking security proofs concrete. In: Oswald, E., Fischlin, M. (eds.) EUROCRYPT 2015. LNCS, vol. 9056, pp. 401–429. Springer, Heidelberg (2015). doi:10.1007/978-3-662-46800-5_16

11. Durvaux, F., Standaert, F.-X., Veyrat-Charvillon, N.: How to certify the leakage of a chip? In: Nguyen, P.Q., Oswald, E. (eds.) EUROCRYPT 2014. LNCS, vol. 8441, pp. 459–476. Springer, Heidelberg (2014). doi:10.1007/978-3-642-55220-5_26

12. Efron, B., Tibshirani, R.J.: An Introduction to the Bootstrap. CRC Press, Boca Raton (1994)

13. Engel, J., Kuhl, D.E., Phelps, M.E., Crandall, P.H.: Comparative localization of foci in partial epilepsy by PCT and EEG. Ann. Neurol. 12(6), 529–537 (1982)

14. Farwell, L.A., Donchin, E.: The truth will out: interrogative polygraphy (lie detection) with event-related brain potentials. Psychophysiology 28(5), 531–547 (1991)

15. Ienca, M.: Hacking the brain: brain-computer interfacing technology and the ethics of neurosecurity. Ethics Inf. Technol. 18(2), 117–129 (2016)

16. Inzlicht, M., McGregor, I., Hirsh, J.B., Nash, K.: Neural markers of religious conviction. Psychol. Sci. 20(3), 385–392 (2009)

17. Kutas, M., Hillyard, S.A.: Reading senseless sentences: brain potentials reflect semantic incongruity. Science 207, 203–205 (1980)

18. Kutas, M., Hillyard, S.A.: Brain potentials during reading reflect word expectancy and semantic association. Nature 307, 161–163 (1984)

19. Lin, C., Wu, R., Liang, S., Chao, W., Chen, Y., Jung, T.: EEG-based drowsiness estimation for safety driving using independent component analysis. IEEE Trans. Circ. Syst. 52–I(12), 2726–2738 (2005)

20. Mangard, S., Oswald, E., Popp, T.: Power Analysis Attacks - Revealing the Secrets of Smart Cards. Springer, New York (2007)

21. Martinovic, I., Davies, D., Frank, M., Perito, D., Ros, T., Song, D.: On the feasibility of side-channel attacks with brain-computer interfaces. In: USENIX Security Symposium, Proceedings, pp. 143–158. USENIX Association (2012)

22. Portas, C.M., Krakow, K., Allen, P., Josephs, O., Armony, J.L., Frith, C.D.: Auditory processing across the sleep-wake cycle: simultaneous EEG and FMRI monitoring in humans. Neuron 28(3), 991–999 (2000)

23. Renauld, M., Standaert, F.-X., Veyrat-Charvillon, N., Kamel, D., Flandre, D.: A formal study of power variability issues and side-channel attacks for nanoscale devices. In: Paterson, K.G. (ed.) EUROCRYPT 2011. LNCS, vol. 6632, pp. 109–128. Springer, Heidelberg (2011). doi:10.1007/978-3-642-20465-4_8

24. Silverman, B.W.: Density Estimation for Statistics and Data Analysis. CRC Press, Boca Raton (1986)

25. Smart, N.P.: Computing on encrypted data. In: Kayaks & Dreadnoughts in a Sea of Crypto, September 2016

26. Standaert, F.-X., Koeune, F., Schindler, W.: How to compare profiled side-channel attacks? In: Abdalla, M., Pointcheval, D., Fouque, P.-A., Vergnaud, D. (eds.) ACNS 2009. LNCS, vol. 5536, pp. 485–498. Springer, Heidelberg (2009). doi:10.1007/978-3-642-01957-9_30

27. Standaert, F.-X., Malkin, T.G., Yung, M.: A unified framework for the analysis of side-channel key recovery attacks. In: Joux, A. (ed.) EUROCRYPT 2009. LNCS, vol. 5479, pp. 443–461. Springer, Heidelberg (2009). doi:10.1007/978-3-642-01001-9_26
28. Veyrat-Charvillon, N., Gérard, B., Renauld, M., Standaert, F.-X.: An optimal key enumeration algorithm and its application to side-channel attacks. In: Knudsen, L.R., Wu, H. (eds.) SAC 2012. LNCS, vol. 7707, pp. 390–406. Springer, Heidelberg (2013). doi:10.1007/978-3-642-35999-6_25

Impacts of Technology Trends on Physical Attacks?

Philippe Maurine[1](✉) and Sylvain Guilley[2,3]

[1] LIRMM, 161 Rue Ada, 34090 Montpellier, France
philippe.maurine@lirmm.fr
[2] Secure-IC S.A.S., 15 Rue Claude Chappe, Bât. B, 35510 Cesson-Sévigné, France
sylvain.guilley@secure-ic.com
[3] LTCI, Télécom ParisTech, Université Paris-Saclay, 75013 Paris, France

Abstract. Chip fabrication technologies evolve at an explosive rate. Notwithstanding, we analyze that attacks on *smartcard* chips are almost not impacted: only the architecture which gets more complex (e.g., the devices transition from mono- to multi-core) and the advanced design solutions (adaptative voltage and frequency scaling, multiple clock domains, asynchronicity, etc.) somehow make attacks slightly more complex. The situation is different for chips tightly integrated in embedded devices, such as *smartphone* chips. Indeed, the chips size and complexity increase drastically, and thus attacks identification phase becomes extremely hard. In addition, the chip targetted by the attacks is usually stacked with other chips (like the memory), which makes access to leakages and injection of faults a challenging task. Therefore, we conclude that there is a clear gain of security in the future to use smartphones as secure elements. Attacks at printed circuit board level associated with signal processing and machine learning could question this conclusion. Also, as a perspective, we notice that new kinds of attacks become possible on smartphones. Those devices being intrinsically connected, the new side-channel and fault injection attacks are realized not physically, but in software (controlled from an external center attack process): such attacks are called microarchitectural cache timing attacks (regarding side-channels) and RowHammer attacks (regarding fault injections). We predict increasing progress in those *cyberattack* threats.

Keywords: CMOS (Complementary Metal-Oxide-Semiconductor) technology · Fabrication evolution · Physical attacks · Side-channel attacks (SCA) · Fault injection attacks (FIA) · Countermeasures · Smartcards · Smartphones

1 Introduction

Physical attacks on cryptographic implementations date back to 1996, i.e., more than twenty years ago. The first side-channel attacks were the *timing attack* [17] (1996) and the *differential power analysis* [18] (1999). Later on, other side-channels had been exploited, such as the electromagnetic (EM) field, which

© Springer International Publishing AG 2017
S. Guilley (Ed.): COSADE 2017, LNCS 10348, pp. 190–206, 2017.
DOI: 10.1007/978-3-319-64647-3_12

allows to capture leakage non-invasively through the plastic packages, and also to narrow down the area of the captured signals. The first *fault injection* attack [4] (1997) consisted in the perturbation of a Rivest-Shamir-Adleman (RSA) computation using the Chinese Remainder Theorem (CRT).

We notice that the first side-channel attack (*timing attack* [17]) has been carried out on a Pentium chip, designed in 350 nm CMOS technology and clocked at 120 MHz. Today, the state-of-the-art processor of the same brand is the core i7 7700, designed in 14 nm CMOS technology and running at 4.20 GHz. This change is truly drastic[1]. This fantastic rate of innovation has been sustained accurately for 50 years[2]. Therefore, a natural question is thus to re-evaluate the potential of physical attacks given so many changes.

Physical attacks on integrated circuits proceed in two steps. First of all, some sensitive signals are either measured (case of *side-channel attacks*) or perturbed (case of *fault injection attacks*). Then, the traces and/or the effect of fault is analyzed, in a view to gain information on the secrets. The first step requires an access to the device. Clearly, the success of the attack depends on the reliability of the first phase, which in turns depends on the way the device is fabricated. As already mentioned, the fabrication technology evolves at a very high pace, for increased performance, cost, and integrability. Therefore, it is important to envision how the attacker potential will evolve. We make a difference between simple chips such as *smartcards* and integrated chips, such as *smartphones*.

In the rest of the paper, we first describe in Sect. 2 the various factors which allow for chip fabrication improvements. Then, in Sect. 3, we analyze how attacks are affected by these trends; our main result is summarized in Table 5 (c.f. Sect. 3.2.5). Emerging attacks for secure chips are discussed in Sect. 4. Finally, conclusions and perspectives are given in Sect. 5.

2   Integrated Circuits: Evolution and Trends

2.1   CMOS Technology Evolution

Gordon Moore is well known as co-founder of Fairchild Semiconductor and Intel corporation, but also owing to its famous "Moore law" [19]. This law predicts that the density of chips increases exponentially with time, namely that it doubles every eighteen months. Said differently, the minimum feature size, typically the transistors gate width, is multiplied by $1/\sqrt{2} \approx 0.707$ every eighteen months. Remarkably, this law has revealed true for more than 50 years. It is unclear today whether the law holds *per se* or whether it is self-realizing. Anyhow, this trend is a strong driver of the electronic industry, and has permitted many applications.

---

[1] Recall that, among all innovative technologies (health, biology, materials, etc.) developed worldwide, electronic chips are one where evolution is the largest and fastest: the number of patents filled every year is the most important (source: WIPO [30, Appendix B]), and the technology generation changes every eighteen months.

[2] Every one and a half year, a new *technological node* is released, where it is possible to integrate twice as more logic.

In practice, Moore law is merely an integration objective. However Dennard et al. [7] explain how to obtain an efficient scaling of MOSFETs (MOS Field Effect Transistors) in a view to integrate them in higher performance circuits.

It is all the more interesting as this explosive integration rate can even be sped up in practice, due to progress of related techniques: for instance, design methodologies and computer aided design (CAD) tools have allowed a better usage of the transistors for a given function.

Still, it is worth mentioning some peculiarities which occurred on the way of Moore/Dennard law. First of all, initially for a scaling $1/\kappa^2$ in density, we could observe an increase of κ of the maximum clock frequency, and a decrease of κ^2 of the power consumption (thence a constant power density ratio). However, starting from 2003 (with the 130 nm technological node), the clock frequency and the power consumption could not manage to scale at the same speed as that of density. This is due to the end of the supply (referred to as vdd) and of the threshold (referred to as vth) voltages shrinking. We recall that:

- vdd determines the power consumption of the chip (it varies as vdd^2), and
- vth determines at which voltage the CMOS gates switch; thus, the speed of the gates slows down when vth increases.

Their evolution with technological nodes is provided in Table 1, where the asymptotic limit $vdd \longrightarrow \approx 1$ V and $vth \longrightarrow \approx 0.3$ V can be clearly seen. Therefore, some problems arises, such as excessive power density. Second, the static power consumption started to become non-negligible compared to dynamic power consumption. Third, the feature size being so nanoscopic, variability issues arose. The reaction to make up for these issues were innovations at the architecture level:

- Power issues have been compensated by the use of clock gating, sleep modes, adaptative clock selection, and adaptative power. Indeed, playing with the vbb, for body bias voltage, it is possible to dynamically trade less speed for less power, and also to reduce static leakage currents.
- Multiplication of elements (e.g., multi-core circuits) allows to compensate for frequency limitation (the throughput is kept increasing at constant speed by increased parallelism).
- Variability due to process variation is mitigated by some redundancy in both the design redundancy (e.g., using spare resources) and adaptive design solutions.

Table 1. Indicative evolution of vdd and vth over 7 technological nodes

Node	250 nm	180 nm	130 nm	90 nm	65 nm	45 nm	28 nm
vdd	2.5 V	1.8 V	1.2 V	1.1 V	1.0 V	1.0 V	1.0 V
vth	0.5 V	0.4 V	0.3 V	0.3 V	0.3 V	0.3 V	0.3 V

Besides, even if the operation frequency is reaching a limit, it is not obvious to keep circuits operate so fast. Therefore, most circuits embed asynchronous clocks. For example the recent processors feature a main clock whose frequency is slightly modulated (thanks to a much slower clock), in order to avoid problems of resonance and electromagnetic compatibility (EMC).

As of today, the next node has a thinness of 7 nm to 5 nm (cf. Fig. 1), which is almost at the atomic scale. Therefore, it can be noticed another evolution of CMOS technologies, namely "More than Moore". This means that a variety of innovations allow to diversify what is feasible in CMOS logic. Example of such *CMOS helpers* are:

- new non-volatile memories (NVM) to make up for FLASH scaling limits and costs,
- 3D stacking of circuits, for a larger density, and also for the overall application to take advantage of various nodes at the same time. Indeed, it is expected an optimization in terms of cost and risk, e.g., due to potential yield issues of monolithic solutions. It is less risky to devise a system based on several chips proved to work in a robust way than with advanced heterogeneous technologies all implemented in a single chip. As a side-effect, the test complexity is also reduced. Eventually, costs are saved owing to a reuse of silicon-proven IPs.

Currently, the technologies in production are *stacked dies* or *package on package*. However, tomorrow, the paradigm will shift from *3D IC packaging* to *3D IC integration* (which is still at research stage). This can consist in chips stacking, by exploiting the *through silicon vias* (TSV) process, or even in monolithic 3D solutions, where field effect transistors (FET) are stacked vertically.

Eventually, CMOS process itself might be questioned in a medium term future. For instance, CMOS might be traded by carbon nanotubes (CNT) or even quantum computing. However, those revolutions, should they occur, are considered out of the scope of this paper, because they would be so disruptive that it is hard to make accurate predictions.

2.2 Today and Tomorrow Secure ICs

We can observe that smartcards, which are single chip in a majority of the cases (since they must be low-cost), face the difficult problem to embed FLASH memory. Indeed FLASH technology requires the generation of voltages greater than *vdd* to write data in it, as it consists in a double gate, in which charges shall be injected permanently. Thus, charge pumps must be integrated, which is complex as they are laid out in analogue logic. Besides, there is an issue related to charge retention: advanced nodes for FLASH are thus less reliable, since data are saved on tiny floating gates. Eventually, FLASH process requires specific (hence more costly) manufacturing process, because double gate transistors need two polysilicon layers. Therefore, they are lagging about 5 to 7 technological nodes behind the state-of-the-art circuits, such as smartphones. This is illustrated in Fig. 1, where the FLASH and pure logic technological specialties are highlighted in red ovals. It clearly appears that advanced technological nodes are most targeting large digital circuits, and not NVM such as FLASH. Indeed, those devices are characterized by the fact their NVM is off-chip, thereby solving the issue of common integration of CMOS logic and FLASH (called eFLASH). From a security point of view, embedded FLASH memories are also vulnerable targets as they become inoperative as soon as the charge pump is inhibited, which happens for instance when the attacker manages to illuminate it strongly with a focused LASER source. So-called *bumping attacks* are also a real-world threat [24].

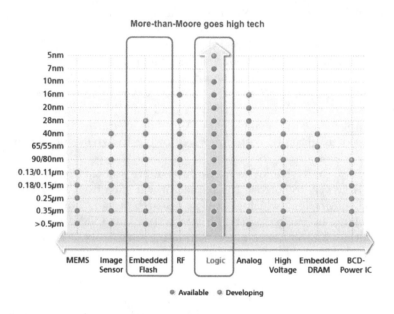

Fig. 1. Different nodes for different markets. Source: TSMC (courtesy of Ed Sperling, Fig. 3 of [25]—with our two "O" annotations)

Owing to the peculiarity of CMOS technologies evolutions (techniques to make up for *vdd* and *vth* limitation in terms of scaling, and "More than Moore" options), smartcards and smartphones secure chips do differ a lot. Typically, smartphone chips implement spacial parallelism (e.g., their architecture is *multicore*) and 3D-stacking, which will have, as we shall see next, a positive impact on their security vis-à-vis physical attacks.

Still, objects that simple as single chip devices still have a usefulness in practice. The reason is the enormous growth in terms of internet-of-things (IoT) devices, for smart applications in retail, building, transportation, energy, health, etc. And let us mention that despite their apparent simplicity (being single chip), they remain all the same very powerful. For instance, a current smartcard microcontroller, such as an STMicroelectronics STM32F4 from 2013, has the same

Table 2. Comparison between a Pentium from year 1993 and a single-chip processor in year 2013.

Brand	Intel	STMicroelectronics
Model	Pentium	STM32F4
Year	1993	2013
Processing power	239 DMIPS @ 133 MHz	255 DMIPS @ 180 MHz
Power efficiency (P/MHz)	75 mW/MHz	40 µW/MHz
Size	3100 K transistors (≈775 K-gates)	1246 K-gates
Minimum feature size (technology node)	800 nm	90 nm
vdd	3.0 V	1.2 V

computing capabilities as a former Intel Pentium (top-class personal computer) processor had twenty years earlier, in 1993, while being much more efficient in terms of power consumption. Refer to Table 2 for quantitative details; the performance is expressed in terms of Dhrystone millions of instructions processed per second (DMIPS, or mega-instructions per second), is similar for both chips. The STM32F4 chip is much more power-efficient than the Pentium, which is a benefit of CMOS down-scaling. Actually, the STM32F4 chip with its on-chip eFLASH memory is 7 technological nodes behind state-of-the-art. Therefore, $20 - 7 \times 1.5 = 12.5$ years of electronic fabrication progress separate the Intel Pentium and the STMicroelectronics STM32F4 chip, which coincides with Moore law:

$$\frac{\text{area}(t_0)}{\text{area}(t_0 + 12.5)} = \left(\frac{\text{size}(t_0)}{\text{size}(t_0 + 12.5)}\right)^2 = \left(\frac{800 \text{ nm}}{90 \text{ nm}}\right)^2 = 2^{12.5/1.5},$$

where $t_0 - 1993$ is the origin date of the oldest technology. Notice that neither Intel Pentium nor STMicroelectronics STM32F4 chips are secure; however, they are both representative of secure architectures (smartcards can be viewed as extremely secure microcontrollers). Eventually, we notice a final difference between smartcard and smartphone types of chips: as of today, a consumer is ready to spend $1 to buy a smart device to monitor its heart while jogging, but is less amenable to spend $1000 (i7 Intel cost) for the similar purpose.

A comparison between features of today's smartcards and smartphones is given in Table 3. Basically, a smartphone processor consists in the assembly of several chips, whereas an archetype smartcard consists in general of only one. A smartphone has several processors, each within its own island, where *vdd*, *vbb* and *frequency* can be chosen independently, and changed dynamically depending on the load and/or the power policy. Eventually, the processors of smartphones are accelerated using cache memory to speed up the access to the main RAM, which is shared among cores.

Table 3. Comparison between features of today's smartcards and smartphones

Features	Smartcards	Smartphones
Number of chips	1	≥ 2 (processing + memory chips + MEMS, etc.)
Number of processors	1 with fixed *vdd*, *vbb* and *frequency*	≥ 4, each with its own configurable *vdd*, *vbb* and *frequency*
NVM	eFLASH	Stacked chips of external memory
Use of cache memory	No	Yes

3   Physical Attacks and Technology Trends

3.1   Current Practice of Physical Attacks

The environment in which state-of-the-art attacks are performed is described in Table 4. We notice that most of the secure chips tested, as of today, consist in single-chip cryptographic modules (as per jargon of [26, Sect. 4.5.2]) of "smartcard" type (e.g., trusted platform modules, secure ICs, etc.). This is mostly the result of a strict security regulation on those objects, for which the highest evaluation assurance levels are demanded (e.g., in terms of Common Criteria [5] certification).

Table 4. List of conditions in which physical attacks are performed as of today

	Side-channel attacks	Fault injection attacks
List of conditions	Access to a leaking signal (power consumption, EM radiation, etc.)	Physical access to the device (laser, EM fault injection, etc.)
	Stability of the leaking signals, in space and time: – constant *vdd*, *vbb*, *frequency*, – constant location of the sensitive calculi	Stability of the targetted signals, in space and time: – constant *vdd*, *vbb*, *frequency*, – constant location of the sensitive calculi
	Moderated clock frequencies, few number of clock domains, asynchronicity	
	Moderated IC complexity (≈ 1 million gates equivalent)	
	Moderated computational noise	
	CMOS 90–65 nm technological nodes	

Clearly, the state-of-the-art of attacks practice (Table 4) is not meant to be rigidly interpreted: the attackers are smart, and adapt to new contexts. For instance, it is demonstrated in [21] an attack on a recent multi-chip system fabricated in 22 nm technology, probably made of hundreds of millions of transistors, and running at a frequency above the gigahertz.

3.2   Adversary's Challenges

We analyze here four factors in electronic circuits fabrication progress which impact the realization of attacks.

3.2.1   "CMOS Scaling" Factor

The effect of Moore's law in the reduction of size of transistors (gates) is clearly in the disadvantage of the attacker. As illustrated in Fig. 2, when the features in circuits shrink:

- **For side-channel attacks**:
  - the algorithmic noise increases,
  - unless the attacker is able to scale down the EM antennæ while conducting local measurements; some minor improvements can be made in this direction, as the radius of the probe (inductance) shall be at least 5 times the width of the metal ($\geqslant 10\,\mu m$). Another option to increase the signal-to-noise ratio (SNR) is merely to collect more traces. Indeed, when the noise is normal and independent from one trace to the other, the SNR increases linearly with the number of collected traces. Alternatively, new side-channels, such as photonic analysis [23] or voltage contrast microscopy [16], can also overcome the decreasing feature size of recent CMOS technologies.
- **For fault injection attacks**:
  - global faults [12] are less selective, since there are more signals (other than the sensitive ones) likely to be faulted,
  - whereas local faults require a scaling of the EM injection antennæ, or body bias probe tip [2], or laser spot diameter, etc. However, we reach here a limit as the section of a laser beam cannot be smaller than its wavelength, which is equal to $\approx 1\,\mu m$ for red light. Still, it is known that, for an attack to succeed, the attacker does not need to have the extremely strong capability to target one gate or one memory cell alone. Besides, it

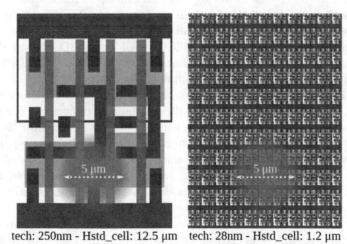

tech: 250nm - Hstd_cell: 12.5 μm   tech: 28nm - Hstd_cell: 1.2 μm

Fig. 2. Comparison of scale between a technology node where standard cells are layouted with a height Hstd_cell = 12.5 μm and a more recent technological node (which is seven nodes apart) where Hstd_cell = 1.2 μm

has been noticed adequately in [1] that the effect of a single *bit flip* can be obtained with a large injection area (wider than the gate carrying the bit to be flipped) all the same. Indeed, assuming a Gaussian profile for the laser beam, the attacker can reduce its intensity so that it effective area is not that at $1/e$ of the power, but at much higher threshold. Furthermore, if the attacker manages to setup an attack path where the bit to fault is surrounded by bits which are unused, then it suffices to fault very coarsely around the intended bit. The collateral effects have no consequence on the success of the attack.

On the contrary, one can notice that the effect of *vdd* and *vth* (recalled in Table 1) induces marginal changes in the current and voltages inside of the gates. Therefore, the *signal* an attacker is able to collect in a side-channel analysis does not weaken significantly. This means that his advantage is almost preserved. In a similar way, the propagation time in gates is also little affected, hence critical paths keep at the same order of magnitude. Thus, global fault attacks (e.g., clock tampering, underfeeding, etc. [12]) continue to work the same (provided the attacker manages to find an experimental fault injection procedure as targetted as possible, e.g., mostly the sensitive application runs, whilst the rest sleeps).

So, to conclude this analysis, one can say that CMOS scaling has either no impact on the physical attacks, or an impact which can be mitigated, typically by scaling down measurement and/or injection antennæ (or probe tips, laser beam focus, etc.).

3.2.2   "Physical Access to Device or Leaking Signals" Factor

Smartcards, by essence, are not concerned by 3D assembly. And even if some rare models implement this technology, one shall keep in mind that in a smart-card, attacks can be done both *frontside* and *backside*. Therefore, the attacker has more freedom to place its probe and/or injection tool. This is not the case for smartphones, since it is hardly possible to desolder the 3D integrated stack of chips and have them work standalone (because the attacker will have hard time to figure out how to plug the power supplies, the clocks, etc., but also because the system boot might be conditioned to the presence of other elements such as peripherals, which would stop the boot process unless connected). For the same reason, it is always possible for a side-channel attacker to monitor the current consumed by a smartcard (because it must be provided externally). However, this is not an option for smartphones, since it is very difficult to deport parts of the smartphone and still have them work (for reasons on signal integrity, in particular, and also because some models might implement anti-tampering techniques). We nonetheless attract the reader's attention to some recent trials of community building on this topic, e.g., through the organization of the http://www.hardwear.io conference. Thus, attention shall be kept on the topic of invasive attacks on complex smart devices such as smartphones.

Still, despite a 3D assembly, it can be imagined that leaking signals are conducted [27], hence can be measured even if the sensitive chip is placed in sandwich between two unrelated chips. Therefore, methodology presented in ISO draft international standard 20085-1 [13] might apply.

The fault injection attacks will be more sensitive to the way the 3D integration is done in practice. As mentioned in Sect. 2.1, such integration is getting tighter and tighter (moving from *stacked dies* to *3D IC packaging/integration*). Therefore, the disassembly required to access the sensitive parts of the chips is getting very challenging. We thus rate such attack at maximum level. However, it shall not be forgotten that novel fault injection on chips assembly might show up (RowHammer, discussed latter in Sect. 4, is a testimony that innovative attacks might be revealed out of the blue). One research direction we would like to point out is the practical study of *conducted perturbation* fault attack [22].

3.2.3   "Architecture and Advanced Design Solutions" Factor

In a view to save energy and better address the tradeoff between power consumption and efficiency, new design strategies are emerging. They include adaptive voltage and frequency scaling (abridged AVFS) which can be activated dynamically. This implies independent clock domains to cooperate (some logic is even fully self-timed, i.e., asynchronous), and charge balance in multicore systems.

The main impact of this trend is that side-channel traces realignment will become difficult. Moreover, in the case of multicore systems, it will become hard to attribute such portion of code to that process (including the one under attack). Maybe side-channel analysis techniques can tolerate these experimental drawbacks (see for instance [6]). But we expect that more genericity will come with a price on the efficiency side. Typically, on mono-threaded systems where *vdd*, *vth* and the *frequency* are fixed, accurate leakage models, namely Hamming weight and Hamming distance are known to match reliably the reality. Such advantage might disappear in the more challenging setup of varying signal amplitude and pace.

Fault attacks can better tolerate asynchronicity. Indeed, a wide array of FIAs need only one single fault to be conclusive on the part of the secret to recover. On the contrary, SCAs, both for "*simple*" [18, Sect. 2] and "*differential*" [18, Sect. 4] analyses, need to accumulate many measurements to cancel out as much noise as possible (and indeed, the noise level is exacerbated in the context of our adaptative device). For instance, a fault attack which consists in skipping a test, can be repeated many times, until the test is eventually successfully skipped, which will statistically happen independently of the AVFS features which randomize the execution pattern of the targetted code. As another example, let us consider differential fault injection attacks on AES. Here, provided the injection is lucky enough (it shall attain only one byte in the antepenultimate round), one fault suffices to recover a full 128-bit key [29]. Obviously, still, FIAs that require many faults (which are not so widespread) encounter the same difficulties as SCA attacks. Concluding, FIAs are less threatened than SCAs in the dynamic environment changes due to smart power-optimizing behaviors of the device.

3.2.4   "Die Size and Complexity" Factor

Regarding smartcards, we notice that their size has been decreasing over time (from more than $10\,\mathrm{mm}^2$ at their inception, to $\approx 1\,\mathrm{mm}^2$ today). The reason is that technological design shrink is dominating, while those objects have become smarter and smarter, hence requiring more logic. From the attacker perspective, this means that the design is more complex. However, the basic building blocks of smartcards have not changed over time: for example, there is still the need for one CPU, however it moved from 8 to 16 bit and then from 16 to 32 bit, over time. Same situation happens for the eFLASH, RAM, ROM, EEPROM memories: they are always part and parcel of a smartcard, however over time, their capacity has been growing, so as to enable more interesting applications. Thus, the attacker can always identify the parts which he intends to measure or fault, and thus complexity does not hurt him.

Smartphones represent a different case. As mentioned in Sect. 3.2.2, owing to 3D stacking of chips, the side-channel measurements and the fault injections can no longer be made local and accurate, all the more so as the application itself is mobile within the chip (e.g., moving from one core to another, so as to balance the load). Therefore, side-channel traces now consist in a patchwork of execution of various unrelated processes, which is in practice very challenging if not impossible to unravel. Fault injection attacks suffer the same problem of sensitive activity "volatility". Tracking for the manipulation of a sensitive data or operation can thus be compared to *searching a needle in a haystack*, thereby making fault injection attacks almost impossible (though probably all the same easier than exhaustive key search).

3.2.5   Summary

Interestingly, CMOS evolution does not make the attacks easier. However, some attack paths are more impacted than others. In order to summarize in one chart the impact of the four considered factors, we provide a qualitative rating using this terminology:

- ☺: no impact at all (the attack remains robust despite technological evolution),
- 😐: small negative impact (the attack is still possible at the price of little more efforts),
- ☹: strong negative impact (the attack is becoming challenging—probably it is no longer a valid attack path),
- 😖: huge negative impact (the attacks becomes almost infeasible).

The impacts of the four factors discussed in Sects. 3.2.1, 3.2.2, 3.2.3 and 3.2.4 are summarized in Table 5, which constitutes the main result of this paper.

Besides, it shall be noticed that smartcards favor the attackers, because the sensitive operations can be directly triggered by APDU (Application Protocol Data Unit, cf. ISO 7816-4 [28]) commands. This eases the attack, compared

Table 5. Summary of the effect of four evolutions in CMOS circuits on physical attacks

Factors	Smartcard		Smartphone	
	SCA	FIA	SCA	FIA
CMOS scaling (Sect. 3.2.1)	☺	☺	☺	☺
Physical access to device or leaking signals (Sect. 3.2.2)	☺	☺	😐	☹
Architecture & advanced design solutions (Sect. 3.2.3)	😐	😐	😐	😐
Die size and complexity (Sect. 3.2.4)	☺	☺	☹	☹

to smartphones, for which the synchronization is a real challenge. Indeed, on smartphones, the access to the API (Application Programming Interface) is less straightforward: there is no direct call from the user, hence it is difficult to master the manipulated data and to control the time/order of executions (which are often based on proprietary mechanisms).

The research efforts required by attackers to overcome the difficulties (denoted by ☺ and ☹ in Table 5) due to CMOS technologies evolutions are listed below:

- To make up for difficult access to the leakage or to the device itself, it is foreseen some advance in terms of *conducted leakage analysis* and *conducted perturbation*;
- Against dynamic behaviour of the chip (AVFS, existence of multiple cores operating in parallel, asynchronicity), we foresee the need for more advanced techniques of *signal processing* and of more flexible side-channel distinguishers;
- Same advances could definitely help advance the power of attacks despite increase of die size and complexity;
- In complement, investments in reverse-engineering (e.g., of 3D stacking structures) would clearly increase the success of fault injection attacks.

4 Logical Side-Channel and Fault Injection Attacks

4.1 Logical Attacks

As mentioned in Table 3, smartphone processors feature *cache* memories, all of them are shared (to some extend) among the cores. Therefore, *microarchitectural attacks* [9] appear to be a nice way to attack the device when other physical counterparts are made difficult due to the four factors presented in Sects. 3.2.1, 3.2.2, 3.2.3 and 3.2.4.

The advantage of such attacks, is that the attacker is a pure software piece of code, which is coming over the top (OTT). It will be "dropped", and then will "land" directly next to the program to attack (victim) where it will be executed. Hence, such attacks allow to circumvent the problem of physical identification of the localization (in the $X-Y$ plane) where the targetted sensitive application runs: the operating system will directly install the attacker the most closely as possible to the victim, since in multitask systems, processors are close one from each other as they depend on the same cache memory.

In other architectures, the *logical* side-channel can arise from an abuse of some monitoring functions. For instance, the integrated sensor in field programmable gates array (FPGA) platforms can be diverted from its intended usage in terms of safety to spy on some IP [20]: it thus behaves as a Trojan horse.

RowHammer attacks [15] are the full cyber counterpart of physical FIA. They share the same advantage as cache attacks: there is no need for the attacker to know the physical layout of the chip(s), nor to have any physical access.

Therefore, we expect much research in those directions. This is illustrated in Fig. 3, regarding the growth of the *remote* threat and other (less successful) *local/physical* analyses.

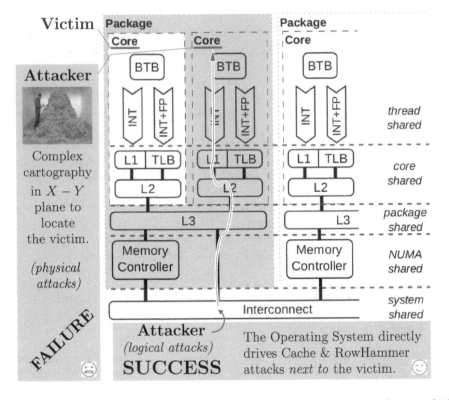

Fig. 3. Contented resources in a multiprocessor system, typical to smartphones, which allows to contrast *physical* cartography with *logical* cache attacks (background image courtesy of Qian Ge et al. [9, Fig. 1, p. 6].)

4.2   Protections Against Logical Attacks

Cache timing and RowHammer attacks demand further focused studies, as there is, as of now, no regulatory incentive to avoid them. Indeed, they are explicitly out of scope of Global Platform TEE Protection Profile [10]. Besides, there is no systematic way to counter such attacks. As an example, cache timing attacks can be made more difficult by the application of some heuristics, such as:

- replacing tests (control flow irregularities) such as d=c?a:b by unconditional code such as m=-(!!c), d=(a&m)^(b&~m),
- trading look-ups in a table T such as y=T[x] by address-independent code such as y=0, for(i=0..#T-1) y^=(-(x==i))&T[i],
- memory access randomization based on oblivious RAM (also known as "ORAM") concept [11],
- obfuscation such as white box cryptography (WBC [3]), etc.

However, even those simple patterns are prone to implementation errors. The article ironically entitled *Make Sure DSA Signing Exponentiations Really are Constant-Time* [8] shows a mistake where the constant-time operations are coded but not called adequately, hence leaving the possibility for an attacker to exploit the code. RowHammer attacks also continue to work because it is possible to access DDR SRAM (Double Data Rate Synchronous Dynamic Random Access Memory) directly (i.e., bypassing the cache memories) at high rates thanks to legacy operations, such as prefetch and clflush. Those instructions allow fast access to the DDR; thereby, paradoxically enough, efficient processors are less secure. Besides, DDR is sensitive to faults because it is very integrated. Hence the practically, as of today, of *cyber-enabled RowHammer attacks*. Notice that error correcting codes (ECC) do not prevent those attacks because their error correction capability is very limited. For instance, 2-bit ECC reduces the success probability only by a factor $2^2 = 4$, hence attacks will require only 4 times more traces to succeed, which is negligible for a determined attacker.

5   Conclusion and Perspectives

We have analyzed the various factors which allow for chip fabrication improvements. Paradoxically, we derive that Moore's law (CMOS minimum feature size decreases over time) does not impact much state-of-the-art attacks. On the contrary, factors such as voltage scaling, designs with multicores and asynchronicity make attacks (slightly) more complex. In the case of smartphones, side-channel and fault injection attacks are very impeded due to the increase of complexity of the chip, and stacking makes it more difficult to access signals needed for side-channel attacks and particularly for fault injection attacks. We conclude that the security level actually increases over time for devices such as smartphones.

Nevertheless, we believe that the mere technological evolution is not going to eradicate the problem of physical attacks. The challenge in front of attackers is now to better process side-channel curves, perform horizontal analysis on a single (or few) curves, develop side-channel specific pattern matching techniques, improve technology to resynchronize and interpret complex curves, etc. In particular, for smartphone devices, the resolution of the timing (required for timing-based side-channel attacks and fault injection triggering) can be enhanced by physical measurements directly on the printed circuit board. In parallel, new attack paths (re)appear, such as timing attacks, hence a paradigm shift in terms of security evaluation.

References

1. Agoyan, M., Dutertre, J.-M., Mirbaha, A.-P., Naccache, D., Ribotta, A.-L., Tria, A.: How to flip a bit? In: 2010 IEEE 16th International On-Line Testing Symposium (IOLTS), pp. 235–239, July 2010
2. Beringuier-Boher, N., Lacruche, M., El-Baze, D., Dutertre, J.-M., Rigaud, J.-B., Maurine, P.: Body biasing injection attacks in practice. In: Palkovic, M., Agosta, G., Barenghi, A., Koren, I., Pelosi, G. (eds.) Proceedings of the Third Workshop on Cryptography and Security in Computing Systems, CS2@HiPEAC, Prague, Czech Republic, 20 January 2016, pp. 49–54. ACM (2016)
3. Beunardeau, M., Connolly, A., Géraud, R., Naccache, D.: White-box cryptography: security in an insecure environment. IEEE Secur. Priv. **14**(5), 88–92 (2016)
4. Boneh, D., DeMillo, R.A., Lipton, R.J.: On the importance of checking cryptographic protocols for faults. In: Fumy, W. (ed.) EUROCRYPT 1997. LNCS, vol. 1233, pp. 37–51. Springer, Heidelberg (1997). doi:10.1007/3-540-69053-0_4
5. Common Criteria Consortium: Common Criteria (aka CC) for Information Technology Security Evaluation (ISO/IEC 15408) (2013). http://www.commoncriteriaportal.org/
6. de Chérisey, É., Guilley, S., Heuser, A., Rioul, O.: On the optimality and practicability of mutual information analysis in some scenarios. In: ArticCrypt (IACR Event), 17–22 July, Longyearbyen, Svalbard, Norway (2016)
7. Dennard, R.H., Gaensslen, F.H., Rideout, V.L., Bassous, E., LeBlanc, A.R.: Design of ion-implanted MOSFET's with very small physical dimensions. IEEE J. Solid-State Circ. **9**(5), 256–268 (1974)
8. García, C.P., Brumley, B.B., Yarom, Y.: Make sure DSA signing exponentiations really are constant-time. In: Weippl, E.R., Katzenbeisser, S., Kruegel, C., Myers, A.C., Halevi, S. (eds.) Proceedings of the 2016 ACM SIGSAC Conference on Computer and Communications Security, Vienna, Austria, 24–28 October 2016, pp. 1639–1650. ACM (2016)
9. Ge, Q., Yarom, Y., Cock, D., Heiser, G.: A survey of microarchitectural timing attacks and countermeasures on contemporary hardware. Cryptology ePrint Archive, Report 2016/613 (2016). http://eprint.iacr.org/2016/613
10. GlobalPlatform Device Committee: TEE Protection Profile Version 1.2.1, GPD_SPE_021, December 2016. https://www.globalplatform.org/specificationform.asp?fid=7831
11. Goldreich, O.: Towards a theory of software protection and simulation by oblivious RAMs. In: Aho, A.V. (ed.) Proceedings of the 19th Annual ACM Symposium on Theory of Computing, 1987, New York, USA, pp. 182–194. ACM (1987)
12. Guilley, S., Danger, J.-L.: Global faults on cryptographic circuits. In: Joye, M., Tunstall, M. (eds.) Fault Analysis in Cryptography. Springer, Heidelberg (2012)

13. ISO/IEC JTC 1/SC 27/WG 3. ISO/IEC CD 20085–1:2017 (E): Information technology - Security techniques – Test tool requirements and test tool calibration methods for use in testing non-invasive attack mitigation techniques in cryptographic modules – Part 1: Test tools and techniques, 25 January 2017

14. Joye, M., Tunstall, M.: Fault Analysis in Cryptography. Springer, Heidelberg (2011). doi:10.1007/978-3-642-29656-7. ISBN 978-3-642-29655-0

15. Kim, Y., Daly, R., Kim, J., Fallin, C., Lee, J.-H., Lee, D., Wilkerson, C., Lai, K., Mutlu, O.: Flipping bits in memory without accessing them: an experimental study of DRAM disturbance errors. In: ACM/IEEE 41st International Symposium on Computer Architecture, ISCA 2014, Minneapolis, MN, USA, 14–18 June 2014, pp. 361–372. IEEE Computer Society (2014)

16. Kison, C., Frinken, J., Paar, C.: Finding the AES bits in the haystack: reverse engineering and SCA using voltage contrast. In: Güneysu, T., Handschuh, H. (eds.) CHES 2015. LNCS, vol. 9293, pp. 641–660. Springer, Heidelberg (2015). doi:10. 1007/978-3-662-48324-4_32

17. Kocher, P.C.: Timing attacks on implementations of Diffie-Hellman, RSA, DSS, and other systems. In: Koblitz, N. (ed.) CRYPTO 1996. LNCS, vol. 1109, pp. 104–113. Springer, Heidelberg (1996). doi:10.1007/3-540-68697-5_9

18. Kocher, P., Jaffe, J., Jun, B.: Differential power analysis. In: Wiener, M. (ed.) CRYPTO 1999. LNCS, vol. 1666, pp. 388–397. Springer, Heidelberg (1999). doi:10. 1007/3-540-48405-1_25

19. Moore, G.E.: Cramming more components onto integrated circuits. In: Readings in Computer Architecture, pp. 56–59. Morgan Kaufmann Publishers Inc., San Francisco (2000)

20. Ngo, X.T., Najm, Z., Bhasin, S., Roy, D.B., Danger, J.-L., Guilley, S., Sensor, I.: A backdoor for hardware trojan insertions? In: 2015 Euromicro Conference on Digital System Design, DSD 2015, Madeira, Portugal, 26–28 August 2015, pp. 415–422. IEEE Computer Society (2015)

21. Saab, S., Rohatgi, P., Hampel, C.: Side-channel protections for cryptographic instruction set extensions. Cryptology ePrint Archive, Report 2016/700 (2016). http://eprint.iacr.org/2016/700

22. Sauvage, L., Guilley, S., Danger, J.-L., Homma, N., Hayashi, Y.-I.: A fault model for conducted intentional electromagnetic interferences. In: 2012 IEEE International Symposium on Electromagnetic Compatibility (EMC), pp. 788–793, 5–10 August 2012, Pittsburgh, PA, USA (2012). http://2012emc.org/. doi:10.1109/ISEMC.2012.6351664

23. Schlösser, A., Nedospasov, D., Krämer, J., Orlic, S., Seifert, J.-P.: Simple photonic emission analysis of AES. J. Cryptogr. Eng. 3(1), 3–15 (2013)

24. Skorobogatov, S.: Flash memory 'bumping' attacks. In: Mangard, S., Standaert, F.-X. (eds.) CHES 2010. LNCS, vol. 6225, pp. 158–172. Springer, Heidelberg (2010). doi:10.1007/978-3-642-15031-9_11

25. Sperling, E.: Reworking Established Nodes, 24 May 2017. Blog article: https://semiengineering.com/reworking-established-nodes/. Accessed 1 June 2017

26. NIST FIPS (Federal Information Processing Standards): Security Requirements for Cryptographic Modules Publication 140-2, 25 May 2001. http://csrc.nist.gov/publications/fips/fips140-2/fips1402.pdf

27. Sugawara, T., Hayashi, Y., Homma, N., Mizuki, T., Aoki, T., Sone, H., Satoh, A.: Mechanism behind information leakage in electromagnetic analysis of cryptographic modules. In: Youm, H.Y., Yung, M. (eds.) WISA 2009. LNCS, vol. 5932, pp. 66–78. Springer, Heidelberg (2009). doi:10.1007/978-3-642-10838-9_6

28. ISO/IEC 7816 (Joint technical committee (JTC) 1/Sub-Committee (SC) 17): Identification cards - Integrated circuit cards. http://www.iso.org/iso/catalogue_detail. htm?csnumber=35168
29. Tunstall, M., Mukhopadhyay, D., Ali, S.: Differential fault analysis of the advanced encryption standard using a single fault. In: Ardagna, C.A., Zhou, J. (eds.) WISTP 2011. LNCS, vol. 6633, pp. 224–233. Springer, Heidelberg (2011). doi:10.1007/ 978-3-642-21040-2_15
30. WIPO (World Intellectual Property Organization): World Patent Report: A Statistical Review - 2008th edn. http://www.wipo.int/ipstats/en/statistics/patents/ wipo_pub_931.html. Accessed 11 June 2017

Low-Cost Setup for Localized Semi-invasive Optical Fault Injection Attacks
How Low Can We Go?

Oscar M. Guillen[1]($^{\boxtimes}$), Michael Gruber[2], and Fabrizio De Santis[2]

[1] Giesecke & Devrient GmbH, Munich, Germany
oscar.guillen@gi-de.com
[2] Chair of Security in Information Technology, Technische Universität München,
Munich, Germany
{m.gruber,desantis}@tum.de

Abstract. Localized semi-invasive optical fault attacks are nowadays considered to be out of reach for attackers with a limited budget. For this reason, they typically receive lower attention and priority during the security analysis of low-cost devices. Indeed, an optical fault injection setup typically requires expensive equipment which includes at least a laser station, a microscope, and a programmable X-Y table, all of which can quickly add up to several thousand euros. Additionally, a careful handling of toxic chemicals in a protected environment is required to decapsulate the chips under test and gain direct access to the die surface. In this work, we present a low-cost fault injection setup which is capable of producing localized faults in modern 8-bit and 32-bit microcontrollers, does not require handling hazardous substances or wearing protective eyeware, and would set back an attacker only a couple hundred euros. Finally, we show that the type of faults which are obtained from such a low-cost setup can be exploited to successfully attack real-world cryptographic implementations, such that of the NSA's Speck lightweight block cipher.

Keywords: Fault injection · Semi-invasive · Optical fault attacks · Backside · Microcontrollers · Embedded devices · Speck

1 Introduction

Integrated Circuit (IC) design is a complex and difficult task, which requires a strict set of conditions to be met. Modern IC's are designed to work within specific operating ranges, defined by the type of applications and environments which they have to endure. Outside of these specification ranges, the correct functionality of the ICs is no longer guaranteed. Faults arise whenever a deviation from the expected operating conditions occurs. Of particular importance to security researchers are the errors produced by faults that can be used to compromise the security of cryptographic devices. The first theoretical work which

© Springer International Publishing AG 2017
S. Guilley (Ed.): COSADE 2017, LNCS 10348, pp. 207–222, 2017.
DOI: 10.1007/978-3-319-64647-3_13

used faults to extract secret information is due to Boneh et al. [4] and was targeted at the RSA public-key cryptosystem. Since then, practical ways to induce faults in secure devices have been extensively researched. A good introduction to the different types of fault attacks can be found in [1]. These attacks can be broadly classified into three categories depending on the type of access that an attacker has to the chip under attack: (1) non-invasive, (2) semi-invasive, and (3) invasive. Non-invasive attacks do not require modifications to the device under attack, i.e. only the chip's external interfaces are used. The most common (and cheap) type of attacks in this category are clock glitches and voltage spikes, e.g. by over- and underfeeding the chip's power supply. On the other side, invasive attacks assume that the attacker has complete access to the internal structures of the chip, thus being able to directly eavesdrop upon buses, set or clear signals and even modify the physical properties of the chip. These are typically the most powerful (but expensive) type of fault attacks. Semi-invasive attacks fall in between these two categories. This latter type of attacks requires the chips to be processed, i.e. (partially) removing the package to have access to the die, but without performing any modifications to the chip itself. Semi-invasive attacks were introduced by Skorobogatov et al. in [15], showing that transient faults could be induced in cryptographic devices through optical probing using lasers and flashguns.

State of the Art. Low-cost setups to perform non-invasive attacks already exist in literature, e.g. the ChipWhisperer [11] and the Generic Implementation Analysis Toolkit (GIAnT)[1]. However, until now there were no similar low-cost setups for *semi-invasive* fault attack exploration. Although the use of low-cost flashguns was already presented in [15], the remaining costs of the setup (which is necessary to perform *localized* attacks) are still considerably high, e.g. high-end microscopes. Also, the faults in [15] were injected in today's obsolete 1300 nm technology. As for current solutions, in 2011, van Woudenberg et al. [17] reported a budget range between $50k and $150k for a laser fault injection setup. In 2015, Breier et al. [5] estimated a price close to €150k for a similar fault injection setup.

Contributions. We present the first *low-cost* setup for semi-invasive *localized* fault attacks that combines a camera flashgun, a motorized X-Y table, and small ball lens to focus the lightbeam. The complete setup runs for a few hundred euros, and is capable of inducing localized faults in modern 8-bit and 32-bit microcontrollers (an ATmega328P AVR and a STM32F030F4P6 ARM Cortex M0) manufactured in 350 nm and 90 nm processes, respectively. It does not require handling hazardous substances or wearing protective eyeware, and can be reproduced in a home laboratory with little effort. Our work is particularly relevant when it comes to evaluating the risks of certain type of physical attacks on security devices, i.e. when the risk depends on the costs of the equipment needed to perform such attacks, like for Common Criteria certifications. Furthermore,

---

[1] https://sourceforge.net/projects/giant/.

although a description of the setup used to conduct the attacks is normally given, exact details are rarely included in literature. Thus comparing different setups and reproducing the exact conditions presented in these works, to estimate the associated risks, are typically not easy tasks. We aim at providing all relevant information so that the exact setup and experiments' conditions can be easily reproduced for the evaluation of attacks and countermeasures in risk assessments. The files used for this project are available online under the MIT license.[2]

Organization. Section 2 provides background information on optical injection setups. Section 3 details our proposed low-cost fault injection station. Section 4 provides a guideline towards an evaluation methodology. Experimental results are discussed in Sects. 5 and 6. Our conclusions are presented in Sect. 7.

2 Background

Fault injection is a class of physical attacks, which aims to modify the behaviour of a device in order to bypass security mechanisms, or cause errors that help to infer secret keys. Integrated circuits are designed to work over a wide range of values for different parameters, including voltage, temperature, and clock frequency among others. The extremes of these parameters receive the name of process corners. ICs created under a specific process are tested to function correctly within a predefined set of corners, which depend on the intended application of the circuit. Modifying the environmental parameters to force ICs to operate outside these corners may cause malfunctions in the circuits, leading to faults which may be exploited by an attacker to compromise their security. In practice, fault injection can be performed using different techniques. These can be classified according to the degree of invasiveness that the attacker operates with respect to the chip:

- **Non-invasive**: The attacker has access to the chip's external interfaces.
- **Semi-invasive**: The attacker is able to partially or completely remove the chip's package and have access to the die's surface.
- **Invasive**: The attacker has complete access to the deep metalization layers.

Non-invasive techniques include over- and under-voltage variations on the power supply (spike attacks), overheating and freezing the chip, and inducing glitches on the clock lines [1]. The variation of these parameters causes the transistors within the ICs to switch at a faster or slower rate than expected. The resulting errors obtained through these attacks can be attributed to setup time violations. Invasive techniques employ the use of expensive tools, that range from micro-probing stations to Focused Ion Beam (FIB) machines. A more interesting class of attacks is known as semi-invasive attacks. This type of attacks allow for more complex fault models than non-invasive attacks, while keeping the equipment

---

[2] https://github.com/open-fi/fault-injector.

costs lower than those needed for invasive attacks. In particular, they allow for *localized* fault injections on the chip's die, which enables an attacker to carry out more powerful and specialized fault attacks.

2.1 Semi-invasive EM Attacks

One of the techniques that can be used to inject faults is Electro-Magnetic (EM) radiation. EM fields can travel through the packaging materials and thus, removing the package of the chip is not necessarily required. However, doing so helps the attacker recognize the features in the IC, making it easier to find the correct point to induce a fault, i.e. by partially removing the encapsulation of the chip, the attacker is able to identify areas of interest where to inject faults (like memories or buses). Moreover, removing grounded metal plates used in some packages may increase the effectiveness of the attacks as they might act as EM shields. Neve et al. [10] describe their experiments with a low-cost setup, using a camera flashgun connected to hand-made coils, to generate EM pulses capable of modifying data values in memories and the address bus. A lower cost alternative comes from Schmidt et al. [12], who used a spark-gap generator from a gas lighter to manually create high frequency sparks instead of magnetic fields. Due to the high charge change caused by the spark gap a strong EM burst is generated, which can be used to temporarily disrupt the device. They were able to affect the program flow as well as the memory contents (SRAM and Flash). Dehbaoui et al. [7] presented a more sophisticated EM fault injection setup, capable of producing pulses with low jitter, wide voltage ranges and high accuracy timing. Their setup is composed of a pulse generator, EM coils, and a motorized X-Y-Z table. However, specific details on the equipment used were not given.

2.2 Semi-invasive Optical Attacks

Optical attacks are a very common type of semi-invasive fault injection attacks. In fact, light radiation on a semiconductor region of an IC can cause currents to form, if the semiconductor's band gap is exceeded. This current is proportional to the intensity of the incident light. This effect is known as the photovoltaic effect. In their seminal paper, Skorobogatov et al. [15] introduced optical fault injections by utilizing lasers and flashguns to produce ionizing radiation to actively control the behavior of ICs. The first known attacks using light were performed with ultraviolet (UV) light, to erase EPROM-like memories. An attack on the security fuses of a microcontroller was described in [14]. Huang [6] later demonstrated that fuses guarded under a metal shield could be still be erased, while keeping most of the program memory intact by placing the microcontroller at an angle and covering the flash memory with a piece of electrical tape. In [13], Schmidt et al. presented an attack that used UV to modify the memory values of flash memories, showing a practical attack on AES when a single bit of the S-box table is modified. In order to erase only the desired bits of information, the parts that do not need to be exposed can be shielded with the help of an opaque material, such as duct tape or using a UV resistant marker. Attacks using white light were also described in [15], where a precision microscope was used to focus the light in a precise manner. With this technique it was possible to flip single bits in SRAM cells. To obtain a more precise beam of light, lasers can be used. Although they can induce the same type of faults as white light, they are typically more effective [15]. In order to focus the laser beam into the vulnerable structures, a precision microscope was used also in this case. To avoid the usage of a costly microscope, Schmidt et al. [12] attack using a laser pointer and focused the light beam using a fiber-optics light guide.

Optical attacks on a chip can be performed either on the frontside, through the metalization layers, or the backside of the die, through the substrate. The two approaches have both advantages, and disadvantages:

- **Frontside attacks**: Red and green (808 nm and 532 nm wavelength) lasers are typically used for attacks from the frontside. Removing the package is usually done using hazardous chemicals, such as nitric acid (HNO_3) and sulfuric acid (H_2SO_4). The chips components can be optically inspected with the aid of a microscope. However, the metalization layers in the chip can reflect or obstruct the light beams. Optical inspection might not be possible with newer technologies due to smaller feature sizes.
- **Backside attacks**: This type of attacks are more successful when conducted using near infrared (1064 nm) lasers, as silicon becomes transparent at these wavelengths. Partial removal of the package to access the backside can be done using mechanical polishing techniques, and is thus safer and easier to perform. Due to the lack of visibility under normal light, correct positioning becomes a more difficult task. However, infrared imaging can be used to overcome this obstacle. Attacking from the backside has the benefit that metalization layers do not interfere with the light beams.

3 A Low-Cost Optical Fault Injection Station

The main control device in our setup is a standard desktop computer (PC), on which different control and configuration scripts are executed. The PC is connected to a motorized X-Y table through the serial port (UART), and controls the position of the device under test (DUT) beneath the flashgun. The DUT is programmed (and debugged) using an external debugger through a Serial Wire Debug (SWD) interface. A STM32F411 Nucleo board is used to reset the DUT, and adjust the trigger timing of the flashgun, and it is connected to the PC via UART. Lastly, the DUT is also connected to the PC via UART to load and retrieve data. Figure 1 shows a block diagram of the proposed setup, while Fig. 2 its actual deployment. In its current form, our fault injection setup can be used to asses the security of devices for which a trigger signal can be output. For devices where this is not possible, our setup could be expanded to allow pattern-based triggering. Low-cost implementations of Sum of Absolute Differences (SAD) triggering have been discussed in [8, p. 44] and [11].

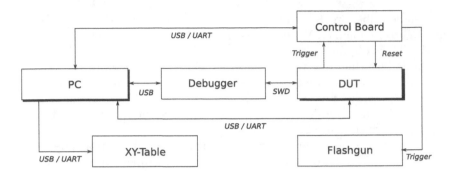

Fig. 1. Block diagram of the fault injection setup.

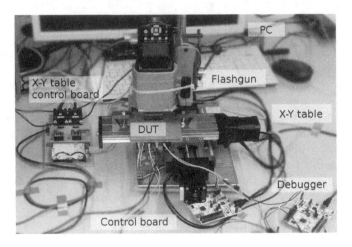

Fig. 2. Picture of the fault injection station.

3.1 Flashgun

Initial experiments were carried out using a flashgun from the brand Yongnuo, specifically the YN560 I. To measure the duration and intensity of the light, an LED was inserted into a hole drilled into a cardboard hood, which was placed on top of the flashgun to avoid dispersion. A quick trigger analysis clearly exhibited a jitter in the activation of the flashgun which emitted light at different times after the trigger. Later experiments were conducted using the YN560 III from the same brand, which resolved the jitter problem. The YN560 III can be bought new for €60 through traditional online retailers. The flashgun was connected directly to an external power supply, building an adapter for this purpose. Figure 3 shows the voltage at the LED, where trigger was captured at $0\,\mu s$. The flashgun emits

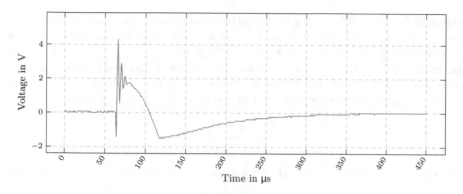

Fig. 3. Voltage across the LED after a trigger has been captured.

light after 64 µs. This delay is constant, and can be easily accounted for during fault injections. After roughly 400 µs the light emission is fully vanished[3].

3.2   Single-Lens Microscope

The idea of using a single ball lens as a microscope was first introduced in the 17th century and is attributed to Anton van Leeuwenhoek, the father of microbiology. He realized that looking through small glass beads would produce a magnification similar to that of a microscope. By improving the quality of his lenses, he achieved greater magnification than that of the compound microscopes of the era, while relaying on a much simpler design. Van Leeuwenhoek's microscopes used a single glass sphere mounted on a brass plate with a small hole. The analyzed objects would be placed on top of a screw on the other side of the plate, and focus would be adjusted with another screw that moved the object closer or farther to the lens. This principle was recently revised in [16], to provide a low-cost alternative to clinical microscopes. The design uses a ball lens coupled to a cellphone, to effectively transform it into a 350× microscope. In the field of physical analysis of ICs a similar technique, called Solid Immersion Lens microscopy, is used to increase the microscope resolution, as the feature sizes shrink at each new technology node [3]. We apply this concept to focus light into the surface of ICs' dies to perform localized fault injection. If we consider the thickness of a lens as two spherical refracting surfaces, then the focal length is given by:

$$\frac{1}{f} = (n - 1)\left[\frac{1}{R_1} - \frac{1}{R_2} + \frac{d(n-1)}{nR_1R_2}\right],\tag{1}$$

where f is the focal length, n is the index of refraction, R the radius of curvature of the concave surface, and d is the lens's thickness. The focal length f is the distance between the center of the lens and the point where the light converges

---

[3] This duration accounts also for the discharging effects inside the LED and various parasitics, i.e. the light is emitted for a much shorter time.

into a focal point. The index of refraction n represents how much the light is bent by the material from which the lens is made. The radius of curvature R is defined as the radius of a sphere's curvature that matches that of the face of the lens. The lens thickness d is the distance between the two faces of the lens, measured at the center. For ball lenses with radius $R > 0$, $R_1 = R$, $R_2 = -R$, and $R = \frac{1}{2}d$; where d is the diameter of the lens. Thus, Eq. 1 simplifies to:

$$\frac{1}{f} = \frac{4(n-1)}{n \cdot d}.$$ (2)

Knowing f, the magnification M of a lens compared to a human eye is given by:

$$M = \frac{250}{f},$$ (3)

where 250 nm is a standard value and is estimated as the closest distance at which the human eye can focus. For our experiments, we used a 1.0 mm diameter, N-BK7 borosilicate-glass ball lens, which was bought online for €25. The index of refraction of the lens is $n = 1.517$. For this specific lens we obtain a focal length[4] of 0.73356 mm and a magnification of 340×. Higher magnifications can be obtained by using lenses with smaller diameters.

3.3   Motorized X-Y Table

A precise positioning table is one of the main components that a localized fault injection setup must have in order to obtain reproducible results [5]. Below the price range of a couple thousand dollars, motorized X-Y tables with small step resolution are hard to find. For this reason, we opted for a low-cost X-Y table designed for fine machining. Namely, a PROXXON KT 70 micro-coordinate table with stepper motors. It is a compact table of 200 mm × 70 mm in size, with 134 mm travel in X and 46 mm in Y direction. To support it, we purchased a MB 200 drilling stand from the same company. To provide PC-controlled automation, the motors were driven using DRV8825 stepper motor controllers, and an Arduino UNO was programmed with Grbl[5], an open source Computer Numerical Control (CNC) controller software. The stepper motors used have a step angle of 1.8° with ±5% accuracy. Given that a complete revolution equals 1 mm displacement, the table's travel resolution is within 5 μm per step, if the motor drivers are configured to use full steps.

4   Evaluation Methodology

In this section, we describe how the devices under test were prepared to perform optical attacks from the backside and how fault characterization was performed.

---

[4] The back focal length is $0.73356 - 0.5 = 0.23356$ mm .

[5] https://github.com/grbl/grbl.

4.1 Device Preparation

We opted for using a mechanical polishing technique, similar to the one discussed in [14]. Figure 4 shows the three steps needed for the preparation of the devices. (a) Sandpaper was used to thin out the epoxy encapsulation of the package. For DIP packages the legs of the chip were first bent out to move them out of the way, other packages such as TSSOP did not require this step. The epoxy was sanded down, using P150 sandpaper (€4). Latex gloves and a protective mask were used as a safety precaution (€20). (b) Once the package's lead frame was reached, we made use of a sharp object to pry it open. (c) The glue used to secure to the chip's die to the lead frame was removed with the use of acetone and a cotton swab. Acetone can be easily bought in pharmacies as nail polish remover for less than €2. Note that the mechanical preparation method used to access the backside of the chip cannot be employed for all type of packages, e.g. in certain BGA packages, the pins are on the backside of the chip, and thus they should be accessed from the frontside.

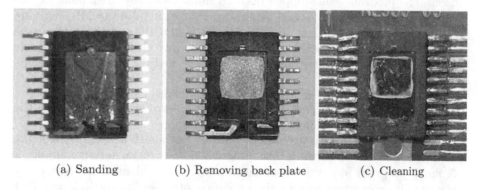

(a) Sanding   (b) Removing back plate   (c) Cleaning

Fig. 4. Preparation steps to perform backside optical fault injection on a microcontroller (STM32F030F4P6).

4.2 Fault Characterization

The prepared chips were placed in hand-made, minimal system boards, that contain the basic components to supply power and clock to the device, a programming interface, and connectivity to the PC through a USB-Serial interface. The chips were turned and soldered upside down to enable access to the die's backside and tested in two different conditions:

– **Global attack:** Using the flashgun directly on top of the exposed die, without any means to focus the light beam.
– **Local attack:** Attaching a ball lens to the end of the flashgun to focus the light beam, and using the X-Y table to scan over the die's surface.

In order to get information about the faulted behavior, three different types of tests were applied:

- **Memory dump:** This test was run to identify changes in the RAM's content after the flash was triggered. For this test, the microcontroller was placed in a reset state before each iteration. After releasing the microcontroller from reset, a user defined pattern (0xFF/0x00) is written into predefined RAM locations, then the flash was triggered. The content of the RAM was then extracted from the microcontroller, through the serial interface, and the downloaded values were compared with the reference values on the PC. Furthermore this test was used as an early indicator for register faults, if the length of the expected ram dump did not match the reference.
- **Register dump:** This test was performed to identify changes in the CPU registers after the flash was triggered. To perform the test, the microcontroller was placed in a reset state before each iteration. After coming out of the reset state, all the accessible CPU registers were cleared to zero. A test program was executed to perform a series of NOP operations for the duration of the flash. The program was stopped once an inline assembly breakpoint at the end of the program was reached. Afterwards, the content of the CPU registers (r0-r12, pc, lr, xPSR) was dumped using OpenOCD[6] and the values were compared with the reference values on the PC. Two checks were performed:
    1. Check whether the program counter (pc) reached the address of the inline assembly breakpoint.
    2. Check whether CPU registers (r0-r12, lr, xPSR) changed their values.
  The reference values to perform this last check were obtained following the same procedure, but without activating the flash.
- **Program execution disturbance:** The last test aimed at identifying faults that caused a disturbance in the program flow (e.g. an instruction skip). A small program was written to include a set of predefined Boolean operations. The correct answer to the execution of these instructions was first recorded. The flashgun was triggered at different time/place instances, to find out the temporal/spatial points which caused a disturbance in the program flow.

Our approach was to first check for global behavior in both microcontrollers. The global tests included, finding disturbances in the program execution, and identifying differences in the content of the RAM. Local behavior was then analyzed. For global attacks finding the temporal instance when an induced fault will give out reproducible results is an easy task, as only the time to inject the fault needs to be taken into consideration. This is more complicated for localized attacks, where the position and the temporal instance play a role on the results obtained. The strategy followed was to first find the spatial location where faults occurred, then restrict the setup to this area to perform temporal tests. Finding the right temporal instances to inject the fault was analyzed using different timings in steps of 12 ns, which is the maximum resolution of the control board running at a frequency of 84 MHz.

---

[6] http://openocd.org/.

5 Analysis and Discussion

We targeted two different microcontrollers, an 8 bit AVR (ATmega328P) and a 32 bit ARM Cortex M0 (STM32F030F4P6) manufactured in 350 nm and 90 nm processes. Each clocked at 2 MHz and 8 MHz, respectively. The results of our evaluation are summarized in Table 1 and detailed in the next paragraphs. To make sure the lens was responsible for the effects being observed, we also tried to induce faults through a small aperture. Using only a small aperture made from aluminum foil, as described in [15], did not result in successful fault injections on the microcontrollers that we tested. We attribute this behavior to the substantially smaller fabrication technology of our targeted chips with respect to those used in [15].

Table 1. Fault characterization table

	ATmega328P		STM32F030F4P6	
	Local	Global	Local	Global
Instruction skip	✗	✓	✓	✗
Register change	✗	✗	✓	✗
RAM change	✓	✗	✗	✗

5.1 8-bit AVR ATmega328P

First, we assessed the behavior of the microcontroller using a global attack, i.e. without magnification lens, and the following setup conditions: The flashgun discharge energy was limited to 1/128 of the maximal energy, while being adjusted to a focal length of 105 mm. The distance between the backside of the chip and the flashgun was adjusted to be 5 cm. Using such a setup configuration, instruction skip faults were observed and could be reproduced with the same outcome in roughly 80% of the injections. Please note that the clocks of the DUT and the control board were *not* synchronized in any manner. We expect that the accuracy can be further improved by synchronizing the clock signals. Using the flashgun on a higher discharge energy setting, i.e. exceeding the discharge energy beyond the aforementioned limit of 1/128, or by reducing the distance between the flashgun and the die, causes the microcontroller to reset immediately. No other type of faults were recognized when conducting a global attack. Localized attacks on the AVR were able to modify the contents of a few RAM cells with 99% accuracy. Note that always the same RAM cells were affected. However, using different ICs resulted in different modification patterns. We attribute these differences to the variations in the fabrication process of RAM cells.

5.2 32-bit ARM Cortex M0 STM32F030F4P6

The STM32 microcontroller showed no usable faults during a global attack. Depending on the intensity of the flash, the microcontroller either worked correctly or would reset. However, the tests using localized injections resulted in much more interesting faults. While the RAM cells could not be disturbed with this setup, the CPU registers did show changes in their values according to the position of the flash over the die. During the examination of the register's fault susceptibility, it was observed that registers {r4, r5, r6, r7, lr} tend to change their contents to random values more often than others. This behavior was confirmed using different samples of the microcontroller. Faults on register's content occurred in roughly 50% of the injections when the spatial and temporal location were fixed, whereby the injected fault always disturbed the same set of registers. During the process of injecting faults, and evaluating the outcome using the register dump test, a spatial dependency was found. This is depicted in the maps of Figs. 5(a), (b), (c) and (d) for the whole chip and different Region of

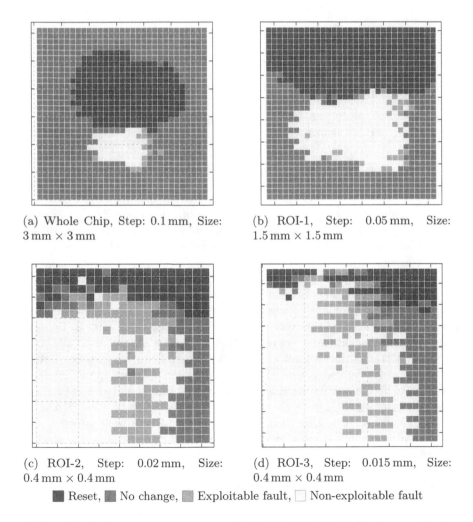

(a) Whole Chip, Step: 0.1 mm, Size: 3 mm × 3 mm

(b) ROI-1, Step: 0.05 mm, Size: 1.5 mm × 1.5 mm

(c) ROI-2, Step: 0.02 mm, Size: 0.4 mm × 0.4 mm

(d) ROI-3, Step: 0.015 mm, Size: 0.4 mm × 0.4 mm

■ Reset, ■ No change, ■ Exploitable fault, □ Non-exploitable fault

Fig. 5. Fault characterization map on STM32F030F4P6. (Color figure online)

Interests (ROIs). The whole die was scanned using a step size of 0.1 mm inside a boundary of dimension 3 mm×3 mm as depicted in Fig. 5(a). Blue marks indicate the locations where no change was observed, green marks indicate that some registers were modified, but the fault was not exploitable (i.e. stuck in Fault Handler/ Reset Handler), red marks indicate the locations that caused a reset, and yellow marks indicate the positions where exploitable faults were found. After performing the first scan, we observed an intermediate area between the locations where resets occur (top) and the locations where unusable faults are found (bottom). In this region usable faults tend to appear more often. The second scan (0.05 mm per step, 1.5 mm × 1.5 mm area) as depicted in Fig. 5(b) shows the aforementioned area in detail. By performing an even more detailed scan (0.02 mm per step, 0.4 mm × 0.4 mm area), we were able to observe the result as depicted in Fig. 5(c). A further decrease in step width to 0.015 mm per step, as depicted in Fig. 5(d) shows the same behavior as in Fig. 5(c), this step size was later one used, to perform very precise scanning in areas of usable faults.

Program flow disturbances could also be observed, however, they would influence more than a single instruction, due to the flashgun's required time to discharge completely using the specified energy amount.

6   Application to Speck

Back in 2013, the NSA introduced the SIMON and SPECK families of lightweight block ciphers [2]. In particular, SPECK is a family of block ciphers based on Feistel-like networks specifically designed for software applications. For the sake of evaluation, we considered SPECK-128-128 which has a block size and a key size of 128-bit. More details on SPECK can be found in the appendix. The attacks were conducted on a unprotected C implementation of SPECK, compiled with the -Os flag on both 8-bit and 32-bit platforms.

6.1   Instruction Skip Attacks

Both microcontrollers were prone to this kind of attack. To provoke an instruction skip on the AVR, the global fault injection setup was used. The last key addition was skipped successfully with the aid of an appropriate timing adjustment, hence leaking the last intermediate secret state entirely. This enables the attacker to directly calculate the last round key with a simple XOR. We were able to observe a similar behaviour when the STM32 was used, with the difference that only some regions of the die can be exposed strong enough to provoke a instruction skip, while the exposure of other regions led immediately to a reset. Therefore, it was necessary to use the localized fault injection setup to successfully mount this kind of attack on the STM32. The whole attack took less than a hour using the STM32 and could be carried out even faster, if the attacker has prior knowledge about the Fault Characterization Map. Provoking an instruction skip on the AVR is even simpler, as there is no spatial dependency, which results in the attack carried out in a matter of minutes.

6.2   Differential Fault Attacks

We tested the Differential Fault Attack (DFA) described in the appendix on the STM32 using the localized setup. The prerequisite for this attack is the occurrence of random faults in the word y^{T-1}. To do so, the faults have to take injected one round earlier, namely in the penultimate round $T-2$. One approach for causing random faults in the word y^{T-1}, is the disturbance of the XOR operation between x^{T-2} and y^{T-2}. Similarly, the processing of the circular left shift of y^{T-2} can be attacked. All the injections vary in their temporal location to gain a sufficiently set of faulty pairs. But the local setup must be adjusted in a manner so that the light beam hits an area which is prone to flipping registers. We took advantage of the previous analysis to fix the Point of Interests (PoIs), i.e. place the injection inside a region of usable faults, as depicted in Fig. 5(d). In total 3×10^3 injections were performed to find the correct temporal locations. Once the timing and spatial point was fixed, then 46 different pairs were collected for a successful attack, some faulty encryptions appeared more than once. This represents ≈ 7 more than the theoretical estimations given in [9]. The whole process took about 1 h, because the injection rate cannot exceed a maximum rate, which is limited by the flashgun's thermal protection. We repeated our attack several times, and were able to recover the entire last round key successfully in all the cases, once the attack location was fixed.

7   Conclusions

We presented a low-cost setup for optical semi-invasive fault attacks on embedded devices. This setup was used to successfully inject localized faults in modern MCUs, including an ARM Cortex M0 microcontroller manufactured in a 90nm process, with high spatial precision and repeatability. The investment for such a setup is much lower than the needed for a professional laser station, and still lower than the one typically needed for side-channel analysis lab equipment, hence making this type of attacks particularly attractive for low-budget attackers. In this regard, semi-invasive fault attacks should have a higher impact during risk assessments than what was previously considered.

Acknowledgements. We thank the anonymous reviewers for their valuable comments and suggestions. This work was performed while Oscar M. Guillen was a research assistant at the Chair of Security in Information Technology of the Technische Universität München.

Appendix: Differential Fault Analysis of Speck

SPECK is a family of block ciphers variable by different block and key sizes. The round function $R(x, y)$ of SPECK has a Feistel-like structure and is described by the following equation:

$$R(x,y) := ((x \ggg \alpha + y) \oplus k, y \lll \beta \oplus (x \ggg \alpha + y) \oplus k),$$

where $\oplus$ denotes a bitwise XOR, $+$ denotes an addition modulo 2^n, $\gg$ α denotes the right circular shift with α bits, $\ll$ β denotes the left circular shift with β bits, x and y are the input n-bit words, and k is the round key.

The last round key k^{T-1} can be recovered by injecting random faults in the word y^{T-1} as proposed by Huo et al. in [9]. The fault propagates through the last round and the pairs of correct and faulty ciphertexts are collected. Then, a system of non linear equations on $\mathbb{F}_2$ is constructed as a set of Differential Equations of Additions (DEAs). Finally, the system of DEAs is solved using a computer algebra system with the aid of Gröbner bases. According to [9], 5–8 pairs are needed on average to solve the system of DEAs, independently of the block size n.

References

1. Bar-El, H., Choukri, H., Naccache, D., Tunstall, M., Whelan, C.: The sorcerer's apprentice guide to fault attacks. IACR Cryptol. ePrint Arch. **2004**, 100 (2004)
2. Beaulieu, R., Shors, D., Smith, J., Treatman-Clark, S., Weeks, B., Wingers, L.: The Simon and speck families of lightweight block ciphers. Cryptology ePrint Archive, Report 2013/404 (2013). https://eprint.iacr.org/2013/404/
3. Boit, C., Schlangen, R., Glowacki, A., Kindereit, U., Kiyan, T., Kerst, U., Lundquist, T., Kasapi, S., Suzuki, H.: Physical IC debug - backside approach and nanoscale challenge. Adv. Radio Sci. **6**, 265–272 (2008)
4. Boneh, D., DeMillo, R.A., Lipton, R.J.: On the importance of checking cryptographic protocols for faults. In: Fumy, W. (ed.) EUROCRYPT 1997. LNCS, vol. 1233, pp. 37–51. Springer, Heidelberg (1997). doi:10.1007/3-540-69053-0_4
5. Breier, J., Jap, D.: Testing feasibility of back-side laser fault injection on a microcontroller. In: Proceedings of the 10th Workshop on Embedded Systems Security, WESS 2015, Amsterdam, The Netherlands, 8 October 2015, p. 5 (2015)
6. Huang, A.B.: Hacking the PIC 18f1320 (2007). https://www.bunniestudios.com/blog/?page_id=40. Accessed 1 Dec 2016
7. Dehbaoui, A., Dutertre, J.-M., Robisson, B., Tria, A.: Electromagnetic transient faults injection on a hardware and a software implementations of AES. In: 2012 Workshop on Fault Diagnosis and Tolerance in Cryptography, Leuven, Belgium, 9 September 2012, pp. 7–15 (2012)
8. Hanft, F.: Entwicklung eines prototypen zur verhaltensanalyse von chipkarten bei fault injection attacks (2016). http://hanft.in/Dokumente/BachelorarbeitHanft.pdf. Accessed 26 Mar 2017
9. Huo, Y., Zhang, F., Feng, X., Wang, L.-P.: Improved differential fault attack on the block cipher speck. In: 2015 Workshop on Fault Diagnosis and Tolerance in Cryptography (FDTC), pp. 28–34. IEEE (2015)
10. Neve, M., Peeters, E., Samyde, D., Quisquater, J.-J.: Memories: a survey of their secure uses in smart cards. In: 2nd International IEEE Security in Storage Workshop (SISW 2003), Information Assurance, The Storage Security Perspective, 31 October 2003, Washington, DC, USA, pp. 62–72 (2003)
11. O'Flynn, C., Chen, Z.D.: ChipWhisperer: an open-source platform for hardware embedded security research. In: Prouff, E. (ed.) COSADE 2014. LNCS, vol. 8622, pp. 243–260. Springer, Cham (2014). doi:10.1007/978-3-319-10175-0_17

12. Schmidt, J.-M., Hutter, M.: Optical and EM fault-attacks on CRT-based RSA: concrete results. In: Posch, K.C., Wolkerstorfer, J. (eds.) Austrian Workshop on Microelectronics - Austrochip 2007, Graz, Austria, 11 October, pp. 61–67. Verlag der Technischen Universität Graz, October 2007. ISBN 978-3-902465-87-0

13. Schmidt, J.-M., Hutter, M., Plos, T.: Optical fault attacks on AES: a threat in violet. In: Naccache, D., Oswald, E. (eds.) Fault Diagnosis and Tolerance in Cryptography - FDTC 2009, 6th International Workshop, Lausanne, Switzerland, 6 September 2009, pp. 13–22. IEEE-CS Press (2009)

14. Skorobogatov, S.P.: Semi-invasive attacks - a new approach to hardware security analysis. Ph.D. thesis, University of Cambridge (2005)

15. Skorobogatov, S.P., Anderson, R.J.: Optical fault induction attacks. In: Kaliski, B.S., Koç, K., Paar, C. (eds.) CHES 2002. LNCS, vol. 2523, pp. 2–12. Springer, Heidelberg (2003). doi:10.1007/3-540-36400-5_2

16. Smith, Z.J., Chu, K., Espenson, A.R., Rahimzadeh, M., Gryshuk, A., Molinaro, M., Dwyre, D.M., Lane, S., Matthews, D., Wachsmann-Hogiu, S.: Cell-phone-based platform for biomedical device development and education applications. PLoS ONE 6(3), 1–11 (2011)

17. Van Woudenberg, J.G., Witteman, M.F., Menarini, F.: Practical optical fault injection on secure microcontrollers. In: 2011 Workshop on Fault Diagnosis and Tolerance in Cryptography, FDTC 2011, Tokyo, Japan, 29 September 2011, pp. 91–99 (2011)

DFA on LS-Designs with a Practical Implementation on SCREAM

Benjamin Lac[1,5]($\boxtimes$), Anne Canteaut[2], Jacques Fournier[3], and Renaud Sirdey[4]

[1] CEA-Tech, Gardanne, France
benjamin.lac@cea.fr
[2] Inria, Paris, France
anne.canteaut@inria.fr
[3] CEA-Leti, Grenoble, France
jacques.fournier@cea.fr
[4] CEA-List, Saclay, France
renaud.sirdey@cea.fr
[5] ENSM-SE, Saint-Étienne, France

Abstract. LS-Designs are a family of SPN-based block ciphers whose linear layer is based on the so-called interleaved construction. They will be dedicated to low-end devices with high performance and low-resource constraints, objects which need to be resistant to physical attacks. In this paper we describe a complete Differential Fault Analysis against LS-Designs and also on other families of SPN-based block ciphers. First we explain how fault attacks can be used against their implementations depending on fault models. Then, we validate the DFA in a practical example on a hardware implementation of SCREAM running on an FPGA. The faults have been injected using electromagnetic pulses during the execution of SCREAM and the faulty ciphertexts have been used to recover the key's bits. Finally, we discuss some countermeasures that could be used to thwart such attacks.

Keywords: Lightweight cryptography · DFA · SPN-based block ciphers · LS-Designs · SCREAM · EM fault attacks

1 Introduction

The advent of the Internet of Things (IoT) has brought the need for new cryptographic primitives to suit the high performance, low power and low resource constraints of IoT devices. Ciphers like AES, which are good enough for embedded devices like smart cards, do not satisfy the constraints of devices like RFID tags or nodes in sensor networks. During the past years, several lightweight block ciphers have been proposed, some are highly efficient software-oriented ciphers like PRIDE [3] or SPECK [5], and some are rather highly efficient hardware-oriented ciphers like PRESENT [10], PRINCE [12] or SIMON [5]. In terms

© Springer International Publishing AG 2017
S. Guilley (Ed.): COSADE 2017, LNCS 10348, pp. 223–247, 2017.
DOI: 10.1007/978-3-319-64647-3_14

of security, these ciphers are mainly designed to resist black-box mathematical attacks. However, since they are used in IoT devices in pervasive environments, we ought to also look at implementation-related attacks. Indeed, resistance against side channel attacks is now considered as a valuable property which should be taken in consideration when designing lightweight ciphers as underlined by the ciphers FIDES [8], PICARO [31] and Zorro [16]. In that respect, several physical attacks have been proposed against lightweight ciphers. One of them is a complete Differential Fault Analysis (DFA) which exploits the design of the linear layer introduced on PRINCE [37] and on PRIDE [1,25]. DFA is a particular physical attack, in which we compare the results of a correct computation to one which has been disturbed at a precise time, in order to infer information about the key bits used in the cipher. In this paper, we propose to extend the DFA described in [25,37] to any SPN-based block cipher, by an in-depth analysis of LS-Designs [18], a family of SPN-based block cipher for which the attack is the most effective. We first present physical attacks and LS-Designs before describing the theoretical DFA using several fault models. These attacks have been validated in practice on a software implementation of PRIDE [25], which follows a construction similar to LS-Designs. In order to validate the practical feasibility of our attack on a hardware implementation, we used electromagnetic pulses to inject faults during the execution of SCREAM running on an FPGA Xilinx Spartan-3E 1600E and we applied our DFA on the obtained corrupted results. SCREAM is the TAE [28] (Tweakable Authenticated Encryption) mode of the block cipher Scream which is an instantiation of LS-designs [20]. It should not be confused with the stream cipher Scream [23]. Then, we detail how to apply the DFA on other families of SPN-based block ciphers: the CUBE family, S-bP structures (i.e. SPN having a bit permutation as a linear layer) and AES-like structures. Finally we discuss countermeasures that can be implemented to thwart such attacks before concluding the paper.

2   Fault Attacks Against Cryptographic Primitives

Fault attacks consist in disturbing the behaviour of the circuit in order to alter the correct execution of the cipher. The faults are injected into the device by various means such as light pulses [36], laser [35], clock glitches [2], spikes on the voltage supply [9] or electromagnetic (EM) perturbations [15]. Some of those techniques, like the laser one, are invasive, requiring the "decapsulation" of the chip using mechanical or chemical means. Laser allows to target one bit in a given register if well manipulated [14]. However it is an expensive means of injection. Other techniques are not invasive such as glitches (power, clock, electromagnetic). Clock and voltage glitches disturb the whole component, and many injections usually have to be made before getting the specific faults required by an attack. EM glitches on the other hand allow to have relatively high spatial and temporal precisions using relatively low-cost equipment [15]. One of the objectives of fault attacks, especially when considering cryptographic primitives, is to perform Differential Fault Analysis (DFA). DFA, originally described in [7,11], consists in retrieving a cryptographic key by comparing the correct ciphertexts with one or more faulty ones. DFA techniques have been described and successfully applied to most of the publicly known ciphers going from symmetric ciphers like the DES [7] or the AES [34] to asymmetric ones like RSA [11] or even more complex schemes like pairing-based systems [26]. In the particular field of lightweight cryptography, DFA have been proposed against ciphers like PRESENT [40], SPECK [39], TRIVIUM [29], PRINCE [37] or PRIDE [25]. DFA techniques are very efficient in retrieving the keys used during a cipher execution, usually requiring a few executions only. It is also quite complex to devise countermeasures against such attacks because of the diversity of the possible injection methods and because the usually deployed countermeasures (like redundancy, error-correcting codes etc.) have serious impacts on performances of the targeted cipher. Therefore, in our approach of analyzing the security of implementations of SPN-based block ciphers, we decided to first focus on their resistance against fault attacks as to identify possible attack paths and devise more efficient countermeasures in order to try to keep the performance characteristics of the original ciphers.

3   LS-Designs

In this paper, bits, bytes, rows and columns are numbered from left to right starting from 1. LS-designs are iterative SPN-based block ciphers composed of r rounds. They were introduced by Grosso et al. [18] in 2014. An LS-design takes as input an n-bit block and uses an n-bit key. The inner state of the cipher, as well as the plaintext, ciphertext, and key, are all represented as $\omega \times c$ bit arrays, with ω the number of rows and c the number of columns such that $n = \omega \cdot c$. The following notation is used for the intermediate values of the state within a round:

I_i the input of the i-th round
X_i the state after the key addition layer of the i-th round
Y_i the state after the substitution layer of the i-th round
Z_i the state after the linear layer of the i-th round
O_i the output of the i-th round

A round $1 \leq i \leq r$ is composed of the following steps:

   i. Add by an XOR the n-bit key K to the state: $X_i = I_i \oplus K$,
  ii. Apply an ω-bit S-box $\mathcal{S}$ to each column of the state (i.e. apply the substitution layer $\mathcal{S}$layer to the state): $Y_i = \mathcal{S}\text{layer}(X_i)$,
 iii. Apply a bijective linear map $\mathcal{L}$, called L-box, operating on c-bit vectors, to each row (i.e. apply the linear layer $\mathcal{L}$layer to the state): $Z_i = \mathcal{L}\text{layer}(Y_i)$,
 iv. Add by an XOR an n-bit round constant $\mathcal{C}t_i$ to the state: $O_i = Z_i \oplus \mathcal{C}t_i$.

An LS-design is parametrized by the choice of r, ω, c, $\mathcal{S}$, $\mathcal{L}$ and the round constants $\mathcal{C}t_i$ for $1 \leq i \leq r$. In order to encrypt a plaintext, the cipher applies the r rounds as previously described, it then performs an XOR between the state

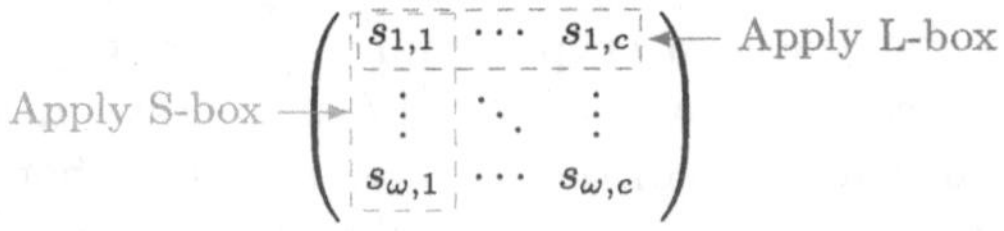

$$\text{Apply S-box} \longrightarrow \begin{pmatrix} s_{1,1} & \cdots & s_{1,c} \\ \vdots & \ddots & \vdots \\ s_{\omega,1} & \cdots & s_{\omega,c} \end{pmatrix} \longleftarrow \text{Apply L-box}$$

Fig. 1. Inner state of an LS-design

and the key. Figure 1 shows the representation of the inner state of an LS-design with an example framed of the input of S-box and the input of L-box.

Robin and Fantomas are two LS-designs proposed by Grosso et al. [18] in 2014 with $\omega = 8$ and $c = 16$. Scream and iScream are a modified version of Robin and Fantomas, referred to as Tweakable LS-Designs, also introduced by Grosso et al. [20] with the same parameters. Those two ciphers are the core of two authenticated encryption schemes submitted to the Caesar competition [13]. It is worth noticing that these block ciphers have been recently broken in the sense that it has been shown that they have a large number of weak keys [38]. However, studying the security offered by the ciphers with respect to other attacks is still of interest. Indeed, the nonlinear-invariant attack can be prevented by changing the round constants, while such a change won't affect the resistance to DFA.

The main difference between Tweakable LS-Designs and LS-Designs lies in the addition of an input parameter, the Tweak, to thwart side-channel attacks if it is used and protected properly. It is a structural countermeasure which, as we shall see, also thwart our attack in this case. Finally, PRIDE, proposed by Albrecht et al. [3] in 2014, has a structure close to the one of an LS-Design with $\omega = 4$ and $c = 16$: it uses one additional key for pre- and post-whitening, uses several L-boxes within the linear layer and has no linear layer on the last round.

4   Applying DFA on LS-Designs

DFA against PRIDE, whose structure is similar to an LS-design, was first introduced in [25]. In this section, we propose a generalisation of fault attacks to any LS-design. First, we introduce the general principle to exploit fault injections. Then, we explain the different strategies depending on the fault model.

4.1   General Principle

Despite some similarities, a DFA against a cryptographic algorithm is different from a classical differential analysis. Indeed, for the latter the differences must be injected on the input of the cipher while for a DFA it can be injected where the attacker wants. The DFA that we propose in this paper also differs from many existing DFA in the sense that, in our attack, the input and output differences of all active S-boxes are known to the attacker. Therefore, there is no need of guessing the input difference as in many other DFA. It consists in corrupting one L-box application (i.e. corrupting a row of the state during the linear layer) in the penultimate round in order to obtain a known difference on the S-boxes inputs in the last round. More precisely, flipping the bit $1 \leq i \leq c$ of the row $1 \leq j \leq \omega$ gives a difference equal to $2^{\omega - j}$ (with a one only at position j) on the input of the i-th S-box in the last round. Moreover, from the knowledge of the correct and the faulty ciphertexts, we can compute the corresponding difference on the S-box output. Figure 2 shows for example the state difference obtained from a flip of the second row before the S-layer.

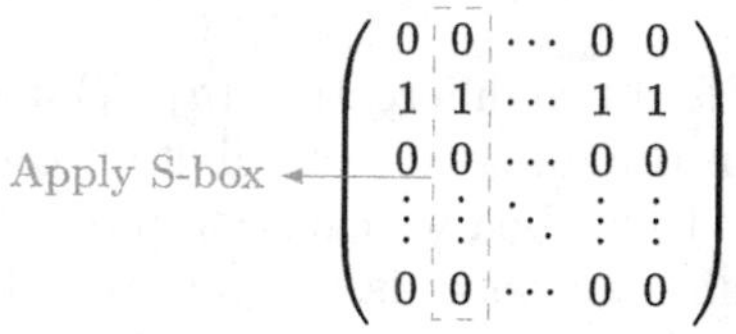

$$
\text{Apply S-box} \leftarrow
\begin{pmatrix}
0 & 0 & \cdots & 0 & 0 \\
1 & 1 & \cdots & 1 & 1 \\
0 & 0 & \cdots & 0 & 0 \\
\vdots & \vdots & \ddots & \vdots & \vdots \\
0 & 0 & \cdots & 0 & 0
\end{pmatrix}
$$

Fig. 2. State difference obtained from a flip of the second row

In this case, we get a difference equal to $2^{\omega - 1}$ on the input of each S-box. Thereby, if we denote ΔX_r (resp. ΔY_r) the difference input (resp. output) of the last substitution layer, we obtain a known differential $(\Delta X_r[i], \Delta Y_r[i])$ on the i-th S-box. Then, we exploit the difference distribution table of the S-box to reduce the number of remaining candidates for the secret key.

Indeed, using the notation introduced in Sect. 3, obtaining information on the key is possible from the following equations:

$$\Delta X_r = \mathcal{S}\mathsf{layer}^{-1}(\mathcal{L}\mathsf{layer}^{-1}(C \oplus K \oplus Ct_r)) \oplus \mathcal{S}\mathsf{layer}^{-1}(\mathcal{L}\mathsf{layer}^{-1}(C^* \oplus K \oplus Ct_r))$$

and

$$\Delta Y_r - \mathcal{L}\mathsf{layer}^{-1}(C \oplus K \oplus Ct_r) \oplus \mathcal{L}\mathsf{layer}^{-1}(C^* \oplus K \oplus Ct_r) = \mathcal{L}\mathsf{layer}^{-1}(\Delta C)$$

where C is the correct ciphertext and C^* the faulty one. We can use these equations for each ω bit word $1 \leq i \leq c$:

$$x = \mathcal{L}\mathsf{layer}^{-1}(C \oplus K \oplus Ct_r)[i] \text{ and } y = \mathcal{L}\mathsf{layer}^{-1}(C^* \oplus K \oplus Ct_r)[i]$$

satisfy

$$x \oplus y = \Delta Y_r[i] = \mathcal{L}\mathsf{layer}^{-1}(\Delta C)[i] \text{ and } \mathcal{S}^{-1}(x) \oplus \mathcal{S}^{-1}(y) = \Delta X_r[i] = 2^{\omega - j}.$$

From the knowledge of a nonzero input difference $x \oplus y$ and of an output difference $\mathcal{S}^{-1}(x) \oplus \mathcal{S}^{-1}(y)$ for the inverse S-box $\mathcal{S}^{-1}$, we reduce the number of possible values for the input x. Moreover, from Proposition 1, we are able to easily find pairs of differentials for the S-box which are simultaneously satisfied for a single element. The proof to this proposition is given in Appendix A.

Proposition 1. *Let $\mathcal{S}$ be an n-bit S-box. Let (a_1, b_1) and (a_2, b_2) be two differentials with $a_1 \neq a_2$ such that the system of two equations*

$$\mathcal{S}(x \oplus a_1) \oplus \mathcal{S}(x) = b_1 \tag{1}$$

$$\mathcal{S}(x \oplus a_2) \oplus \mathcal{S}(x) = b_2 \tag{2}$$

has at least two solutions. Then, each of the three Eqs. (1), (2) and

$$\mathcal{S}(x \oplus a_1 \oplus a_2) \oplus \mathcal{S}(x) = b_1 \oplus b_2$$

has at least four solutions.

In other words, if we can find two differentials (a_1, b_1) and (a_2, b_2) such that one out of the three entries in the difference distribution table (a_1, b_1), (a_2, b_2) and $(a_1 \oplus a_2, b_1 \oplus b_2)$ is equal to 2, then we can guarantee that the input satisfying these two differentials simultaneously is unique. Note that if one of the three equations has no solution then the system of two Eqs. (1), (2) has no solution.

4.2 Ideal Fault Model

The ideal fault model consists simply in a first step in finding two output differences $b_1 = \Delta X_r^1 = 2^{\omega - i_1}$ and $b_2 = \Delta X_r^2 = 2^{\omega - i_2}$ with $1 \leq i_1 < i_2 \leq \omega$ such that (a_1, b_1) and (a_2, b_2) are simultaneously satisfied for a single element for all a_1 and a_2. After, it is sufficient to flip the row i_1 then the row i_2 of the state during the penultimate linear layer with two successive fault injections in order to retrieve the complete secret key. Table 1 gives as example the pairs of differentials which are simultaneously satisfied for a single element in the case of the S-boxes involved in some LS-designs.

Table 1. Exploitable differential pairs

Cipher	Pair
PRIDE	$(a_1, 0 \times 1)$, $(a_2, 0 \times 8)$
Robin	$(a_1, 0 \times 01)$, $(a_2, 0 \times 40)$
Fantomas	$(a_1, 0 \times 01)$, $(a_2, 0 \times 80)$
Scream	$(a_1, 0 \times 01)$, $(a_2, 0 \times 02)$
iScream	$(a_1, 0 \times 02)$, $(a_2, 0 \times 80)$

Note: Applying DFA on Scream and iScream is possible only if the attacker can execute encryption (or decryption) twice consecutively with the same tweak.

4.3 Random Fault Model

We call random fault model one where we have one chance out of two to flip each bit of a desired word. It is close to what is obtained in practice with electromagnetic pulses where it is possible to target a precise word (more precisely a specific instruction) but the injected faults follow a random distribution.

In order to achieve the attack, we must flip all the bits of two c-bit words in the ideal fault model used in the preceding part. However, we can see that flipping one bit provides an active S-box, it is therefore enough to flip all the bits of the desired c-bit words non simultaneously using as many faults as necessary. Accordingly, as in the ideal model, it consists first in finding two output differences $b_1 = \Delta X_r^1 = 2^{\omega-i_1}$ and $b_2 = \Delta X_r^2 = 2^{\omega-i_2}$ with $1 \leq i_1 < i_2 \leq \omega$ such that (a_1, b_1) and (a_2, b_2) are simultaneously satisfied for a single element for all a_1 and a_2. Let A_1 (resp. A_2) be the average number of remaining candidates for a key byte from an active S-box obtained from a fault on row i_1 (resp. i_2). Let m_1 (resp. m_2) denote the number of obtained faults on row i_1 (resp. i_2).

Then, the number of remaining candidates for the key from the faults and from the knowledge of ΔY_r for each fault is

$$N = \left(\frac{2^\omega}{2^{m_1+m_2}} + \sum_{i=1}^{m_1} \frac{A_1}{2^{i+m_2}} + \sum_{i=1}^{m_2} \frac{A_2}{2^{i+m_1}} + \left(\sum_{i=1}^{m_1} \frac{1}{2^i} \right) \left(\sum_{i=1}^{m_2} \frac{1}{2^i} \right) \right)^c .$$

Indeed, when m faults have been injected on one word, the probability to obtain no difference on a byte is equal to $1/2^m$ and the probability to obtain at least one is equal to $\sum_{i=1}^{m} 1/2^i = 1 - 1/2^m = (2^m - 1)/2^m$. Moreover, if we get no difference with all the faults (on the first and on the second word) then we still have 2^ω candidates for the corresponding byte. On the other hand, if we get only one difference we obtain A_1 or A_2 candidates. Finally, if we get two differences we retrieve the correct value. We then deduce that

$$N = \left(\frac{2^\omega + A_1(2^{m_1} - 1) + A_2(2^{m_2} - 1) + (2^{m_1} - 1)(2^{m_2} - 1)}{2^{m_1+m_2}} \right)^c$$

$$= \left(\frac{2^\omega - A_1 - A_2 + 1}{2^{m_1+m_2}} + \frac{A_1 - 1}{2^{m_2}} + \frac{A_2 - 1}{2^{m_1}} + 1 \right)^c .$$

4.4 Properties that Make the Attack Effective

Our attack mainly exploits the following 2 properties of the building-blocks of most SPN-based block ciphers. It is worth noticing that these two properties come from the fact that the building-blocks of the cipher have been designed for maximizing the resistance of the cipher against differential and linear cryptanalysis.

The Design of the Linear Layer. Indeed, the motivation for LS-Designs, as well as the more general interleaved construction [3], is to guarantee a large number of active S-boxes in any differential or linear attack (see e.g. Theorem 1 in [3] and Page 23 in [18]). Indeed, in LS-Designs, flipping all bits of any row at the input of the linear layer activates all S-boxes in the next round. Indeed, by construction, the c bits of any row go to different S-boxes. It follows that flipping one row of the penultimate round allows the attacker to recover information on the whole subkey used in the last round. The same situation occurs in the interleaved construction. In other words, the optimal diffusion offered by those constructions enables the attacker to recover the whole last-round subkey. This would not be the case if the linear layer was weaker with respect to the classical diffusion criteria.

The Differential Properties of the S-box, Which Avoids the Existence of Differentials of High Probability Over a Large Number of Rounds. The counterpart of this property required for resistance against classical differential cryptanalysis is that the number of inputs which satisfy two valid differentials simultaneously is usually reduced to a single element. In the context of a DFA, this property enables the attacker to drastically reduce the number of key candidates. In many cases, two faults are enough to obtain a single candidate for the secret key.

5   Practical Implementation of the DFA on SCREAM

5.1   The TAE Mode SCREAM

The TAE (Tweakable Authenticated Encryption) mode is a mode of operation introduced by Liskov et al. [27,28]. In order to encrypt a message M, the TAE mode splits it into m blocks of size n such that the last block is padded if necessary, and in this case it is added by an XOR to its output. Then, it applies on each block a tweakable block cipher

$$E : \{0,1\}^k \times \{0,1\}^t \times \{0,1\}^n \to \{0,1\}^n$$

using a key K of size k and a tweak T of size t. It uses the same key K to encrypt each block and different tweaks produced from the same nonce N (the recommended size for the nonce in SCREAM is 11 bytes). Finally, a tag is produced from the checksum of all blocks.

SCREAM optionally proposes to authenticate blocks of associated data with the message. It uses the tweakable block cipher Scream and the tweaks are produced from the same nonce concatenated with a block counter. Scream is an iterative block cipher composed of N_s steps, each of them made of N_r rounds, denoted Tweakable LS-Design and introduced by Grosso et al. [20] in 2014. The recommended parameters for a related-key security are $N_s = 12$ and $N_r = 2$. It takes as inputs a 128-bit block, a 128-bit key K and a 128-bit tweak $T = t_0 \| t_1$. The tweak is used as a "lightweight key schedule": the output of the step s is added by an XOR to a subkey equal to

$$\begin{aligned} K \oplus (t_0 \| t_1) & \quad \text{if } s = 3i, \\ K \oplus (t_0 \oplus t_1 \| t_0) & \quad \text{if } s = 3i + 1, \\ K \oplus (t_1 \| t_0 \oplus t_1) & \quad \text{if } s = 3i + 2. \end{aligned}$$

Each round is composed of the following steps: it applies a nonlinear layer composed of 8-bit S-boxes, then it adds by an XOR a round constant and finally it applies a 16-bit L-boxes layer. Specifications on these components are given in

Appendix of [20]. In order to encrypt a plaintext P, Scream adds by an XOR the first subkey to P (with $s = 0$), then it applies the N_s steps described before. The tweak, or the nonce in the case of the TAE mode, is a protocol-level countermeasure against side-channel analysis added directly into the design. It must be changed at each execution in order to be effective. In this case, we will see it also protects against our DFA. However, attacking a Tweakable LS-Design using a fixed tweak is equivalent to attacking an LS-Design, what interests us in our case to validate our DFA.

5.2   The DFA on SCREAM

Briefly, we recall the general principle of our attack on SCREAM. Firstly, we consider that the same nonce is used at each execution, i.e. each block uses the same tweak at each encryption. As mentioned, that provides a structure equivalent to an LS-Design. Then, flipping one row before the S-box layer in the last round during one execution of Scream allows to obtain known differentials on each S-box. The best case is to flip, from two faults, the last and the penultimate row since in this case the number of inputs which satisfy the obtained differentials $(a_1, 0 \times 01)$ and $(a_2, 0 \times 02)$ simultaneously is reduced to a single element. It can be verified by testing all intersections of the obtained sets from each possible pair of differentials. Thereby, an attacker can use these differentials for retrieving the key from the following equations, for each byte $1 \leq i \leq 16$:

$$x = \mathcal{L}\mathsf{layer}^{-1}(C \oplus K \oplus T)[i] \oplus Ct_{23}[i] \text{ and } y = \mathcal{L}\mathsf{layer}^{-1}(C^* \oplus K \oplus T)[i] \oplus Ct_{23}[i]$$

satisfy

$$x \oplus y = a_1 \text{ (resp. } a_2) \text{ and } \mathcal{S}^{-1}(x) \oplus \mathcal{S}^{-1}(y) = 0 \times 01 \text{ (resp. } 0 \times 02).$$

Finally, in case of the ideal fault model, an attacker can retrieve the complete secret key from two faults only (flip two complete rows) since she knows C, T and $Ct_{23} - 50577$ (defined in [20]). In case of the random fault model, any differential $(a_1, 0 \times 01)$ allows to obtain $A_1 \approx 2.286$ candidates for a key byte and any differential $(a_2, 0 \times 02)$ allows to obtain $A_2 \approx 2.639$ candidates. Moreover, the inner state of Scream is represented as $\omega \times c$ bit arrays, with $\omega = 8$ the number of rows and $c = 16$ the number of columns. Therefore, the average number of remaining candidates for the key from m_1 (resp. m_2) random faults on the last (resp. penultimate) row is approximately:

$$\left(\frac{252.075}{2^{m_1+m_2}} + \frac{1.286}{2^{m_2}} + \frac{1.639}{2^{m_1}} + 1 \right)^{16}$$

Table 2 gives the average number of remaining candidates for some values of m_1 and m_2. We note that 6 faults or more are enough to retrieve the key since in this case the attacker obtain less than 2^{40} remaining candidates.

Table 2. Number of remaining candidates for K

		m_1			
		1	2	3	4	5
m_2	1	$2^{96.5}$	$2^{81.2}$	$2^{66.4}$	$2^{52.6}$	$2^{40.6}$
	2	$2^{81.1}$	$2^{66.1}$	$2^{51.8}$	$2^{39.1}$	$2^{28.4}$
	3	$2^{66.2}$	$2^{51.7}$	$2^{38.5}$	$2^{27.2}$	$2^{18.5}$
	4	$2^{52.3}$	$2^{38.8}$	$2^{27.1}$	$2^{17.9}$	$2^{11.3}$
	5	$2^{39.9}$	$2^{27.9}$	$2^{18.2}$	$2^{11.2}$	$2^{6.7}$

5.3 Practical Implementation of the DFA

In order to test the feasibility of our attack on a hardware implementation, we have implemented and run the 128-bit reference VHDL code of SCREAM, given in [19], on an FPGA Xilinx Spartan-3E 1600E manufactured with advanced 90 nm process technology using a frequency of 50 MHz. The FPGA die was composed of components CRYPTO, UART and FSM. The CRYPTO component contained the whole reference code of SCREAM, the UART allowed us to send data from the computer to the chip and the FSM allowed to define all the internal states. The input parameters used at each encryption was:

i. Nonce (11 bytes): $0 \times$ e9e6f9281b86c8470ba120,
ii. Key: $0 \times$ 2ff6963dd72462ab67d5da22c0e264ae,
iii. Associated data (2 blocks): $0 \times$ 5c0e6a47bc146679d2d64aca577463679782953 401 57eb9d2581bfbb14a0cb39,
iv. Data (3 blocks): $0 \times$ 6b36f33ff882e432861448a61183583b0df1f908593481535 b6eb bc6abfc07ae22cd50a331678301fd8535690335dcbe.

The correct ciphertext was $0 \times$ c9018ef2804f85e0de4d6519593a3e5ed83c22bdc 8b2db2229e6801071cdea6785856feac83bbe335c6bcb2f5f6d81a6 and the tag was $0 \times$ 84670ef3aaba9ee5d7358858c65c41ed. In practice, an attacker can target any block of data to carry out the attack, she must just target the last linear layer of the execution of the running Scream. In our case, we injected all the faults during the last Scream execution. Therefore, we will give only the value of the last block for each fault, the correct value without fault being $0 \times$ 85856feac83bbe335c6bcb2f5f6d81a6. First we performed a simple electromagnetic analysis of one SCREAM execution to identify the last round. We needed actually to only know the total duration of the encryptions in order to deduce the temporal position of the last round - which is always feasible in practice, even if there is an added noise. Figure 3 shows the obtained curves from the simple electromagnetic analysis.

So as to conduct the attack, we used electromagnetic pulses because with this approach we did not need to decapsulate the chip and we were able to inject faults at precise enough instants. The set-up we used is quite similar to the one described in [15]. In our case, the duration of the full encryption was approximately 1200 ns, i.e. 240 ns for one Scream encryption and 10 ns for each

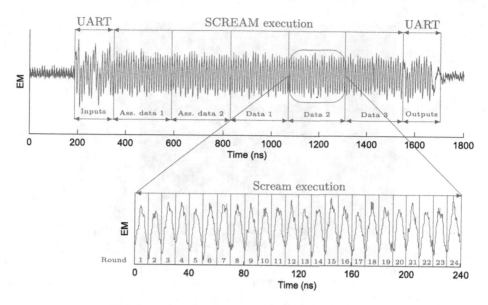

Fig. 3. Simple electromagnetic analysis of SCREAM

round. A pulse duration, from our pulses generator, could be 6 ns at minimum, which is almost half of a round. However, some signals are only used by the linear layer and the pulse can affect only few of them, which allowed us to obtain the desired fault model. First, we have done a cartography of the obtained faults on the full chip of size 19×19 mm. We injected pulses on 100 positions distributed on a 10×10 grid. On each, we tested 11 different temporal positions, 4 different voltages and we injected 2 pulses, i.e. a total of 88 shots by spatial position. On the 8800 total injections, we obtained 465 faults of which at most 88 to one spatial position. Figure 4 shows the faults distribution on the chip. Then, we targeted the sensitive area of the chip - which probably corresponds to the FPGA die - and we injected a total of 69250 pulses. We obtained a total of 2482 faults, among which 937 were different. For each fault, we calculated the value of the difference output on the last substitution layer and we verified if each byte could have been obtained by the same difference in input equal to 2^j with $0 \leq j \leq 7$.

A total of 36 different faults complied with this property. The obtained faults as well as our knowledge about the differences values around the last substitution layer are given in Appendix B. Finally, we obtained $6144 \approx 2^{12,6}$ candidates for $\mathcal{L}\mathsf{layer}^{-1}(C \oplus K \oplus T) \oplus Ct_{23}$ and we retrieve K by testing all from the knowledge of C, T and Ct_{23}. The obtained number of remaining candidates does not correspond to theoretical analysis because EM pulses not allow to target a chosen row.

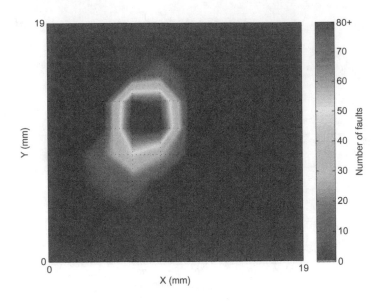

Fig. 4. Cartography of the obtained faults on the full chip

6 Application on Other SPN-based Block Ciphers

6.1 Application on the CUBE Family

A cipher belonging to the CUBE family (called CUBE) is an iterative SPN-based
block cipher composed of r rounds whose concept was introduced by Berger
et al. [6] in 2015. It takes as input an n-bit block and uses an n-bit or a $2n$-bit
key with a key schedule defined in [6]. The inner state of the cipher, as well as
the plaintext, ciphertext, and key, are all represented as a $\omega \times \omega \times \omega$ cube. The
cube is filled beginning with its least significant bit at position $(1, 1, 1)$ according
the reference (X, Y, Z). A round $1 \le i \le r$ is then composed of the following
steps:

i.   Add by an XOR an n-bit subkey SK_i to the state,
ii. Apply an ω-bit S-box S to each row of X,
iii. Apply a quasi-involutive Feistel-MDS transformation [33], denoted M, on ω
     words of size ω bits, for each plane (i, Y, Z), $1 \le i \le \omega$,
iv. Rotate the axes (X, Y, Z) as (Z, X, Y).

An instance is parametrized by the choice of r, ω, S and M. In order to encrypt
a plaintext, the cipher applies the r rounds as previously described, it then
performs an XOR between the state and a last subkey SK_r. Figure 5 shows the
representation of the inner state of a CUBE instance with $\omega = 4$ illustrating an
example of the input of S and the input of M.

CUBE proposed in [6] is an example with $\omega = 4$. The decryption process of
PRESENT-80 is also an example with $\omega = 4$ but without the quasi-involutive

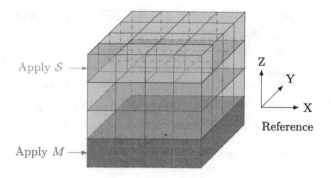

Fig. 5. Inner state of a CUBE

Feistel-MDS transformation. The DFA consists in this case in flipping a $\omega \times \omega$ plane in X before the axes rotation, i.e. during the quasi-involutive Feistel-MDS transformation, in order to obtain known differences at the inputs of all the S-boxes on the last round. Indeed, flipping the plane $1 \leq k \leq \omega$ in X before the axes rotation allows to obtain differences equal to 1 only on the plane k in Z after it, i.e. differences equal to 2^{k-1} on each of the S-box inputs. Thereby, an attacker can run the DFA on the last round to retrieve the last subkey and repeat the attack on the previous rounds until she recovers enough key information-bits to retrieve the complete key from the key schedule.

6.2   Application on S-BP Structures

An S-bP structure is an iterative SPN-based block cipher with a bit permutation layer. It takes as input an n-bit block, uses an n-bit key with a key schedule, a ω-bit S-box and a bitwise permutation layer which diffuses each S-box output to different S-boxes input. Such a cipher with r rounds is called an S-bP(n, ω, r) structure according to these parameters. A round consists of the following steps:

i. Add by an XOR the current n-bit subkey to the state,
ii. Divide the state into n/ω words and apply the ω-bit S-box to each word,
iii. Apply the bitwise permutation layer.

PRESENT-80 is an S-bP$(64, 4, 31)$ structure introduced by Bogdanov et al. [10] in 2007. PRINTCIPHER is an S-bP$(32, 3, 48)$ structure proposed by Knudsen et al. [24] in 2010. The DFA consists in this case in flipping the output of one S-box in the penultimate round to obtain known differences at input of several S-boxes on the last round thanks to the design of the bitwise permutation layer. Figure 6 shows the diffusion of a difference obtained from a flip on the output of a 4-bit S-box before the permutation layer of an S-bP structure, which allows to obtain 4 differences equals to 0×8 at the input of the next substitution layer. It is the case for a flip of the first S-box output on a round of PRESENT-80 for example.

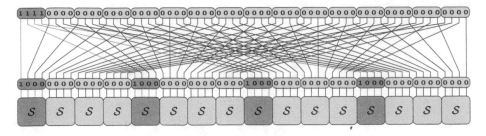

Fig. 6. Propagation of a difference obtained by a flip of a nibble before the permutation layer of PRESENT-80

In case of PRESENT-80, the differentials $(a_1, 0 \times 1)$ and $(a_2, 0 \times 8)$ are simultaneously satisfied for a single element for all a_1 and a_2. Using this attack, the attacker can retrieve the complete key from the key schedule once she gets the last two subkeys from faults on the nibbles 0, 3, 4, 7, 8, 11, 12 then 15 before the permutation layer in the last two rounds. Note that the attack on the decryption needs only two faults to retrieve a subkey: on the first and the last 16-bit word.

6.3 Application on AES-like Structures

An AES-like structure is an iterative SPN-based block cipher composed of r rounds. It takes as input an n-bit block and uses an n-bit key with a key schedule. The inner state of the cipher is represented as an $l \times c$ ω-bit array, with l the number of rows and c the number of columns. The size of the input and the key is $n = \omega \cdot l \cdot c$. A round is composed of the following steps:

  i. Add by an XOR an n-bit round constant to the state,
 ii. Add by an XOR the current n-bit subkey to the state,
iii. Apply an ω-bit S-box to each cell of the state,
 iv. Apply shifts of the cells on each row of the state,
  v. Apply a matrix multiplication transformation to each column of the state,

Midori64 is an AES-like structure, introduced by Banik et al. [4] in 2015, with $\omega = 4$, $l = 4$ and $c = 4$. KLEIN-64 is another example, introduced by Gong et al. [17] in 2012, with $\omega = 4$, $l = 8$ and $c = 2$. LED, proposed by Guo et al. [22] in 2011, is also an example with $\omega = 4$, $l = 4$ and $c = 4$. The DFA consists in this case in flipping a row before the matrices application (step v), i.e. flipping a c-bit word of the state, in order to obtain known differences at the input of each S-box. Unlike other families of SPN-based block ciphers, the AES-like structures do not use a bitwise permutation (which is transparent in the case of LS-designs but which is however exploitable). The bitwise permutation allows us to obtain differences equal to $0 \times 1, 0 \times 2, 0 \times 4$ or 0×8 at input of each S-box which allows a highly efficient exploitation of the obtained faults. In the case of AES-like structures, the differences obtained mainly depend on the matrices used, making these designs more resistant to our attack, although it remains relevant. For example, flip the first row then the third row of the state after the substitution layer of the AES-like structure LED allows to respectively obtain the following state differences before the substitution layer in the next round:

$$\begin{pmatrix} 0 \times 9 & 0 \times 9 & 0 \times 9 & 0 \times 9 \\ 0 \times 1 & 0 \times 1 & 0 \times 1 & 0 \times 1 \\ 0 \times 3 & 0 \times 3 & 0 \times 3 & 0 \times 3 \\ 0 \times d & 0 \times d & 0 \times d & 0 \times d \end{pmatrix} \text{ and } \begin{pmatrix} 0 \times d & 0 \times d & 0 \times d & 0 \times d \\ 0 \times 6 & 0 \times 6 & 0 \times 6 & 0 \times 6 \\ 0 \times c & 0 \times c & 0 \times c & 0 \times c \\ 0 \times a & 0 \times a & 0 \times a & 0 \times a \end{pmatrix}$$

Moreover, pairs of differentials $\{(a_1, 0 \times 9), (a_2, 0 \times d)\}$, $\{(a_1, 0 \times 1), (a_2, 0 \times 6)\}$, $\{(a_1, 0 \times 3), (a_2, 0 \times c)\}$ and $\{(a_1, 0 \times d), (a_2, 0 \times a)\}$ on the inverse S-box of LED guarantee that the input satisfying these two simultaneously is unique. However, if the attacker does not control the value of the fault, it is more difficult to exploit it because she cannot predict the previous state difference before the last substitution layer.

7 Countermeasures

An LS-design and more generally a block cipher is always used following a well-defined mode of operation. We will show that a number of such modes intrinsically protect against our attack. Then, we will present and briefly analyze possible countermeasures to thwart DFA: masking and the so-called Internal Redundancy Countermeasure (IRC) which we propose as a new kind of countermeasure.

7.1 Modes of Operation

In order to encrypt data, a block cipher is always used with a mode of operation. It turns out that some well-known modes - standardized and already used in practice - thwart our DFA. Therefore, in this usage context, it is not necessary to add a countermeasure to protect the cipher. It is the case for the modes which use a random initialization vector, denoted IV, to encrypt data, as for example the OFB mode, which is illustrated in Fig. 7.

This mode applies the cipher $\mathcal{E}_K$ directly on the IV and thus manipulate the plaintext only to add it to the obtained output. The IV changes at each execution and cannot be controlled by the attacker to comply with the correct

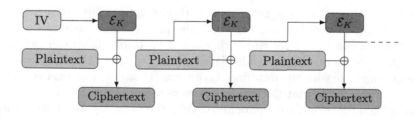

Fig. 7. OFB mode encryption

use of these modes of operation. Therefore, an attacker cannot by construction have two executions of the cipher with the same input and so she cannot apply a DFA since it is a necessary condition. It is also the case for the CTR mode which combines a unique nonce and a counter which is incremented at each execution of the algorithm, therefore guaranteeing the latter is never run twice on the same plaintext input. Some modes of operation like the CBC mode add the IV directly to the plaintext P and apply the cipher on the result. In this case, the IV must be unpredictable by the attacker in advance, otherwise the attacker can mount an adaptive plaintext attack by using two pairs of plaintext and IV, (IV_1, P_1) and (IV_2, P_2) such that $P_1 \oplus IV_1 = P_2 \oplus IV_2$.

Now, we will provide two countermeasures to thwart DFA when the mode of operation leaves the cipher unprotected against it, for example when used to derive a keyed one-way function for authentication.

7.2 Masking

Description: A countermeasure proposed by Guilley et al. in [21] is to add a random mask to the message to prevent two consecutive executions of the same plaintext. More precisely, in its original description, it consists in generating a random mask different at each execution, to XOR it to the plaintext and to return the corresponding ciphertext with the mask. However, in our case, the mask generator must be unpredictable, otherwise an attacker can choose an adaptive plaintext as described in the previous section.

Another technique - which we have already hinted at - is to use a tweakable cipher. The tweak can be considered like a mask which is added at each step and must be changed at each execution. This countermeasure, originally proposed for avoiding side-channel attacks, also protects against DFA since, once again, an attacker cannot obtain two executions on the same plaintext.

In order to guard only against DFA, we propose to use only one mask (different for each execution of the cipher) which must be added to the state SM in the middle of the encryption $\mathcal{E}_K$. More precisely, if $\mathcal{E}_K$ is composed of r rounds and takes an n-bit block as input, the countermeasure consists in computing

$$\mathcal{E}_K^{(1)}(\mathcal{E}_K^{(0)}(\text{Plaintext}) \oplus \text{RV})$$

where RV is an n-bit random value and $\mathcal{E}_K^{(0)}$ (resp. $\mathcal{E}_K^{(1)}$) corresponds to the first $\lceil r/2 \rceil$ rounds (resp. last $\lfloor r/2 \rfloor$ rounds) of the cipher. The decryption $\mathcal{D}_K = \mathcal{D}_K^{(1)} \circ \mathcal{D}_K^{(0)}$ must be synchronized with encryption (like for a tweakable cipher). In that respect, we can use the same process as a mode of operation which synchronizes the IV for encryption and decryption and which can therefore be expected to be already available in existing systems. Figure 8 illustrates the countermeasure with the introduced notation. Then, the mask generator can be public if we assume that the attacker does not have access to the encryption and decryption functions, both parametrized by the same key and mask. This is a necessary condition for each masking we have presented, otherwise the attacker can lead the DFA on the decryption since she knows the correct plaintext.

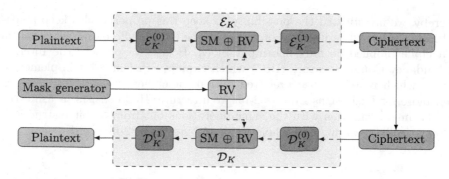

Fig. 8. Masking countermeasure

Indeed, to mount a DFA on the encryption, an attacker must obtain a correct ciphertext $C = \mathcal{E}_K^{(1)}(\mathcal{E}_K^{(0)}(P_1) \oplus RV_1)$ and a faulty ciphertext $C^* = \mathcal{E}_K^{(1)}(\mathcal{E}_K^{(0)}(P_2) \oplus RV_2)$ such that the inputs of $\mathcal{E}_K^{(1)}$ are the same in both computations, i.e.

$$\mathcal{E}_K^{(0)}(P_1) \oplus RV_1 = \mathcal{E}_K^{(0)}(P_2) \oplus RV_2. \tag{3}$$

Similarly, if the attack is mounted against the decryption function, the inputs of $\mathcal{E}_K^{(0)}$ must be the same in both computations. There are two strategies for finding two pairs of inputs which satisfy (3). The first one consists in using a generic algorithm (without exploiting any specific property of the cipher). From the birthday paradox, this requires $2^{n/2}$ encryptions where n is the block size. In our case, the attacker has then to perform 2^{32} fault injections, which is infeasible in practice. A second strategy consists in exploiting some differential properties of $\mathcal{E}_K^{(0)}$. But this is again infeasible if $\mathcal{E}_K^{(0)}$ does not have any differential of probabilty higher than 2^{32}. Therefore, the mask generator can be a simple LFSR implemented in hardware which must not be modifiable by the attacker.

Cost: The cost depends on the choice of the random mask generation. A simple LFSR implemented in hardware has a low cost with respect to IoT constraints.

7.3 Internal Redundancy Countermeasure

Description: Recently, a countermeasure based on Intra-Instruction Redundancy [30] was proposed to thwart fault attacks. It uses a bit-sliced implementation of a given cipher applied to 15 blocks of data interleaved with 15 blocks of redundancy and 2 blocks of references in order to fit with a 32-bit architecture. The blocks of references are constant plaintexts for which the corresponding ciphertexts are known. Unfortunately, this countermeasure imposes to use - in most cases - a less efficient implementation of the cipher due to the Boolean circuit transformation overhead necessary for bit-slicing [32] and need to encrypt data blocks 15 by 15 from n encryption for an n-bit input. However, using reference blocks as part of a countermeasure is very effective against skip instruction.

240   B. Lac et al.

Thereby, we investigated the possibility to keep this property, also keep a spatial redundancy, while using a conventional (i.e. non-bitsliced) implementation of a cipher applied on only one input block. Hence, we propose the Internal Redundancy Countermeasure (IRC) which exploits efficient 8-bit implementations - which is usually the preferred option for ciphers used in IoT devices, even more for LS-Designs - on a 32-bit architecture. IRC consists in using the original implementation with the same operations but from 32-bit instructions systematically operating as a whole on the 4 bytes of a 32-bit word.

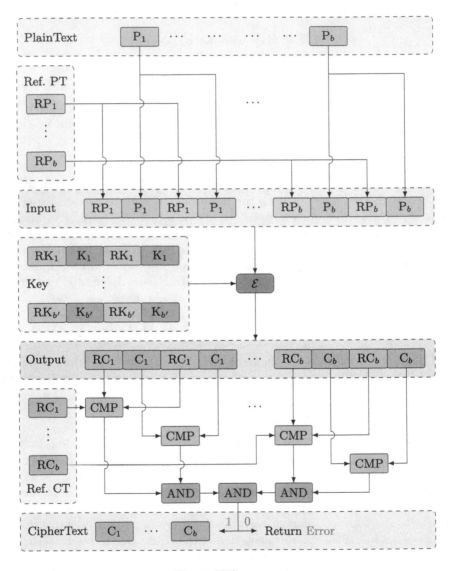

Fig. 9. IRC process

Let $\mathcal{E}$ denote a cipher which takes as input a b-byte plaintext $P = P_1 \cdots P_b$, uses a b'-byte key $K = K_1 \cdots K_{b'}$ and produces a b-byte ciphertext $C = C_1 \cdots C_b$. IRC uses a b-byte reference plaintext $RP = RP_1 \cdots RP_b$, a b'-byte reference key $RK = RK_1 \cdots RK_{b'}$ and a b-byte reference ciphertext $RC = RC_1 \cdots RC_b$. Figure 9 shows one encryption protected by IRC.

Firstly, for each $i \in \{1, \cdots, b\}$, IRC stores in a 32-bit word the byte P_i concatenated with RP_i, P_i and RP_i. IRC also stores, for each $i \in \{1, \cdots, b'\}$, the byte K_i concatenated with RK_i, K_i and RK_i. Then, it applies the cipher by means of a single stream of 32-bit instructions to obtain, for each $i \in \{1, \cdots, b\}$, the byte C_i concatenated with RC_i, C_i and RC_i. Finally, it makes comparisons between redundant bytes and with the reference ciphertexts.

The resulting structure intrinsically protects against fault attacks. Indeed, if the fault is an instruction skip, which corresponds to a random fault for our DFA, the value of the reference ciphertext will be different, so the fault will be detected and the system trapped. Now, if the fault directly affects the value contained in a 32-bits word manipulated by the cipher, it must not affect the first and the third bytes and it must have the same impact on the second and the last bytes. This last case is extremely difficult to control in practice.

Cost: LS-Designs mainly use bitwise operators like logical AND, OR, exclusive OR, shift etc. and also nonlinear operators over $\mathbb{F}_2^8$ like addition, multiplication, modulo etc. operating on bytes. Thereby, IRC simply uses bitwise operators on 32-bit - with sometimes minor changes like the shift for which it must use one mask - and use masks to implement nonlinear operators using few additional 32-bit instructions systematically operating as a whole on the 4 bytes of a 32-bit word to ensure the unicity of the instruction stream. Finally, IRC can use SIMD instructions, depending on the targeted device, to replace some nonlinear operations. Therefore, we obtain performances close to those on an 8-bit architecture while having a structure that intrinsically protects against DFA.

8   Conclusion

In this paper we propose a general method for differential fault analysis on any block cipher based on LS-designs and other families of SPN with similar structures. Such an approach had already been used against a software implementation of PRIDE [25], whose design is close to the one of an LS-Design. But our generalisation has allowed us to successfully perform such an attack against a hardware implementation of SCREAM [20], using the TLS-Design Scream with a fixed tweak. Faults were injected using electromagnetic pulses, which constitutes a low-cost means of injection. We believe that the resistance against DFA is important for LS-Designs, which are expected to be largely deployed in low-resource connected devices. Finally, we propose some countermeasures to thwart such attacks while keeping the efficiency of the ciphers for IoT devices, especially the so-called Internal Redundancy Countermeasure which we propose as a new kind of countermeasure.

Acknowledgement. Benjamin Lac's research work is partly supported by the French DGA-MRIS scholarship.

A   Differential Properties of S-boxes

A.1   Proof of Proposition 1

Proof (of Proposition 1). Let $\mathcal{D}(a, b)$ denote the set of solutions of the equation

$$\mathcal{S}(x \oplus a) \oplus \mathcal{S}(x) = b.$$

Let us consider (a_1, b_1) and (a_2, b_2) be two differentials with $a_1 \neq a_2$ such that

$$\#\mathcal{D}(a_1, b_1) \cap \mathcal{D}(a_2, b_2) \geq 2.$$

It is clear that any element x in $\mathcal{D}(a_1, b_1) \cap \mathcal{D}(a_2, b_2)$ is a solution of

$$\mathcal{S}(x \oplus a_2) \oplus \mathcal{S}(x \oplus a_1) = b_1 \oplus b_2,$$

i.e., $x \oplus a_1 \in \mathcal{D}(a_1 \oplus a_2, b_1 \oplus b_2)$ and $x \oplus a_2 \in \mathcal{D}(a_1 \oplus a_2, b_1 \oplus b_2)$.
Let $\{x, x \oplus a_4\} \subseteq \mathcal{D}(a_1, b_1) \cap \mathcal{D}(a_2, b_2)$ for some $a_4 \neq 0$. Then

$$\{x, x \oplus a_1, x \oplus a_4, x \oplus a_1 \oplus a_4\} \subseteq \mathcal{D}(a_1, b_1),$$

$$\{x, x \oplus a_2, x \oplus a_4, x \oplus a_2 \oplus a_4\} \subseteq \mathcal{D}(a_2, b_2),$$

$$\{x \oplus a_1, x \oplus a_2, x \oplus a_1 \oplus a_4, x \oplus a_2 \oplus a_4\} \subseteq \mathcal{D}(a_1 \oplus a_2, b_1 \oplus b_2).$$

Since $a_1 \neq a_2$, we just must prove that if $a_4 = a_1$ or $a_4 = a_2$ or $a_4 = a_1 \oplus a_2$ then $\mathcal{D}(a_1, b_1)$, $\mathcal{D}(a_2, b_2)$ and $\mathcal{D}(a_1 \oplus a_2, b_1 \oplus b_2)$ have each at least 4 elements. If $a_4 = a_1$ then $x \oplus a_1 \in \mathcal{D}(a_2, b_2)$ imply

$$\mathcal{S}(x \oplus a_1 \oplus a_2) \oplus \mathcal{S}(x \oplus a_1) = b_2 = \mathcal{S}(x \oplus a_2) \oplus \mathcal{S}(x)$$

implying that

$$\mathcal{S}(x \oplus a_1) \oplus \mathcal{S}(x) = \mathcal{S}(x \oplus a_2 \oplus a_1) \oplus \mathcal{S}(x \oplus a_2).$$

Thus $x \oplus a_2 \in \mathcal{D}(a_1, b_1)$ and $\mathcal{D}(a_1, b_1)$, $\mathcal{D}(a_2, b_2)$ and $\mathcal{D}(a_1 \oplus a_2, b_1 \oplus b_2)$ contain $\{x, x \oplus a_1, x \oplus a_2, x \oplus a_1 \oplus a_2\}$. It's identical if $a_4 = a_2$ following the same reasoning. Now, $a_4 = a_1 \oplus a_2$ implies that $x \oplus a_2 \oplus a_4 = x \oplus a_1$ belongs to $\mathcal{D}(a_2, b_2)$, i.e., $x \oplus a_1 \in \mathcal{D}(a_1, b_1) \cap \mathcal{D}(a_2, b_2)$. Therefore, $x \oplus a_1$, $x \oplus a_2$, x and $x \oplus a_1 \oplus a_2$ all belong to $\mathcal{D}(a_1 \oplus a_2, b_1 \oplus b_2)$ and $\#\mathcal{D}(a_1 \oplus a_2, b_1 \oplus b_2) \geq 4$.   □

B   Different Exploitable Faults Obtained on SCREAM

Table 3 provides the 36 different exploitable faults obtained on the reference hardware implementation of SCREAM as well as our knowledge about the differences values around the last substitution layer for each fault, i.e. the value of the output difference ΔY_{24} and the possible values for each input byte difference, denoted ΔIn (which must be the same for all bytes). Among these faults, only 7 displayed in red gave as much information as all faults.

The faults were sorted according to the value of ΔIn. As we can see, some faults have several possibilities for the value of ΔIn. However, its correct value can be retrieve from the others faults. Indeed, the fault 16 for example at the same difference output on the 9-th byte as the fault 15. Thereby, the correct value of ΔIn for the fault 16 is 0×01. Like this, we can retrieve the correct value of ΔIn for each fault - the last fault can not have been obtained from $\Delta In = 0 \times 08$ since the correct value for the last nibble is this case is 0xe9. Finally, by

Table 3. Exploitable faults obtained on SCREAM

No	Value of the last block	Value of ΔY_{24}	Value of ΔIn
1	0x04eb0f2430df5f9301047fae10109f6d	0x003347ca19002a00d95d000000548a00	0x01
2	0x907766fd6347e9904999306543889d43	0x003300ca000000d5d900fd0000000000	0x01
3	0xc0414fd2280187f719af457254f68a3d	0x000000ca00002a00d900fd002a0000c2	0x01
4	0xc0416fea2801fbf719af2b151aa9c462	0x0000000000000000d900fd0000000000	0x01
5	0xc041e0232801280619afea8354f68a3d	0x000000ca00000000d900fd002a0000c2	0x01
6	0xc098d36128d80ce3760cea8308179d43	0x000047ca000000d5d95dfd002a000000	0x01
7	0xdb313ac52801e99002df306554f68a3d	0x003300ca00000000d900fd00000000c2	0x01
8	0xdb3154a2280187f702df5e0254f68a3d	0x003300ca00002a00d900fd002a0000c2	0x01
9	0xdb3166fd2801e99002df306508ced605	0x003300ca00000000d900fd0000000000	0x01
10	0xdb31fb532801280602dff1f354f68a3d	0x003300ca00000000d900fd002a0000c2	0x01
11	0xdb31fb5345d645d102dff1f33921e7ea	0x003300ca00000000d900fdcc2a0000c2	0x01
12	0xdbe8c81128d80cc36d7cf1f308179d43	0x003347ca000000d5d95dfd002a000000	0x01
13	0xe47dd3610c3d280652e9ea832cf29d43	0x000047ca000000d5d900fd002a000000	0x01
14	0xff0d00870c3de990400030652cf20d43	0x003347ca000000d5d900fd0000000000	0x01
15	0xff0dc8110c3d28064999f1f32cf29d43	0x003347ca000000d5d900fd002a000000	0x01
16	0x207b6feac83b1bcdf995cb2ffa932458	0x0000000000000000d900000000000000	0x01, 0x02 or 0x20
17	0x207b7d8dc83b09aaf995cb2fe8f4363f	0x000000ca00000000d900000000000000	0x01 or 0x02
18	0x3b0b66fdc83b09aac2c5d05fc8f4363f	0x003300ca00000000d900000000000000	0x01 or 0x02
19	0x3b0b749ac83b1bcde2e5d05ffa932458	0x0033000000000000d900000000000000	0x01 or 0x02
20	0x9ef5749ac83b0fb9f69161d55f6d302c	0xb8330000000000000000000000000000	0x01, 0x02 or 0x04
21	0xea8ab50cc83bba5082eea0435f6d302c	0xb8330001900000000000002a000000	0x01 or 0x02
22	0x8585749ac83bbe335c6bd05f441d81a6	0x0062000000000000000000000000000	0x02
23	0x97e266fdda5cac545c6bd05f567a81a6	0x0062004f0000000000000000000000	0x02
24	0x97e286c7da5cac545c6bd05f567a619c	0x0062004f00000000820000000000	0x02
25	0x85858fd0c83bbe335c6bcb2f5f6d619c	0x000000000000000820000000000	0x01, 0x02 or 0x10
26	0xeaff0090da5cd1494e0ccb2f227081a6	0x00004b540000000000000000000000000	0x04
27	0x65bfae7cc83b9f9f7dc72b15bf57619c	0x0000000000000000000f9001a000000	0x08
28	0x880d56417b40b585a3ee75dc74f12425	0xb9880000000001a0058f9001aee00e9	0x08
29	0xd435ae7cc83b2e15cc4d9a9fbf57d016	0xb90000000000000000000f9001a000000	0x08
30	0xd435ae7cc83b3565cc4d9a9fbf57cb66	0xb98000000000000000f9001a000000	0x08
31	0xd9bdd4df0cbe9dbb23dbe8a7aa07b049	0x0000003a00000000000000bc000046e9	0x08
32	0xd9bdd4df6169f06c4e0c8570aa07dd9e	0x0000003a00000000000000000046e9	0x08
33	0x340f6feac83b14c9ede17aa55f6d2b5c	0xb9880000400f00000000000000000000	0x02 or 0x08
34	0x65bf6feac83b5e09bc512b15bf57619c	0x00000000000000000000f90000000000	0x04 or 0x08
35	0xd9bd7d8dc83bf06c4e0c85700355dd9e	0x0000003a00000000000000000000e9	0x02 or 0x08
36	0x8585c6b8522b4b595c6bcb2f30171bb6	0x0000c40000000000000000000008e8c	0x08 or 0x80

intersection of the obtained output differences for each value of ΔIn, we obtain the following exploitable differentials on the inverse S-boxes:

(0 × b83347ca19002ad5d95dfdcc2a548ac2, 0 × 0101010101010001010101010101010101),
(0 × 0062004f0000000000000820000000000, 0 × 0002000200000000000000020000000000),
(0 × 00004b540000000000000000000000000, 0 × 0000040400000000000000000000000000),
(0 × b988003a0000001a0058f9bc1aee46e9, 0 × 0808000800000008000808080808080808),
(0 × 0000c40000000000000000000008e8c, 0 × 0000800000000000000000000008080).

Finally, Table 4 gives for each $1 \leq i \leq 16$, the value of the output byte difference $\Delta Y_{24}[i]$ that the 7 faults which give as much information as all faults have allowed us to know for each obtained input difference $\Delta X_{24}[i] = 2^j$ denoted ΔIn (i.e. the intersection of the obtained output differences). Table 4 also gives the byte candidates obtained for $\mathcal{L}\text{layer}^{-1}(C \oplus K \oplus T)[i] \oplus Ct_{23}[i]$ (equal to $0 \times 030e2eef32dbfbcbdb3f4859d1e49e97$). The symbol $\emptyset$ means that the faults did not provide any information about the byte (i.e. the 256 values are possible).

Table 4. Values of $\Delta Y_{24}[i]$, $1 \leq i \leq 16$, for each obtained input difference ΔIn and bytes candidates obtained for $\mathcal{L}\text{layer}^{-1}(C \oplus K \oplus T)[i] \oplus Ct_{23}[i]$

		ΔIn			$\mathcal{L}\text{layer}^{-1}(C \oplus K \oplus T)[i] \oplus Ct_{23}[i]$
		0x01	0x04	0x08	
$\Delta Y_{24}[i]$	$i = 1$	0xb8	$\emptyset$	0xb9	0x03
	$i = 2$	0x33	$\emptyset$	0x88	0x0e
	$i = 3$	0x47	0x4b	$\emptyset$	0x2e
	$i = 4$	0xca	0x54	0x3a	0xef
	$i = 5$	0x19	$\emptyset$	$\emptyset$	0x2b, 0x32, 0x4f, 0x56, 0x65 or 0x7c
	$i = 6$	$\emptyset$	$\emptyset$	$\emptyset$	$\emptyset$
	$i = 7$	0x2a	$\emptyset$	$\emptyset$	0xd1 or 0xfb
	$i = 8$	0xd5	$\emptyset$	0x1a	0xcb
	$i = 9$	0xd9	$\emptyset$	$\emptyset$	0x02 or 0xdb
	$i = 10$	0x5d	$\emptyset$	0x58	0x3f
	$i = 11$	0xfd	$\emptyset$	0xf9	0x48
	$i = 12$	0xcc	$\emptyset$	0xbc	0x59
	$i = 13$	0x2a	$\emptyset$	0x1a	0xd1
	$i = 14$	0x54	$\emptyset$	0xee	0xe4
	$i = 15$	0x8a	$\emptyset$	0x46	0x9e
	$i = 16$	0xc2	$\emptyset$	0xe9	0x97

References

1. Adomnicai, A., Lac, B., Canteaut, A., Fournier, J.J., Masson, L., Sirdey, R., Tria, A.: On the importance of considering physical attacks when implementing lightweight cryptography. In: NIST Lightweight Cryptography Workshop 2016, Gaithersburg, Maryland, October 2016

2. Agoyan, M., Dutertre, J.-M., Naccache, D., Robisson, B., Tria, A.: When clocks fail: on critical paths and clock faults. In: Gollmann, D., Lanet, J.-L., Iguchi-Cartigny, J. (eds.) CARDIS 2010. LNCS, vol. 6035, pp. 182–193. Springer, Heidelberg (2010). doi:10.1007/978-3-642-12510-2_13

3. Albrecht, M.R., Driessen, B., Kavun, E.B., Leander, G., Paar, C., Yalçın, T.: Block ciphers – focus on the linear layer (feat. PRIDE). In: Garay, J.A., Gennaro, R. (eds.) CRYPTO 2014. LNCS, vol. 8616, pp. 57–76. Springer, Heidelberg (2014). doi:10.1007/978-3-662-44371-2_4

4. Banik, S., Bogdanov, A., Isobe, T., Shibutani, K., Hiwatari, H., Akishita, T., Regazzoni, F.: Midori: a block cipher for low energy. In: Iwata, T., Cheon, J.H. (eds.) ASIACRYPT 2015. LNCS, vol. 9453, pp. 411–436. Springer, Heidelberg (2015). doi:10.1007/978-3-662-48800-3_17

5. Beaulieu, R., Shors, D., Smith, J., Treatman-Clark, S., Weeks, B., Wingers, L.: SIMON and SPECK: block ciphers for the internet of things. Cryptology ePrint Archive, Report 2015/585 (2015). http://eprint.iacr.org/2015/585

6. Berger, T.P., Francq, J., Minier, M.: CUBE cipher: a family of quasi-involutive block ciphers easy to mask. In: El Hajji, S., Nitaj, A., Carlet, C., Souidi, E.M. (eds.) C2SI 2015. LNCS, vol. 9084, pp. 89–105. Springer, Cham (2015). doi:10.1007/978-3-319-18681-8_8

7. Biham, E., Shamir, A.: Differential fault analysis of secret key cryptosystems. In: Kaliski, B.S. (ed.) CRYPTO 1997. LNCS, vol. 1294, pp. 513 525. Springer, Heidelberg (1997). doi:10.1007/BFb0052259

8. Bilgin, B., Bogdanov, A., Knežević, M., Mendel, F., Wang, Q.: FIDES: lightweight authenticated cipher with side-channel resistance for constrained hardware. In: Bertoni, G., Coron, J.-S. (eds.) CHES 2013. LNCS, vol. 8086, pp. 142–158. Springer, Heidelberg (2013). doi:10.1007/978-3-642-40349-1_9

9. Blömer, J., Seifert, J.-P.: Fault based cryptanalysis of the advanced encryption standard (AES). In: Wright, R.N. (ed.) FC 2003. LNCS, vol. 2742, pp. 162–181. Springer, Heidelberg (2003). doi:10.1007/978-3-540-45126-6_12

10. Bogdanov, A., Knudsen, L.R., Leander, G., Paar, C., Poschmann, A., Robshaw, M.J.B., Seurin, Y., Vikkelsoe, C.: PRESENT: an ultra-lightweight block cipher. In: Paillier, P., Verbauwhede, I. (eds.) CHES 2007. LNCS, vol. 4727, pp. 450–466. Springer, Heidelberg (2007). doi:10.1007/978-3-540-74735-2_31

11. Boneh, D., DeMillo, R.A., Lipton, R.J.: On the importance of checking cryptographic protocols for faults. In: Fumy, W. (ed.) EUROCRYPT 1997. LNCS, vol. 1233, pp. 37–51. Springer, Heidelberg (1997). doi:10.1007/3-540-69053-0_4

12. Borghoff, J., Canteaut, A., Güneysu, T., Kavun, E.B., Knezevic, M., Knudsen, L.R., Leander, G., Nikov, V., Paar, C., Rechberger, C., Rombouts, P., Thomsen, S.S., Yalçın, T.: PRINCE – a low-latency block cipher for pervasive computing applications. In: Wang, X., Sako, K. (eds.) ASIACRYPT 2012. LNCS, vol. 7658, pp. 208–225. Springer, Heidelberg (2012). doi:10.1007/978-3-642-34961-4_14

13. CAESAR: Competition for Authenticated Encryption (2014). https://competitions.cr.yp.to/caesar.html

14. Courbon, F., Loubet-Moundi, P., Fournier, J.J.A., Tria, A.: Adjusting laser injections for fully controlled faults. In: Prouff, E. (ed.) COSADE 2014. LNCS, vol. 8622, pp. 229–242. Springer, Cham (2014). doi:10.1007/978-3-319-10175-0_16

15. Dehbaoui, A., Dutertre, J., Robisson, B., Tria, A.: Electromagnetic transient faults injection on a hardware and a software implementations of AES. In: Bertoni, G., Gierlichs, B. (eds.) FDTC 2012, pp. 7–15. IEEE Computer Society, Leuven, 9 September 2012

16. Gérard, B., Grosso, V., Naya-Plasencia, M., Standaert, F.: Block ciphers that are easier to mask: How far can we go? Cryptology ePrint Archive, Report 2013/369 (2013). http://eprint.iacr.org/2013/369

17. Gong, Z., Nikova, S., Law, Y.W.: KLEIN: a new family of lightweight block ciphers. In: Juels, A., Paar, C. (eds.) RFIDSec 2011. LNCS, vol. 7055, pp. 1–18. Springer, Heidelberg (2012). doi:10.1007/978-3-642-25286-0_1

18. Grosso, V., Leurent, G., Standaert, F.-X., Varıcı, K.: LS-designs: bitslice encryption for efficient masked software implementations. In: Cid, C., Rechberger, C. (eds.) FSE 2014. LNCS, vol. 8540, pp. 18–37. Springer, Heidelberg (2015). doi:10.1007/978-3-662-46706-0_2

19. Grosso, V., Leurent, G., Standaert, F.X., Varıcı, K., Journault, A., Durvaux, F., Gaspar, L., Kerckhof, S.: Implementations of the SCREAM authenticated encryption algorithm. https://perso.uclouvain.be/fstandae/SCREAM

20. Grosso, V., Leurent, G., Standaert, F.X., Varıcı, K., Journault, A., Durvaux, F., Gaspar, L., Kerckhof, S.: SCREAM, side-channel resistant authenticated encryption with masking. https://competitions.cr.yp.to/round2/screamv3.pdf, submission to the CAESAR competition

21. Guilley, S., Sauvage, L., Danger, J., Selmane, N.: Fault injection resilience. In: Breveglieri, L., Joye, M., Koren, I., Naccache, D., Verbauwhede, I. (eds.) FDTC 2010, pp. 51–65. IEEE Computer Society, Santa Barbara, 21 August 2010. http://dx.doi.org/10.1109/FDTC.2010.15

22. Guo, J., Peyrin, T., Poschmann, A., Robshaw, M.: The LED block cipher. In: Preneel, B., Takagi, T. (eds.) CHES 2011. LNCS, vol. 6917, pp. 326–341. Springer, Heidelberg (2011). doi:10.1007/978-3-642-23951-9_22

23. Halevi, S., Coppersmith, D., Jutla, C.: Scream: a software-efficient stream cipher. In: Daemen, J., Rijmen, V. (eds.) FSE 2002. LNCS, vol. 2365, pp. 195–209. Springer, Heidelberg (2002). doi:10.1007/3-540-45661-9_15

24. Knudsen, L., Leander, G., Poschmann, A., Robshaw, M.J.B.: PRINTCIPHER: a block cipher for IC-printing. In: Mangard, S., Standaert, F.-X. (eds.) CHES 2010. LNCS, vol. 6225, pp. 16–32. Springer, Heidelberg (2010). doi:10.1007/978-3-642-15031-9_2

25. Lac, B., Beunardeau, M., Canteaut, A., Fournier, J.J.A., Sirdey, R.: A first DFA on PRIDE: from theory to practice. In: Cuppens, F., Cuppens, N., Lanet, J.-L., Legay, A. (eds.) CRiSIS 2016. LNCS, vol. 10158, pp. 214–238. Springer, Cham (2017). doi:10.1007/978-3-319-54876-0_17

26. Lashermes, R., Fournier, J., Goubin, L.: Inverting the final exponentiation of tate pairings on ordinary elliptic curves using faults. In: Bertoni, G., Coron, J.-S. (eds.) CHES 2013. LNCS, vol. 8086, pp. 365–382. Springer, Heidelberg (2013). doi:10.1007/978-3-642-40349-1_21

27. Liskov, M., Rivest, R.L., Wagner, D.: Tweakable block ciphers. In: Yung, M. (ed.) CRYPTO 2002. LNCS, vol. 2442, pp. 31–46. Springer, Heidelberg (2002). doi:10.1007/3-540-45708-9_3

28. Liskov, M., Rivest, R.L., Wagner, D.: Tweakable block ciphers. J. Cryptol. **24**(3), 588–613 (2011). doi:10.1007/s00145-010-9073-y

29. Mohamed, M.S.E., Bulygin, S., Buchmann, J.: Using SAT solving to improve differential fault analysis of trivium. In: Kim, T., Adeli, H., Robles, R.J., Balitanas, M. (eds.) ISA 2011. CCIS, vol. 200, pp. 62–71. Springer, Heidelberg (2011). doi:10.1007/978-3-642-23141-4_7

30. Patrick, C., Yuce, B., Ghalaty, N., Schaumont, P.: Lightweight fault attack resistance in software using intra-instruction redundancy. In: 23rd Conference on Selected Areas in Cryptography (2016)

31. Piret, G., Roche, T., Carlet, C.: PICARO – a block cipher allowing efficient higher-order side-channel resistance. In: Bao, F., Samarati, P., Zhou, J. (eds.) ACNS 2012. LNCS, vol. 7341, pp. 311–328. Springer, Heidelberg (2012). doi:10.1007/978-3-642-31284-7_19

32. Pornin, T.: Implantation et optimisation des primitives cryptographiques. Ph.D. thesis, Université Paris 7 (2001)

33. Sajadieh, M., Dakhilalian, M., Mala, H., Sepehrdad, P.: Recursive diffusion layers for block ciphers and hash functions. In: Canteaut, A. (ed.) FSE 2012. LNCS, vol. 7549, pp. 385–401. Springer, Heidelberg (2012). doi:10.1007/978-3-642-34047-5_22

34. Sakiyama, K., Li, Y., Iwamoto, M., Ohta, K.: Information-theoretic approach to optimal differential fault analysis. IEEE Trans. Inf. Forensics Secur. **7**(1), 109–120 (2012)

35. Skorobogatov, S.: Semi-invasive attacks - a new approach to hardware security analysis. Technical report 630, University of Cambridge, April 2005

36. Skorobogatov, S.P., Anderson, R.J.: Optical fault induction attacks. In: Kaliski, B.S., Koç, K., Paar, C. (eds.) CHES 2002. LNCS, vol. 2523, pp. 2–12. Springer, Heidelberg (2003). doi:10.1007/3-540-36400-5_2

37. Song, L., Hu, L.: Differential fault attack on the PRINCE block cipher. Cryptology ePrint Archive, Report 2013/043 (2013). http://eprint.iacr.org/2013/043

38. Todo, Y., Leander, G., Sasaki, Y.: Nonlinear invariant attack. In: Cheon, J.H., Takagi, T. (eds.) ASIACRYPT 2016. LNCS, vol. 10032, pp. 3–33. Springer, Heidelberg (2016). doi:10.1007/978-3-662-53890-6_1

39. Tupsamudre, H., Bisht, S., Mukhopadhyay, D.: Differential fault analysis on the families of SIMON and SPECK ciphers. Cryptology ePrint Archive, Report 2014/267 (2014). http://eprint.iacr.org/2014/267

40. Zhao, X., Wang, T., Guo, S.: Improved side channel cube attacks on PRESENT. Cryptology ePrint Archive, Report 2011/165 (2011). http://eprint.iacr.org/2011/165

Multiple-Valued Debiasing for Physically Unclonable Functions and Its Application to Fuzzy Extractors

Manami Suzuki[✉], Rei Ueno, Naofumi Homma, and Takafumi Aoki

Tohoku University, Aramaki Aza Aoba 6–6–05, Aoba-ku, Sendai-shi 980-8579, Japan
manami@aoki.ecei.tohoku.ac.jp

Abstract. This paper proposes a new debiasing method for a stable and efficient extraction of uniform random binary responses from physically unclonable functions (PUFs). The proposed method handles multiple-valued (i.e., ternary) responses from PUF responses, including unstable response bits, and stably extracts uniform random-bit responses from them. In this paper, we evaluate the stability and effectiveness of the proposed method with two experiments with simulated and actual responses of latch PUFs implemented on an FPGA. We demonstrate that the proposed method can obtain longer debiased random-bit responses than the conventional method. In addition, we apply the proposed method to the construction of a fuzzy extractor (FE), and show the advantages of the proposed FE in terms of response length and authentication success rate in an experimental evaluation.

Keywords: PUF · Fuzzy extractors · Latch PUF · Debiasing · Multiple-valued logic

1 Introduction

Authentication technologies for LSI are now essential for efficient distributions, traceability, and counterfeit detection. In particular, counterfeiting prevention is in high demand, as the amount of counterfeit products increases due to the advancement of analytical and manufacturing technologies on LSI. Physical unclonable functions (PUFs) are expected to construct a more robust authentication technology that can prevent the counterfeiting of LSI products [7]. PUFs usually generate random responses uniquely determined by process variations in their own LSIs, such as the drive capabilities of two NAND gates and the difference between their wire lengths. Applications of PUFs include key generation and storage for secure authentication.

One important feature of PUF responses is their stability, indicated by the consistency of PUF response bits obtained from multiple observation trials. However, it is difficult to manufacture a PUF that can always generate stable responses under a wide variety of circumstances [8] because every PUF exploits

© Springer International Publishing AG 2017
S. Guilley (Ed.): COSADE 2017, LNCS 10348, pp. 248–263, 2017.
DOI: 10.1007/978-3-319-64647-3_15

uncontrollable physical variations in LSIs. A fuzzy extractor (FE) is a possible solution for stable PUF-based authentication owing to its error correction scheme [8]. However, a less-stable PUF requires stronger error correction from the FE, decreasing the amount of values available for authentication [8].

Another important feature of PUF responses is their unpredictability, which affords them their resistance to counterfeiting. One necessary condition for unpredictability is the uniformity of responses, which can be quantified by the bias between the appearance rates of 0 and 1 in a PUF response. The smaller the bias, the closer to balance the appearance rates are. When the bias of a PUF response is large, entropy leakage occurs from the helper data in the FE [3]. Ideally, a non-biased PUF should be implemented; however, doing so is often difficult because the PUF response is dependent on small and uncontrollable process variations in LSIs. To reduce bias in PUF responses, a debiasing method based on a deterministic randomness extractor [9] was proposed in [8]. This debiasing method enables the suppression of information leakage from the helper data at the expense of decreasing the resulting response length to extract secret information.

As described above, lower stability and uniformity decrease the entropy of PUFs. Even in unstable PUFs, though, we can see two types of locations (cells), whose values are stable and unstable, respectively, in generating repeated responses. That is, PUFs commonly include two types of cells: one has a constant output value of 0 or 1 and the other has an output value flipped randomly[1]. Using the characteristics of PUFs, [12] took a ternary approach, which assigned the third "random value" to the inconstant output value in addition to the values of 0 and 1. PUF stability and unpredictability were successfully enhanced by the ternary approach; however, application of the ternary (or, more generally, multiple-valued [12]) responses was limited since there were neither debiasing methods nor FEs available for such ternary responses.

In order to take full advantage of the ternary approach described above, we propose a ternary debiasing method that extracts uniform random-bit responses from ternary PUF responses. With the proposed debiasing method, we also present an extended FE structure that can handle ternary PUF responses. Since the proposed debiasing method outputs binary values, we can apply a conventional FE to the latter part of the extended FE. The stability and effectiveness of the proposed debiasing method are demonstrated with two experiments with simulations and latch PUFs (L-PUFs) implemented on an FPGA. We demonstrate that the proposed method can obtain debiased random-bit responses that are longer than those of conventional debiasing methods. The effectiveness of the proposed FE is evaluated in terms of response length and authentication success rate.

---

[1] Note that the information about these locations is secret information.

2   Related Work

2.1   Latch PUFs and Fuzzy Extractors

This paper focuses on an L-PUF [10,11], because it can be implemented on both an ASIC and an FPGA, and available for generating ternary responses. Figure 1 shows an overview of an L-PUF consisting of n RS latches, in which the input is a set signal and the output is an n-bit response. When the set signal changes from 0 to 1, the RS latch temporarily goes to a metastable state and finally to a stable state. Depending on the individual physical variation, such as the drive capabilities of two NAND gates and the wire lengths between them, many RS latches have a high probability of terminating at one specific state that outputs 0 or 1. As a result, the L-PUF generates unique responses for each LSI device.

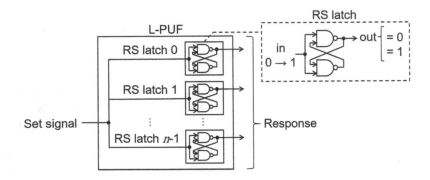

Fig. 1. L-PUF block diagram.

However, some RS latches do not terminate at one specific state; their responses change each time the input signal is received. Accordingly, we can classify the outputs of RS latches into three types: (i) always 0, (ii) always 1, and (iii) either 0 or 1 (a random number) for repeated input signals. In this paper, we call RS latches with output types (i) and (ii) constant and type (iii) random. Random latches have been considered to be noise. There are two conventional approaches to suppressing noise in response generation: one is detecting random latches in advance and removing from responses in authentication and another is the use of error correction code (ECC) [2,6] to correct inconsistencies caused by random latches in the response.

In particular, FEs are commonly used for the latter approach. Figure 2 shows the block diagram of an FE [1] composed of an enrollment unit and a reconstruction unit. The enrollment unit consists of an ECC encoder and a universal hash function [4,5], and generates secret information (key) K and helper data W from an n-bit PUF response X and a m-bit $(m < n)$ random seed S. The reconstruction unit consists of an ECC decoder and the same universal hash function as the enrollment unit, and generates secret information (key) K from a regenerated

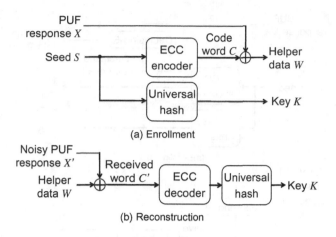

Fig. 2. FE block diagram.

(noisy) PUF response X' and the helper data W. W is stored in nonvolatile memory in this authentication scheme. It is known that if a PUF has enough uniformity, entropy leak does not occur even if an attacker observes W [3].

The drawback of the conventional approaches is that they result in responses with greatly reduced bit length. To solve this problem, [12] proposed a multiple-valued approach that assigns the third, random value to the response bit of a random latch. This method can improve the stability, and therefore increase the entropy, of PUF responses in comparison with the conventional binary approach. However, multiple-valued responses have not been combined with FEs until now.

2.2 Debiasing

Debiasing was proposed to increase the uniformity of PUF responses [8]. Note that when a PUF response has significant bias, entropy leakage from the helper data can occur even if the PUF response is stable. Here, let p_0 and p_1 be the occurrence probability of 0 and 1 in the responses, respectively. $p_0 = 0.5$ for the case of an ideal unbiased PUF. In [8], a typical FE can generate secret information from PUF responses without any leakage from the helper data if the PUF satisfies the condition $0.418 \leq p_0 \leq 0.582$. The basic idea of debiasing is to extract unbiased responses from biased PUF responses by a deterministic randomness extractor. The classic deterministic randomness extractor (CDRE) [9] proposed by von Neumann was used in the conventional work [8]. The CDRE handles a PUF response with a pair of two consecutive bits. A pair is discarded if both the bits are equal, whereas the first bit of the pair is retained as a debiased bit if the bits are different. The resulting occurrence probability of 0 and 1 in a debiased result Y are equal, since both occurrence probabilities are $p_0 p_1$, respectively.

Figure 3 shows an FE block diagram with debiasing. Debiasing-E generates a debiased response Y and debiasing data D from a PUF response X with the

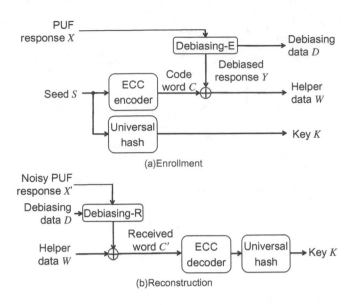

Fig. 3. FE block diagram with debiasing.

CDRE. The debiasing data D has a length of $n/2$ bits, indicating where bits are discarded. Debiasing-R generates a debiased response Y' from a (noisy) PUF response X' and debiasing data D. Table 1 shows the input-output relations of debiasing using the CDRE for enrollment and reconstruction, where x_i, x'_i, y_i, y'_i, and d_i are the ith ($0 \leq i \leq n/2-1$) bits of X, X', Y, Y', and D, respectively[2].

Table 1. Debiasing using CDRE

Enrollment			Reconstruction		
Input	Output		Input		Output
$x_{2i}x_{2i+1}$	y_i	d_i	$x'_{2i}x'_{2i+1}$	d_i	y'_i
0 0	discard	0	0 -	1	0
0 1	0	1	1 -	1	1
1 0	1	1	- -	0	discard
1 1	discard	0			

Figure 4 shows an example of debiasing using the CDRE. The method generates an unbiased response Y and debiasing data D from a biased response X using the CDRE for enrollment. If two bits in a pair are not equal, the bit of the

---

[2] Y and Y' do not have a precisely $n/2$-bit length. However, when $x_{2i} = x_{2i+1}$ (or $x'_{2i} = x'_{2i+1}$), the output y_i (or y'_i) is considered to have a "discard" value in order to simplify the formulation.

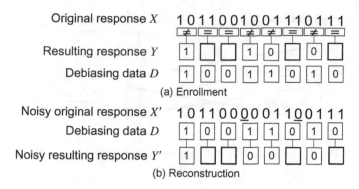

(a) Enrollment

(b) Reconstruction

Fig. 4. Example of debiasing using CDRE.

debiasing data is 1; otherwise, it is 0. In reconstruction, the noisy response Y' is reconstructed based on D. In Fig. 4, an underlined bit indicates an error bit compared with the original response bit in the enrollment. The use of debiasing decreases the resulting response length. The expected entropy of Y (or Y') is limited to $np_0p_1 \times (1 - p_r)$, where n and p_r are a response length of X (or X') and an occurrence probability of a random latch, respectively. Note here that the responses of random latches do not contribute to the entropy of Y, since these responses are considered noise. As the bias of X increases, the resulting response length decreases, suggesting that the PUF response length should be increased to generate sufficient length for a biased PUF. If the implementation cost size to increase the response length is nontrivial, a more efficient debiasing method with lower overhead is necessary.

3   Multiple-Valued Debiasing

We propose an effective debiasing method applicable to ternary PUF responses. The proposed debiasing generates a binary output and debiasing data from a ternary response given by the enrollment method in [12]. It generates a (noisy) debiased response from a (noisy) ternary response and the debiasing data in reconstruction. The basic idea of the proposed debiasing method is to handle an output of random latch r as an erasure bit and apply an ECC based on the Hamming erasure distance. While the false detection of a random latch causes an error in the multiple-valued response, the proposed debiasing generates a binary output from a ternary input with error correction.

Figure 5 shows the error pattern of response bits in reconstruction, where $p_{s \to s'}$ indicates the probability that a value s at enrollment is changed to s' at reconstruction. Note that correct reconstruction with no errors does not appear in Fig. 5. The main reason for such errors is the false detection of random latches, and therefore the probabilities $p_{0 \to r}$, $p_{1 \to r}$, $p_{r \to 0}$, and $p_{r \to 1}$ are large, whereas the probabilities $p_{0 \to 1}$ and $p_{1 \to 0}$ would be very small. Under this assumption, we can correct the error by generating y_i' based on minimum distance decoding, that

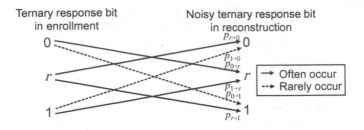

Fig. 5. Error patterns of response bits in reconstruction.

the distance between r and 0 and between r and 1 is smaller than the distance between 0 and 1. One such distance is the Hamming erasure distance, which is taking account of an erasure channel, used in the proposed method.

Table 2 shows the input-output relations of the proposed debiasing method in enrollment and reconstruction, where y_i is the ith bit of a resulting response Y, d_i is the ith bit of debiasing data D, and t_i and t'_i are the ith bits of a ternary response in enrollment T and a (noisy) ternary response in reconstruction T', respectively. The proposed method handles ternary responses with a pair of consecutive digits, and then outputs 0 or 1 or discards it by the value of a pair.

Figure 6 shows an example of debiasing using the proposed method, where 0 and 1 at the ternary response are the outputs of the constant latch, and ternary response is the output of the random latch. If $t_{2i} = t_{2i+1}$ the value of D is $d_i = 0$, otherwise $d_i = 1$. $y_i = 0$ when (t_{2i}, t_{2i+1}) is $(0, 1)$, $(r, 1)$, or $(0, r)$, while $y_i = 1$ when (t_{2i}, t_{2i+1}) is $(1, 0)$, $(r, 0)$, or $(1, r)$ according to minimum distance decoding, assuming that r is an erasure bit in binary code. Let p_0, p_1, and p_r be the occurrence probabilities of 0, 1, and r, respectively. Here,

Table 2. Proposed debiasing

Enrollment			Reconstruction		
Input	Output		Input		Output
$t_{2i}t_{2i+1}$	y_i	d_i	$t'_{2i}t'_{2i+1}$	d_i	y'_i
0 0	discard	0	0 -	1	0
1 1	discard	0	1 -	1	1
$r\ r$	discard	0	$r\ r$	1	1
0 1	0	1	r 0	1	1
r 1	0	1	r 1	1	0
0 r	0	1	- -	0	discard
1 0	1	1			
r 0	1	1			
1 r	1	1			

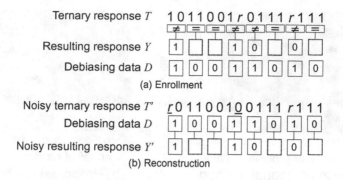

Fig. 6. Example of proposed debiasing.

the occurrence probabilities of both 0 and 1 in Y are equal to $p_1 p_0 + p_1 p_r + p_r p_0$. In the reconstruction, the proposed method discards bits based on D, and generates y'_i from (t_{2i}, t'_{2i+1}) based on the same minimum distance decoding as in the enrollment. Note that the method can detect an error when $d_i = 1$ and $t_{2i} = t_{2i+1}$, but it assigns the value of 1 to the error, since it cannot correct it at this stage. The expected value of the resulting response's bit length in the proposed method is $n \times (p_0 p_1 + p_r p_1 + p_0 p_r)$. Our method can exploit random latches for increasing the entropy of PUFs, whereas the conventional method cannot.

4 Performance Evaluation

4.1 Simulation Experiment

We first evaluate the proposed debiasing method with ternary responses randomly generated, assuming L-PUF responses with 1,024 RS latches. In the simulation, we change the bias of the responses and the number of random latches.

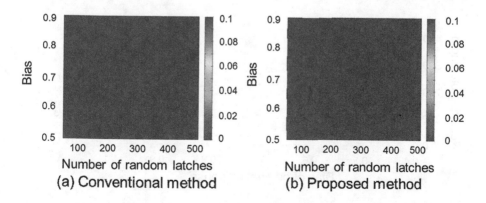

Fig. 7. Average bias of resulting response after debiasing.

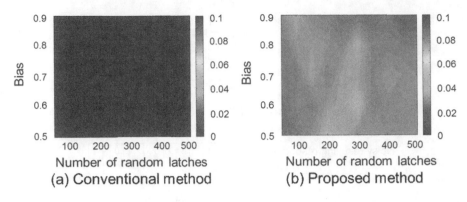

Fig. 8. Worst-case bias of the resulting response after debiasing. (Color figure online)

We then compare the resulting uniformity (i.e., bias value) and output response length of our method with those of the conventional method in [8].

Figures 7 and 8 show the average and worst bias of the resulting responses, respectively. Here, the color maps in (a) and (b) indicate the results of the conventional and proposed methods, respectively. The vertical axis indicates the initial bias of response p_0 (p_0 indicates the probability of 0 from all the constant latches) and the horizontal axis indicates the number of random latches in the response. If the resulting bias p_0' after debiasing is close to the ideal value (i.e., $|p_0' - 0.5| = 0$), the map turns set to deep blue. If not, it is set to deep red. The reddest point is $|p_0' - 0.5| = 0.1$. Note that an FE should meet the condition that $|p_0' - 0.5| \leq 0.082$ to generate secret information securely. The results of Figs. 7 and 8 show that while both methods were, on average, satisfied with the condition of the resulting biases, the conventional method did not satisfy this condition on the worst-case biases in the experiment. However, we did confirm that the proposed method satisfied this condition even in the worst case. This is because the use of ternary response could increases the entropy of PUF response [12] in comparison with binary response.

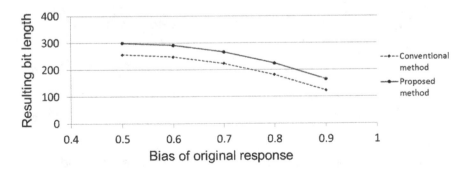

Fig. 9. Resulting bit length after debiasing for different original biases.

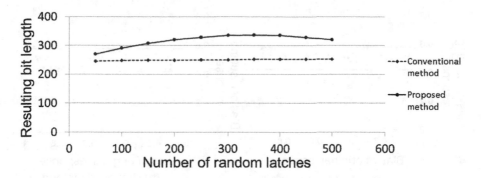

Fig. 10. Resulting bit length after debiasing for different numbers of random latches.

Figures 9 and 10 show the bit lengths of the resulting responses generated by the conventional and proposed debiasing methods for different biases and numbers of random latches, respectively. In Fig. 9, both methods decreased the resulting bit length as bias increased. In Fig. 10, our proposed method obtained the longest bit length when the number of random latches was 300–400, because the entropy of the ternary response is largest when the number of random latches is 1/3 of all latches (i.e., 1,024). Note here that the resulting bit length of the conventional method was unrelated to the number of random latches under the experimental condition. These results show that the bit length generated by the proposed method is larger than that of the conventional method for both evaluations.

4.2 Experiment with FPGA Implementation

The proposed debiasing was then evaluated with actual responses obtained from L-PUFs implemented on Xilinx Spartan6 FPGAs (XC6SLX150). We implemented 30 L-PUFs, and each of them consisted of 1,024 RS latches. The constant and random latches were determined using 256 challenges. If the output bits were always 0 (or 1) for all challenges, the latches were considered to be constant. The remaining latches were considered to be random.

Figure 11 shows the relation between the bias of the original response (T or X) and the resulting response Y obtained from (a) the conventional and (b) the proposed method, where the vertical axis indicates the bias of the resulting response and the horizontal axis indicates the bias of the original (input) response. Note here that obtained 30-bit positions are plotted in the figure. In this experiment, both the conventional and proposed methods can reduce the response biases significantly, with no significant difference between the two, because the percentage of random latches is at most 10% of all the latches in this experiment. Figure 12 shows the bit lengths of the output responses by the conventional and proposed methods.

The results also show that both methods decrease the resulting bit lengths for large biases, as in the simulation; however, the proposed method can generate

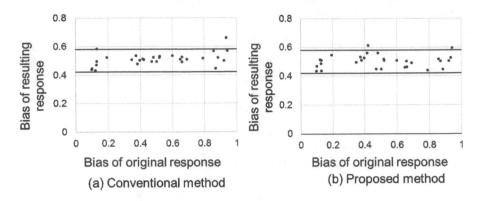

Fig. 11. Biases of resulting responses.

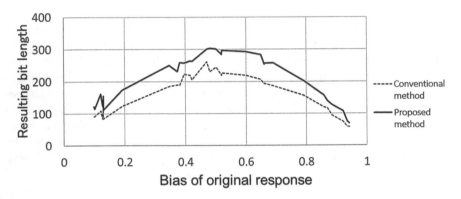

Fig. 12. Resulting bit length after debiasing for different original biases given from L-PUFs.

a larger resulting bit length than the conventional method. Note here that the number of random latches in the L-PUF is dependent on the technology and physical layout. [12] states that the appropriate percentage of random latches in an L-PUF is 30%. In such cases, the proposed method would be at a more significant advantage.

5   FEs Based on the Multiple-Valued PUF

In this section, we present a new FE structure for multiple-valued responses [12] and the proposed debiasing. We first describe the modeling for ternary PUF responses to determine the design parameters, and then construct the proposed FE and evaluate its performance in terms of resulting response lengths and authentication failure rates.

5.1 Modeling for Ternary PUF Responses

To design an efficient FE using the proposed method, we first investigated the incorrect detection and miss rates of random latches. In this paper, for detecting random latches, we use the detection method as above and in [12], which performs challenges m times to identify the value of a PUF response. If the output bits are always 0 (or 1) for all challenges, the latches were considered constant; otherwise, they were considered to be random. If m is sufficiently large, we can say that the miss rate of a random latch is negligibly small in enrollment. Hence, we consider incorrect detections and misses of a random latch in reconstruction. We calculate an incorrect detection and miss probability in reconstruction. We assume here that a probability of outputting 1 (or 0) every challenge for the i-th cell ($0 \le i \le n - 1$) is equal to P_i. The probability that the ith cell is constant is $P_i^m + (1 - P_i^m)$ when challenges are performed m times. Here, if m is large enough, it is assumed that the smaller of P_i^m and $(1 - P_i^m)$ is negligibly small (when $P_i = 0.5$, it is 0.5^m). For simplicity, let P_i' be $|P_i - 0.5| + 0.5$, and the probabilities of a constant and random latch are P_i^m and $1 - P_i^m$, respectively. Hence, if the number of challenges in enrollment is equal to that in reconstruction, an incorrect detection or miss by a random latch occurs on the ith cell with probability $P_i^m (1 - P_i^m)$ in reconstruction. As a result, in the entire L-PUF, the detection failure rate is $1 - \{1 - (P_i')^m \{1 - (P_i')^m\}\}^n$. Note here that the detection failure rate is calculated theoretically and therefore does not consider bit errors in the PUF response due to noise. With the above calculation, we investigate the histogram of random latches with a sufficient amount of challenges, and then determine the most efficient m and ECC in terms of the detection failure rate of random latches.

In this paper, we implemented an L-PUF with 1,024 RS latches on 3 FPGA (Xilinx Spartan6 XC6SLX150) boards as an example, and calculated the histogram and detection failure rate of a random latch. Figure 13 shows the number of detected random latches for different numbers of challenges (i.e., m).

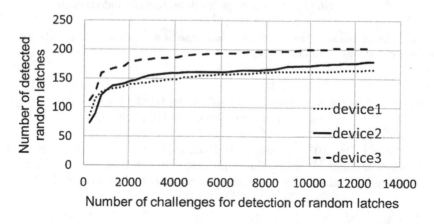

Fig. 13. Number of detected random latches for different numbers of challenges.

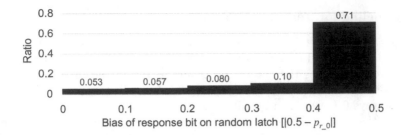

Fig. 14. Occurrence probability of a random latch.

The curve first increases and converges at around 10,000. As an example, we determine P'_i generated by 10,000 challenges in the experiment. Figure 14 show a histogram of P'_i on random latches. In this experiment, 535 random latches are obtained from $1,024 \times 3$ latches. It is clear from Fig. 14 that the number of random latches whose P'_i is close to 0.5 is much smaller than that whose P'_i is close to 0 or 1. This means that the above assumption that if m is large enough, the smaller of P^m_i and $(1 - P^m_i)$ is negligibly small for a simple calculationis met in most cells.

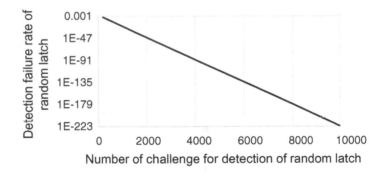

Fig. 15. Relation between the detection failure rate of a random latch and the number of challenges to detect random latches.

Figure 15 shows a detection failure rate calculated by the above histogram for the L-PUFs. We confirm that the detection failure rate of random latches exponentially decreases when the number of challenges increases. The detection failure rate is 1.21×10^{-223} when m is 10,000. Note that the detection failure rate does not account for noise-related bit errors in the PUF response. Thus, an ECC in the proposed FE can be selected to cope with the calculated bit errors.

5.2   FE Using Proposed Method and Its Evaluation

Figure 16 shows the block diagram of the proposed FE consisting of the (a) enrollment and (b) reconstruction units. In enrollment, the proposed FE generates secret information K, helper data W, and debiasing data D from random seed S and PUF response X. In reconstruction, the FE regenerates the secret information K from a noisy PUF response X' and the debiasing data D.

We evaluated the proposed FE with the simulated responses described above and compared it with those of the conventional FE [8]. The number of challenges for detecting the three types of RS latches (i.e., m) was 100, and the detection failure rate of random latches was 0.004. For comparison, we used connected code for the ECC in the FE, a (24,12)-Golay code for the outer code, and a (8,1)-repetition code for the inner code, as in [8].

Table 3 shows the authentication results for various biases and random latch rates, where the bias ranges from 0 to 0.3 and the random latch rate ranges from 0 to 0.4. Here, the length of an original PUF response is $|X|$, the resulting bit length after debiasing is $|Y|$, and the efficiency is $|Y|/|X|$. The authentication failure rate is given as P_{fail}. We confirmed from Table 3 that the authentication success rate of the proposed method was 100%, even when the number of random latches increased, while that of the conventional method was less than 20% at the lowest. The major reasons for this success are the higher stability and improved error correction capability with the multiple-valued approach

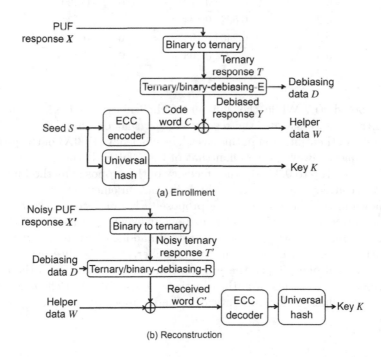

Fig. 16. FE block diagram with multiple-valued debiasing.

Table 3. Comparison of debiasing results by conventional and proposed FEs

Bias	Random latch	Conventional method		Proposed method	
		P_{fall}	Efficiency	P_{fall}	Efficiency
0	0	0	0.252	0	0.245
	0.1	0	0.250	0	0.300
	0.2	0	0.246	0	0.320
	0.3	0.003	0.250	0	0.331
	0.4	0.316	0.248	0	0.330
0.1	0	0	0.240	0	0.240
	0.1	0	0.236	0	0.286
	0.2	0	0.237	0	0.312
	0.3	0.013	0.243	0	0.328
	0.4	0.434	0.240	0	0.327
0.2	0	0	0.204	0	0.209
	0.1	0	0.212	0	0.259
	0.2	0	0.216	0	0.299
	0.3	0.010	0.222	0	0.310
	0.4	0.327	0.229	0	0.314
0.3	0	0	0.157	0	0.162
	0.1	0	0.172	0	0.220
	0.2	0.002	0.184	0	0.264
	0.3	0.240	0.195	0	0.287
	0.4	0.829	0.203	0	0.300

in the proposed FE. While the number of challenges (m) has an impact on authentification success rate, the results suggest that one hundred is sufficient for successful authentification in this case. The efficiency $|Y|/|X|$ of the proposed method is almost always larger than that of the CDRE. If the ratio of random latches increases (up to 0.3–0.4), the efficiency of the proposed method increases, whereas the conventional CDRE has little impact on efficiency.

As shown in the above results, the proposed FE can decrease error bits that should be corrected by ECC, and hence avoid using a costly strong ECC to achieve a certain authentication success rate. In addition, for extracting a given entropy, the proposed FE can reduce the number of RS latches in comparison with conventional ones. Thus, the proposed method is superior to the conventional method [8] in terms of authentication success rate and efficiency, which leads to cost reduction for authentication and tamper-resistant cryptographic modules.

6   Conclusion

We proposed a new debiasing method for a stable and efficient extraction of uniform binary responses from PUFs. We also showed the stability and effectiveness of the proposed method with two simulation experiments and actual responses obtained from latch-PUFs implemented on an FPGA. We also applied the proposed method to the construction of an FE, and showed the advantages of the proposed FE in terms of response length and authentication success rates through experimental evaluation. Future work remains for designing a proper ECC utilizing the capability of the proposed method. Additionally, we will evaluate the proposed method for other PUFs in the future.

Acknowledgment. This work has been supported by JSPS KAKENHI Grants No. 16K12436 and No. 16J05711.

References

1. Dodis, Y., Reyzin, M., Smith, A.: Fuzzy extractors: how to generate strong keys from biometrics and other noisy data. SIAM J. Comput. **38**(1), 97–139 (2008)
2. Guajardo, J., Kumar, S.S., Schrijen, G.-J., Tuyls, P.: FPGA intrinsic PUFs and their use for IP protection. In: Paillier, P., Verbauwhede, I. (eds.) CHES 2007. LNCS, vol. 4727, pp. 63–80. Springer, Heidelberg (2007). doi:10.1007/978-3-540-74735-2_5
3. Koeberl, P., Li, J., Rajan, A., Wu, W.: Entropy loss in puf-based key generation schemes : the repetition code pitfall, pp. 44–49 (2014)
4. Lily, C.: Recommendation for key derivation using pseudorandom functions(revised) (2009)
5. Lily, C.: Recommendation for key derivation through extraction-then-expansion (2011)
6. Lim, D., Lee, J., Gassend, B., Suh, G., van Dijk, M., Devadas, S.: Extracting secret keys from integrated circuits. IEEE Trans. Very Large Scale Integer VLSI Syst. **13**(10), 1200–1205 (2005)
7. Maes, R.: Physically Unclonable Functions. Springer, Heidelberg (2013)
8. Maes, R., Leest, V., Sluis, E., Willems, F.: Secure key generation from biased PUFs. In: Güneysu, T., Handschuh, H. (eds.) CHES 2015. LNCS, vol. 9293, pp. 517–534. Springer, Heidelberg (2015). doi:10.1007/978-3-662-48324-4_26
9. von Neumann, J.: Various techniques used in connection with random digits. Appl. Math. Ser. **12**, 36–38 (1951). National Bureau of Standards, USA
10. Su, Y., Holleman, J., Otis, B.: A 1.6pj/bit 96% stable chip-id generating circuit using process variations. In: IEEE International Solid-State Circuits Conference (ISSCC2007), pp. 406–611 (2007)
11. Su, Y., Holleman, J., Otis, B.: A digital 1.6pj/bit chip identification circuit using process variations. IEEE J. Solid-State Circ. **43**(1), 69–77 (2008)
12. Yamamoto, D., Sakiyama, K., Iwamoto, M., Ohta, K., Takenaka, M., Itoh, K.: Variety enhancement of PUF responce using the locations of random outputting RS latches. Crypt. Eng. **3**(4), 197–211 (2013)

Getting the Most Out of Leakage Detection
Statistical Tools and Measurement Setups Hand in Hand

Santos Merino del Pozo$^{(\boxtimes)}$ and François-Xavier Standaert

ICTEAM/ELEN/Crypto Group, Université catholique de Louvain,
Louvain-la-Neuve, Belgium
santos.merino@uclouvain.be

Abstract. In this work, we provide a concrete investigation of the gains that can be obtained by combining good measurement setups and efficient leakage detection tests to speed up evaluation times. For this purpose, we first analyze the quality of various measurement setups. Then, we highlight the positive impact of a recent proposal for efficient leakage detection, based on the analysis of a (few) pair(s) of plaintexts. Finally, we show that the combination of our best setups and detection tools allows detecting leakages for a noisy threshold implementation of the block cipher PRESENT after an intensive measurement phase, while either worse setups or less efficient detection tests would not succeed in detecting these leakages. Overall, our results show that a combination of good setups and fast leakage detection can turn security evaluation times from days to hours (for first-order secure implementations) and even from weeks to days (for higher-order secure implementations).

1 Introduction

State-of-the-Art. The concrete evaluation of cryptographic hardware and software against *side-channel analysis* (SCA) attacks is a complex and expensive process. This is especially true in the case of implementations protected with (combinations of) countermeasures, for which cryptographic engineers aim to ensure that the leakages are noisy and carry little sensitive information. In this context, minimizing the evaluation time (which we assume proportional to the evaluation cost) generally benefits from a combination of three ingredients:

1. Obtaining good measurements, with high signal and minimum noise.
2. Simplifying the evaluation goals, e.g. from key recovery to simpler detections.
3. Optimizing the distinguishers, in particular their data and time complexity.

Despite its practical relevance, the first problem is rarely the primary focus in the literature. Yet, several recent papers have highlighted the significant impact of good setups for the evaluation of masked implementations [12], especially when it comes to devices running at higher frequencies [2] or for the investigation of static leakages [11]. The second problem typically illustrates the tradeoff between the evaluation time and the accuracy of the conclusions in SCA. That is, the ultimate goal of an evaluator is to obtain accurate evaluations of the worst-case security

© Springer International Publishing AG 2017
S. Guilley (Ed.): COSADE 2017, LNCS 10348, pp. 264–281, 2017.
DOI: 10.1007/978-3-319-64647-3_16

level of an implementation. Yet, in view of the higher cost of such worst-case analyzes, an increasingly popular approach in SCA evaluations is to start with simpler leakage detection tests, i.e., the *test vector leakage assessment* (TVLA) methodology introduced by Goodwill et al. [9]. In this case, the goal is not to estimate a key recovery success rate, but to detect whether the leakages depend on the data manipulated by the device: an arguably easier task. Following, such leakage detections can become the sole goal of the evaluation (which is then limited to qualitative conclusions: see [18] for a recent discussion), or serve as a preliminary step for more advanced (quantitative) investigations. Eventually, the last problem has been the daily bread of researchers in SCA for the last fifteen years, with various proposals of distinguishers optimized for efficiency or genericity, in profiled or non-profiled attack settings, e.g. [5,6].

Our Contributions. In this paper, we investigate the extent to which a combination of good measurement setups and state-of-the-art leakage detection tools can speed up the evaluation time for cryptographic implementations.

For this purpose, we start by comparing various measurement setups with TVLA, taking advantage of the optimizations by Schneider et al. [17]. Our experiments allow us to exhibit the significant impact of bad measurements (i.e. with limited signal and a lot of noise), which can lead to incorrectly conclude about the (in)security of a threshold implementation (TI) of the PRESENT block cipher.

Next, we study the gains that can be obtained by exploiting a more recent proposal of TVLA, based on a partition of the measurements in two classes corresponding to two fixed plaintexts, next denoted as a *fixed vs. fixed* TVLA [8], rather than a partition of the measurements in two classes where one class corresponds to a fixed plaintext and the other class to random plaintexts, as originally proposed in [9] and next denoted as the *fixed vs. random* TVLA. We show that the fixed vs. fixed test allows a consistent reduction of the data complexity needed for successful detections, with different factors depending on the signal and noise of the target implementation.

Eventually, we pushed the investigation of our first-order TI in a very high noise regime, typically corresponding to signal-to-noise ratios below 0.01, where TIs are supposed to lead to high security levels. This experiment allows us to exhibit a concrete case where 100 million measurements (corresponding to one day of sampling) were not sufficient to spot any leakage with the fixed vs. random test, whereas the fixed vs. fixed test leads to clear detections. It also leads to an interesting change of the most informative statistical moment in the leakage traces, from the third-order moment in low noise contexts to the second-order one in high noise contexts, as predicted by theory [7].

Overall, these examples highlight that as the security level of an implementation increases, the impact of the gains due to a good measurement setup and selection of optimized statistical tools is magnified. Put very simply, reducing the evaluation time from 100 s to 1 s is a gain by a much larger factor than reducing the evaluation time from 5 days to 1 day. Yet the second improvement is much more relevant for evaluation laboratories for which time and cost are absolute (rather than relative) metrics. The sound combination of state-of-the-art tools in this work therefore contribute to this important goal of minimizing evaluation time (and cost).

2   Preliminaries

In order to make the paper self-contained, in this section we introduce notations and provide background information about the statistical tools and measurement setups considered throughout the paper.

2.1   Notations

We use capital letters for random variables and lower-case letters for their realization. We denote vectors and matrices with bold notations and sets with calligraphic ones.

We refer to the targeted cryptographic device storing a private key as the *device under test* (DUT). In order to assess the vulnerability of the DUT to (higher-order) SCA attacks, we consider an evaluator with full knowledge of the implementation and the ability to measure the current across the DUT. We denote by $\mathcal{T}$ the set comprising m so-called SCA traces $t_{i \in \{1,\ldots,m\}}$, each of them made of n time samples $t_i^{j \in \{1,\ldots,n\}}$, collected while the DUT is fed with the associated plaintexts $x_{i \in \{1,\ldots,m\}}$.

2.2   Leakage Assessment Methodology

The two versions of TVLA proposed in [9] (i.e., *specific* and *non-specific*) aim to detect the existence of (possibly) exploitable leakages at a certain statistical moment on the DUT. Following the guidelines in [17], since our DUT is equipped with a masking countermeasure (i.e., a first-order TI), we start our practical investigations by considering the non-specific fixed vs. random TVLA. Though this tool does not bring information about the hardness of mounting successful attacks against the DUT, it comes in handy when examining the existence of leakages at higher-order moments (e.g., during prototyping) where, in comparison with higher-order attacks, the number of required measurements can be significantly reduced. To this end, the evaluator records two sets of measurements $\mathcal{T}_0$ and $\mathcal{T}_1$, respectively associated to fixed and randomly generated inputs (i.e., fixed vs. random), while keeping the device key constant. The test does not require any prior knowledge of the DUT or assumption about how it leaks (i.e., non-specific). More recently, with the aim of speeding up the detection of leakages, the so-called non-specific fixed vs. fixed TVLA has been proposed by Durvaux et al. [8] as a tweak of the aforementioned fixed vs. random version. To that end, $\mathcal{T}_0$ and $\mathcal{T}_1$ are now associated to two fixed plaintexts (i.e., fixed vs. fixed), an approach that has been shown to improve the signal, remove the algorithmic noise intrinsic to the set of SCA traces associated with random plaintexts, and thereby reduce the measurement complexity of security evaluations. We should note that in order to guarantee a non-deterministic internal state of the DUT at the beginning of a new measurement, and therefore to avoid false-positives, in both methodologies SCA traces must be collected in a randomly-interleaved fashion. Then the two sets of traces are compared by computing the Welch's (two-tailed) t-test in a univariate fashion (i.e., individually at each time sample $j \in \{1, \ldots, n\}$):

$$ t = \frac{\mu(\mathcal{T}_0) - \mu(\mathcal{T}_1)}{\sqrt{\frac{\sigma^2(\mathcal{T}_0)}{|\mathcal{T}_0|} + \frac{\sigma^2(\mathcal{T}_1)}{|\mathcal{T}_1|}}}, $$

where $|.|$ is the sample size. The result determines if the samples have been drawn from the same population, i.e., the *null hypothesis*. A typical significance threshold to reject the null hypothesis in SCA evaluations, and therefore to pinpoint the existence of first-order leakages in the DUT, is $|t| \geq 4.5$. To enable the detection of leakages at higher orders, SCA traces need to be preprocessed accordingly (we omit the details but refer an interested reader to [17]).

2.3 Measurement Setups

In the following sections, the platform employed to conduct our practical experiments is a SAKURA-G [1] featuring two Spartan-6 FPGAs (*target* and *control*) built in a 45 nm technology and which has been especially designed for research on hardware security, e.g., SCA attacks. The board provides three built-in attack points (i.e., the two heads of a resistor and the output of an embedded amplifier) to measure the voltage drop over the $1\,\Omega$ shunt resistor placed in the Vdd path of the target FPGA that, by means of the corresponding voltage regulator, was supplied at 1.2 V. Since running the DUT at high frequencies, e.g., 24 MHz, is known to harden the detection (and exploitation) of SCA leakages (i.e., due to the intrinsic windowing effect), we clocked the target FPGA at 3 MHz.

In all the experiments, SCA traces were collected by means of a Teledyne Lecroy HRO66Zi WaveRunner 12-bit digital oscilloscope (DSO) at a sampling rate of 500 MS/s and a bandwidth limit of 20 MHz to reduce the environmental noise. Besides, a passive probe (i.e., a SMA-to-BNC coaxial cable) that avoids the additional noise induced by, e.g., active components in differential probes, was used in all the experiments as well. Since practical investigations in this work involved the analysis of millions of SCA leakages, our acquisition framework was designed following the guidelines in [17], allowing us to perform millions of measurements per hour by exploiting the sequence mode of the employed DSO. It should be noted that the UART communication channel between the PC and the control FPGA also contributes to increase the noise level in the SCA traces. Therefore, we made sure that before triggering the DSO, the UART channel was closed and remained in such state until the completion of the measurement.

Typically, when measuring the dynamic power consumption, it is advantageous to remove the DC shift that, from the evaluator's perspective, can be seen as an additional source of noise. To this end, the most straightforward way is to perform current measurements using the AC coupling mode of the DSO (from now on termed *setup 1*). Further, quantization noise due to a low peak-to-peak signal amplitude can exacerbate the measurement complexity of TVLA. To cope

Table 1. Summary of the different setups considered in this work.

	Coupling	Amplifier	Low Pass	High Pass	Pass Band
setup 1	AC 1 MΩ	✗	✗	✗	N/A
setup 2	DC 50 Ω	✓	✗	✗	N/A
setup 3	DC 50 Ω	✓	✓	✓	0.1 to 5MHz
setup 4	DC 50 Ω	✓	✓	✓	1.2 to 5MHz

with it, amplifiers, such as the ADI AD8000 embedded on the SAKURA-G board (in the following called *setup 2*), can be employed to increase the amplitude of the signal. In fact, security evaluators may use more elaborate setups featuring a combination of AC amplifiers, DC blockers and/or hardware filters to maximize the signal (e.g., when targeting ultra low-power designs) and reduce the noise (e.g., for masking schemes whose security level relies on having sufficiently noisy leakages). To assess their benefits (but also the repercussions of capacitive elements), we considered two more setups. In the so-called *setup 3*, we employed an AC amplifier (i.e., ZFL-1000LN+ from Mini-Circuits), a DC to 5 MHz filter (i.e., SLP-5+ from Mini-Circuits) and a DC blocker (i.e., BLK-89-S+ from Mini-Circuits), whereas in *setup 4* the DC blocker was replaced by a 1.2 to 800 MHz filter (i.e., ZFHP-1R2+ from Mini-Circuits) that together with the SLP-5+ formed a band pass filter around 3 MHz (the clock frequency of the DUT). A high-level overview of all the aforementioned setups is given in Table 1.

3   Case Studies

After describing the tools employed in our analysis, in this section we provide an overview of the two (hardware-oriented) countermeasures deployed in our designs. More precisely, we start by describing a fully-serialized architecture of a first-order TI of PRESENT. Next we detail the Gaussian noise engine that we implemented on our target FPGA.

3.1   Threshold Implementations

In this work we considered the first-order TI technique introduced in [15], which was designed to prevent any first-order leakage even in a glitchy hardware implementation. Following the principles of multi-party computation and secret sharing, the sensitive variables and functions are implemented using at least $s \geq d+1$ shares, where d is the algebraic degree of the targeted function. As in any other (e.g., Boolean) masking scheme, the intermediate value x is represented using a vector of s shares $\boldsymbol{x} = (x_1, \ldots, x_s)$ such that $x = \bigoplus_{i=1}^{s} x_i$. While a linear function can be easily applied on each share with s instances in parallel, the implementation of non-linear functions is not a trivial task. TI implements a target non-linear function f as a vector of component functions $\boldsymbol{f} = (f_1, \ldots, f_s)$ such that every component function f_i, where $i \in \{1, \ldots, s\}$, is independent of at least one input share. Such a property is referred as the *non-completeness* property and it is the main contribution of [15]. When the sharing is correct, so-called *correctness* property, and the input shares are uniformly distributed, so-called *uniformity* property (and which indeed are standard properties in masking schemes), then the non-completeness property provides probable security against first-order SCA (even in the presence of glitches). It is noteworthy that, if the outputs of the component functions are used as inputs in next parts of the implementation, which is usually the case in symmetric-key algorithms, then the uniformity of the shared functions and their outputs must be carefully examined. Whenever this property is not satisfied, different techniques to repair the problem have already been proposed in the literature (see e.g., [13,16]).

First-Order TI of PRESENT-80. For our practical evaluations, we implemented a serialized architecture of PRESENT-80, and more concretely, *profile 2* as given in [16], where the 80-bit key is not represented in a shared form. First, the plaintext (resp. the secret key) is loaded in parallel mode into the corresponding state (resp. key) register which is made of 16 (resp. 20) 4-bit wide registers, and that also behaves as a shift register. During the 16 first clock cycles within a encryption round, the state and key registers provide the shared Sbox with the corresponding 4-bit chunks (so-called nibbles). Eventually, when the 16 Sbox computations are done, the PLayer and Key Schedule are performed during the last clock cycle. Following the minimum settings of a first-order TI, the datapath is represented using a 3-share Boolean masking, so all state (i.e., shift) registers and PLayer instances have to be tripled.

For the TI representation of the PRESENT Sbox $S(x)$, we exploited the decomposition proposed by Moradi et al. [14]. In their work, the Sbox is decomposed into two quadratic bijections $\mathcal{Q}_{294} \times \mathcal{Q}_{299}$ such that $S(x) = \mathcal{A}_3 \circ \mathcal{Q}_{299} \circ \mathcal{A}_2 \circ \mathcal{Q}_{294} \circ \mathcal{A}_1$ where $\mathcal{A}_1$, $\mathcal{A}_2$ and $\mathcal{A}_3$ are affine functions. As it can be seen in Fig. 1, where we provide a graphical representation of the resulting shared Sbox, three intermediate registers (i.e., one per share) must be placed between the component functions of the two quadratic bijections. By doing so, it is possible to disallow the propagation of glitches, which is required to ensure that the

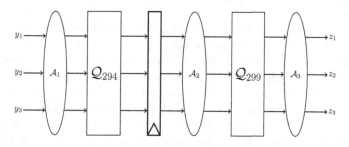

Fig. 1. Uniform first-order TI of the PRESENT Sbox in [14]

non-completeness property is satisfied. As a result, each Sbox lookup now takes two clock cycles. For more details we refer the interested reader to [14].

We used Xilinx ISE version 14.7 for design synthesis, implementation and configuration of the board. Following the recommendations in [3] to satisfy the non-completeness property, we used the KEEP HIERARCHY constraint when generating the bitstream of the crypto module. Note that this is needed to make sure that the assumption of component functions leaking independently is not violated (which might lead to undesired first-order leakages).

3.2 Gaussian Noise Engine

In [10], the authors investigated different FPGA-dedicated techniques to achieve maximum levels of noise in SCA traces (i.e., to hide data-dependent leakages) by configuring unused available logic. Xilinx FPGAs contain n-to-m Look-Up Tables (LUTs) that, besides being used as a Boolean function generators, can also be configured as 16-bit shift registers. This so-called Shift Register LUT (SRL) mode is exploited by the authors of [10] to create r cycling registers made of s LUTs in SRL mode that are initialized with the pattern 01...0101. The enable signal of each cycling register is driven by a PRNG, in our case, a LFSR with enough period to record up to 100 million traces. Only when both the clock and enable signals are high, the power consumption will increase due to the additional bit flips in the registers. The r and s parameters are used by the designer to set respectively the variance and amplitude of the noise. Concretely, we implemented a noise engine made up of $r = 16$ cycling registers, each of them consisting of $s = 100$ LUTs in SRL mode. Since this module did not have an output, we prevented the synthesizer from removing this unconnected component by using SAVE NET FLAG and KEEP constraints. Further, in order to not introduce algorithmic noise in the measurements, the PRNG (needed for mask generation and TVLA) was implemented on the control FPGA as the realization of AES-128 in CTR mode.

4 Comparing Setups with CRI's Fixed vs. Random TVLA

In this section we evaluate the ability of the aforementioned measurement setups (see Sect. 2.3) to ease the detection of (higher-order) SCA leakages.

In the last years, the non-specific fixed vs. random TVLA has emerged as a very popular technique for the SCA evaluation of cryptographic devices. For this reason, and because recent academic works had used it to assess the security level of higher-order TIs [4], we based our preliminary investigations on this technique.

As explained in the previous section, our DUT features a fully-serialized architecture with small combinatorial circuits, and so with negligible algorithmic noise. Therefore a relatively small number of measurements was expected to suffice for our purposes, so we performed the analysis up to third-orders using a set of 1 million SCA traces collected with fixed and random plaintexts in

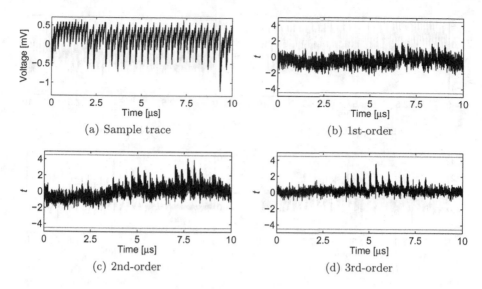

Fig. 2. Setup 1, fixed vs. random TVLA using 1 million traces

arbitrarily interleaved order. Note that, because we controlled the PRNG, we systematically repeated this process for each setup.

A sample power trace (comprising the first encryption round of the targeted design) recorded with *setup 1* is shown by Fig. 2(a). In this setting, even limiting the bandwidth of the DSO to 20 MHz and using its maximum vertical accuracy (i.e., 1 mV/div), the traces were very noisy due to the small amplitude of the signal. First, in order to verify the correct behavior of the measurement framework and the DUT, we turned the PRNGs (in charge of mask generation) off. As illustrated by Fig. 15(a) in Appendix A, TVLA reported clear first-order leakages using 10 000 traces. As expected, when masks were enabled there was no easy to detect first-order leakage using up to 1 million traces, see Fig. 2(b). Further, extending the analysis to second- (security in combinational logic) and third-orders (security in the memory elements) showed that such a number of measurements was not high enough to detect second- (Fig. 2(c)) and third-order leakages (Fig. 2(d)). By using traces collected from the amplified output provided by the SAKURA-G board (i.e., *setup 2*), and so with less quantization noise due to the higher peak-to-peak signal (see Fig. 3(a)), made the detection of third-order leakages feasible (Fig. 3(d)). Yet, second-order ones still remained undetectable (Fig. 3(c)). Despite leading to a greater signal amplitude, similar results were obtained by considering *setup 3* (see Fig. 4). The strong windowing effect induced by such setup is obvious. This is due to the DC blocker and the AC amplifier that, according to the authors of [12], makes consecutive power peaks overlap. Hence, such a behavior was expected for *setup 4* as well. As we can see in Fig. 5(a), the inclusion of the high pass filter changed the polarity of the signal. Interestingly, in this case the existence of third-order leakages was pin-

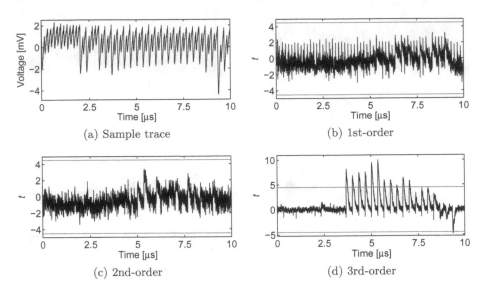

(a) Sample trace

(b) 1st-order

(c) 2nd-order

(d) 3rd-order

Fig. 3. Setup 2, fixed vs. random test TVLA using 1 million traces

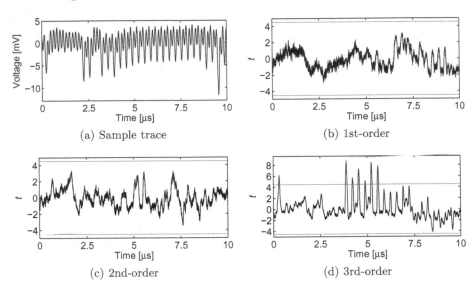

(a) Sample trace

(b) 1st-order

(c) 2nd-order

(d) 3rd-order

Fig. 4. Setup 3, fixed vs. random TVLA using 1 million traces

pointed with a greater level of confidence than before (see Fig. 5(d)). Moreover, and despite being featuring and AC amplifier, the aforementioned windowing effect became negligible, as shown by Fig. 15(d) in Appendix A.

Figure 6 summarizes the results of these experiments. Figure 6(a) shows that, due to the register-oriented architecture of the DUT, the non-specific fixed vs. random TVLA cannot spot second-order leakages with up to 1 million SCA traces

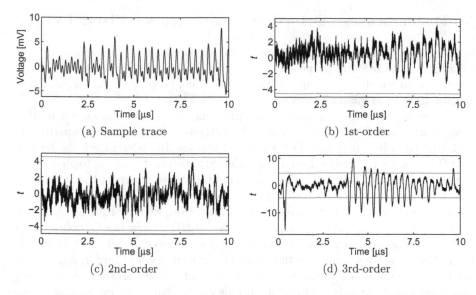

Fig. 5. Setup 4, fixed vs. random TVLA using 1 million traces

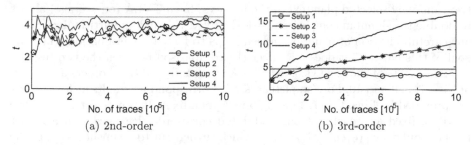

Fig. 6. Fixed vs. random TVLA in function of the number of traces (absolute values)

with any of the setups. On the other hand, Fig. 6(b) highlights the importance of having signals with enough peak-to-peak amplitude, as shown by the results of *setup 1* where the quantization noise led to unsuccessful results. To conclude with this section, we should note that in comparison with *setup 2* and *3*, *setup 4* reduced by a factor of ≈3 the measurement complexity for detecting third-order leakages.

As a side-note, we finally mention that preprocessing the traces with digital filters can be used to get rid of (a part of) the noise in the measurements. By contrast, it cannot be used to improve the signal as enabled by the analog amplifiers in our setups. Furthermore, in our experiments such a preprocessing turned out to be only marginally useful for the best setups, confirming that good measurements significantly simplify the statistical evaluation of leaking devices.

5   Faster Leakage Detection with Fixed vs. Fixed TVLA

In the last section, we have shown the impact of choosing fine-tuned setup. Now, we focus on the recently proposed non-specific fixed vs. fixed TVLA to evaluate the extent to which it can improve the speed of convergence of the original non specific fixed vs. random TVLA.

Following the same approach as in previous section, we collected 1 million measurements with each setup. The only difference with previous experiments is the partitioning, now based on two fixed classes. In order to get the best of both TVLAs, we carefully selected the plaintexts such that they led to an optimal signal. For the fixed vs. fixed TVLA, this amounts to select inputs that maximize the Hamming distance (HD) between each two consecutive values in the register (that we both control). For the fixed vs. random case, we only control one of the values, and therefore can only guarantee a halved HD on average. Of course, such an approach assumes a strong adversary with full knowledge and control over the DUT. However, we should note that this is a natural assumption for, e.g., secure hardware designers assessing the security level of a cryptographic device they have engineered. At this point we should also note that, even in the case where the two fixed plaintexts cannot be carefully selected (e.g., if the evaluator uses two randomly selected plaintexts), the fixed vs. fixed TVLA (on average) allows to double the signal in comparison with the fixed vs. random partitioning [8]. So our careful selection of plaintext indeed helps to fasten the evaluations (i.e., the goal of this paper) but does not affect the comparison between the two tests.

The results of the fixed vs. fixed TVLA at second- and third-orders are shown in Figs. 7, 8, 9 and 10. For completeness, the corresponding results at first-orders are provided by Figs. 16, 17, 18 and 19 in Appendix A. It is noteworthy that the fixed vs. fixed approach not only exhibited third-order leakages with a higher level of confidence (even with *setup 1*, which turned out to be unsuccessful in the previous section), but it also enabled the detection of second-order leakages by means of *setup 2*, *setup 3* and *setup 4*. Hence, these results further confirm the negative impact of having signals with limited peak-to-peak amplitude, which can be mitigated thanks to the fixed vs. fixed TVLA.

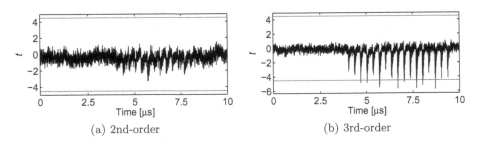

(a) 2nd-order   (b) 3rd-order

Fig. 7. Setup 1, fixed vs. fixed TVLA using 1 million traces

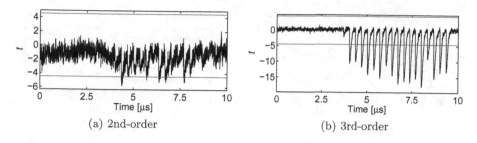

(a) 2nd-order (b) 3rd-order

Fig. 8. Setup 2, fixed vs. fixed TVLA using 1 million traces

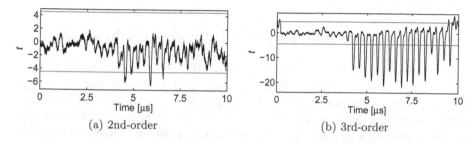

(a) 2nd-order (b) 3rd-order

Fig. 9. Setup 3, fixed vs. fixed TVLA using 1 million traces

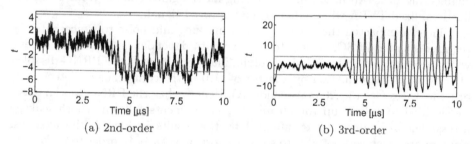

(a) 2nd-order (b) 3rd-order

Fig. 10. Setup 4, fixed vs. fixed TVLA using 1 million traces

Finally, in order to compare the speed of convergence of both tests, we summarize the aforementioned results in Fig. 11 as a function of the number of traces. As clear from Fig. 11(a), the improvements at second-orders are significant, and they also serve to reaffirm *setup 4* as the best candidate among the different setups considered in this work. Besides, we should also highlight the results obtained for the third-order moments in Fig. 11(b), where we can notice a reduction by a factor ≈ 4 on the number of required leakages for detections with *setup 2* and *setup 3*.

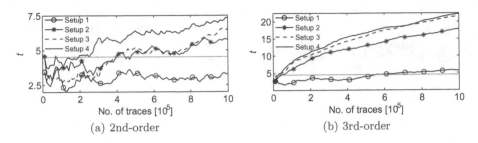

Fig. 11. Fixed vs. fixed TVLA in function of the number of traces (absolute values)

6   Noisy Implementations and Intensive Evaluations

The last empirical results have shown the significant gains that we can obtain by relying on the non-specific fixed vs. fixed TVLA. Yet, it remains largely unclear whether such gains are due to increasing the signal or reducing the noise. Motivated by this question, in this section we move towards a more challenging scenario featuring a combination of masking and noise. In this new context the noise is synchronous with the crypto core, and so filtering it out looks a hard engineering task. All in all, in this setting all the gains coming from using the fixed vs. fixed TVLA will be due to the improved signal.

Since we expected the combination of masking and noise to exacerbate the required number of SCA traces to observe data dependencies at higher orders, we only considered *setup 4*. Indeed, preliminary results with the PRNG disabled already showed a reduction by a factor of ≈ 10 in the confidence to detect first-order leakages (see Fig. 20 in Appendix A). Hence, when the PRNG was enabled, we decided to collect up to 100 million measurements following both leakage assessment techniques. As exemplified by the results in Fig. 12, the fixed vs. random approach was not able to spot any sort of leakage. In order to make sure that our results were not biased by a poor choice of the fixed class, we tested different plaintexts that also led to analogous results. By contrast, Fig. 13 shows that relying on the fixed vs. fixed TVLA, such amount of measurements suffices to detect second-order leakages. Indeed, results in Fig. 14 reveal that ≈ 60 million SCA traces can spot second-order leakages.

These results are well in line with theoretical expectations, which state that lower-order moments become more informative when the level of noise is sufficiently high [7]. They also confirm the effectiveness of generating additive noise on top of a masked implementation to harden the detection of higher-order leakages. More importantly, they show that even when the noise cannot be canceled out, by increasing the signal (i.e., with the fixed vs. fixed TVLA) the number of traces required to spot a higher-order leakage can be significantly reduced. In such a context, where millions of traces must be acquired, even small gain factors can save a lot of time and storage requirements to security evaluators.

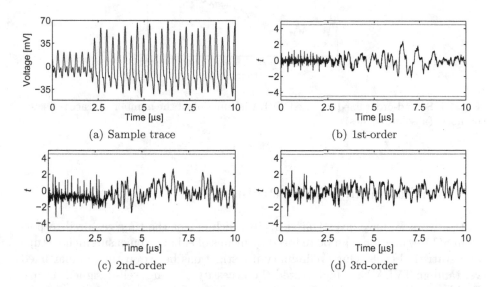

Fig. 12. Setup 4, fixed vs. random TVLA using 100 million traces effected by noise

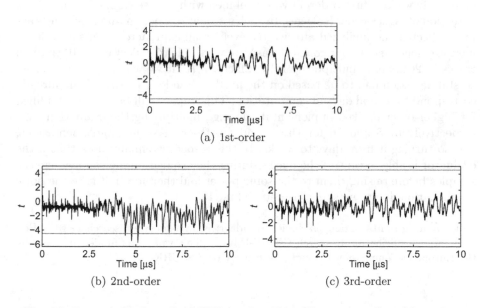

Fig. 13. Setup 4, fixed vs. fixed TVLA using 100 million traces effected by noise

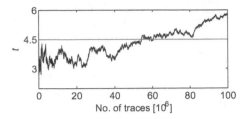

Fig. 14. Second-order fixed vs. fixed TVLA in function of the number of traces effected by noise (absolute values)

7   Conclusion and Open Problems

In this work we have investigated up to which extent the time required to perform SCA evaluations can be reduced by means of enhanced measurement setups and statistical tools. Our preliminary investigations based on the popular fixed vs. random TVLA have highlighted the necessity of using well-designed setups. Besides, we have presented a fair comparison between it and its more recent counterpart, the so-called fixed vs. fixed TVLA. Our results have shown the benefits of the latter technique for reducing the measurement complexity of TVLA and therefore its time and storage requirements. In this regard, we refer to our last case study where a first-order TI was combined with a noise engine to harden the detection of higher-order leakages. In such a scenario, where millions of traces are first collected and analyzed afterwards, even small gain factors can save critical measurement time. In our concrete case, we were able to collect up to 100 million traces in 20 h. So, a multiplication by a factor 4 (that we would expect if a successful detection had to be based on the fixed vs. random TVLA) would already correspond to several days of computation. This impact will be further amplified for higher-order masked implementations, e.g., multiplying the evaluation time respectively by 8 and 16 for third- and fourth-order secure implementations, and so turning it from days to weeks. In this respect we finally note that if the evaluator is able to control the masks during the acquisition, he can average the samples before raising them to the some power and therefore mitigate the noise amplification due to masking, as detailed in [18].

Acknowledgments. François-Xavier Standaert is an associate researcher of the Belgian Fund for Scientific Research (FNRS-F.R.S.). This work was funded in parts by the European Commission through the project REASSURE.

A   Additional Figures

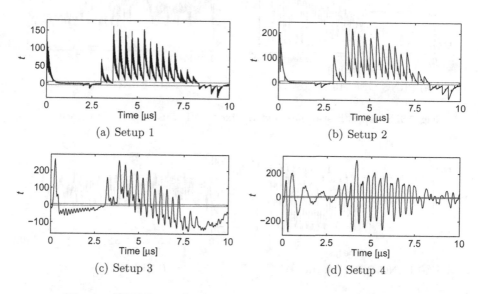

(a) Setup 1

(b) Setup 2

(c) Setup 3

(d) Setup 4

Fig. 15. (PRNG OFF) fixed vs. random TVLA using 10k traces

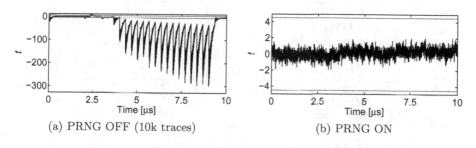

(a) PRNG OFF (10k traces)

(b) PRNG ON

Fig. 16. Setup 1, first-order fixed vs. fixed TVLA using 1 million traces

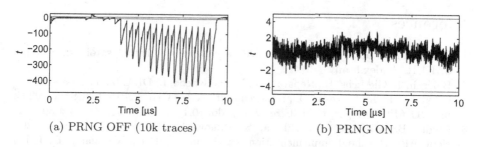

(a) PRNG OFF (10k traces)

(b) PRNG ON

Fig. 17. Setup 2, first-order fixed vs. fixed TVLA using 1 million traces

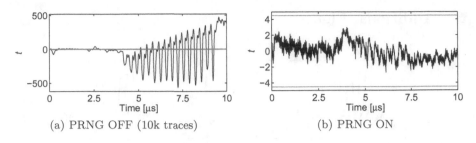

(a) PRNG OFF (10k traces)   (b) PRNG ON

Fig. 18. Setup 3, first-order fixed vs. fixed TVLA using 1 million traces

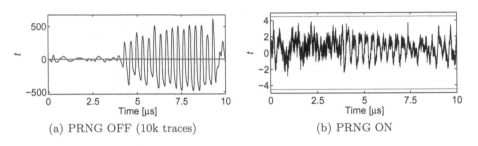

(a) PRNG OFF (10k traces)   (b) PRNG ON

Fig. 19. Setup 4, first-order fixed vs. fixed TVLA using 1 million traces

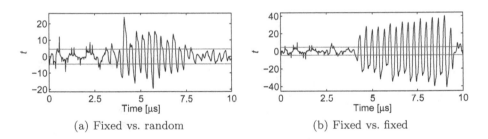

(a) Fixed vs. random   (b) Fixed vs. fixed

Fig. 20. (PRNG OFF) setup 4, TVLA using 10 k traces effected by noise

References

1. Side-channel attack user reference architecture. http://satoh.cs.uec.ac.jp/SAKURA/index.html
2. Balasch, J., Gierlichs, B., Reparaz, O., Verbauwhede, I.: DPA, bitslicing and masking at 1 GHz. In: Güneysu, T., Handschuh, H. (eds.) CHES 2015. LNCS, vol. 9293, pp. 599–619. Springer, Heidelberg (2015). doi:10.1007/978-3-662-48324-4_30
3. Bilgin, B., Gierlichs, B., Nikova, S., Nikov, V., Rijmen, V.: A more efficient AES threshold implementation. In: Pointcheval, D., Vergnaud, D. (eds.) AFRICACRYPT 2014. LNCS, vol. 8469, pp. 267–284. Springer, Cham (2014). doi:10.1007/978-3-319-06734-6_17

4. Bilgin, B., Gierlichs, B., Nikova, S., Nikov, V., Rijmen, V.: Higher-order threshold implementations. In: Sarkar, P., Iwata, T. (eds.) ASIACRYPT 2014. LNCS, vol. 8874, pp. 326–343. Springer, Heidelberg (2014). doi:10.1007/978-3-662-45608-8_18

5. Brier, E., Clavier, C., Olivier, F.: Correlation power analysis with a leakage model. In: Joye, M., Quisquater, J.-J. (eds.) CHES 2004. LNCS, vol. 3156, pp. 16–29. Springer, Heidelberg (2004). doi:10.1007/978-3-540-28632-5_2

6. Chari, S., Rao, J.R., Rohatgi, P.: Template attacks. In: Kaliski, B.S., Koç, K., Paar, C. (eds.) CHES 2002. LNCS, vol. 2523, pp. 13–28. Springer, Heidelberg (2003). doi:10.1007/3-540-36400-5_3

7. Duc, A., Faust, S., Standaert, F.-X.: Making masking security proofs concrete. In: Oswald, E., Fischlin, M. (eds.) EUROCRYPT 2015. LNCS, vol. 9056, pp. 401–429. Springer, Heidelberg (2015). doi:10.1007/978-3-662-46800-5_16

8. Durvaux, F., Standaert, F.-X.: From improved leakage detection to the detection of points of interests in leakage traces. In: Fischlin, M., Coron, J.-S. (eds.) EUROCRYPT 2016. LNCS, vol. 9665, pp. 240–262. Springer, Heidelberg (2016). doi:10.1007/978-3-662-49890-3_10

9. Goodwill, G., Jun, B., Jaffe, J., Rohatgi, P.: A testing methodology for side channel resistance validation. In: NIST Non-invasive Attack Testing Workshop (2011)

10. Güneysu, T., Moradi, A.: Generic side-channel countermeasures for reconfigurable devices. In: Preneel, B., Takagi, T. (eds.) CHES 2011. LNCS, vol. 6917, pp. 33–48. Springer, Heidelberg (2011). doi:10.1007/978-3-642-23951-9_3

11. Del Pozo, S.M., Standaert, F.-X., Kamel, D., Moradi, A.: Side-channel attacks from static power: when should we care? In: DATE 2015, pp. 145–150. ACM (2015)

12. Moradi, A., Mischke, O.: On the simplicity of converting leakages from multivariate to univariate. In: Bertoni, G., Coron, J.-S. (eds.) CHES 2013. LNCS, vol. 8086, pp. 1–20. Springer, Heidelberg (2013). doi:10.1007/978-3-642-40349-1_1

13. Moradi, A., Poschmann, A., Ling, S., Paar, C., Wang, H.: Pushing the limits: a very compact and a threshold implementation of AES. In: Paterson, K.G. (ed.) EUROCRYPT 2011. LNCS, vol. 6632, pp. 69–88. Springer, Heidelberg (2011). doi:10.1007/978-3-642-20465-4_6

14. Moradi, A., Wild, A.: Assessment of hiding the higher-order leakages in hardware. In: Güneysu, T., Handschuh, H. (eds.) CHES 2015. LNCS, vol. 9293, pp. 453–474. Springer, Heidelberg (2015). doi:10.1007/978-3-662-48324-4_23

15. Nikova, S., Rijmen, V., Schläffer, M.: Secure hardware implementation of nonlinear functions in the presence of glitches. J. Cryptol. 24(2), 292–321 (2011)

16. Poschmann, A., Moradi, A., Khoo, K., Lim, C., Wang, H., Ling, S.: Side-channel resistant crypto for less than 2, 300 GE. J. Cryptol. 24(2), 322–345 (2011)

17. Schneider, T., Moradi, A.: Leakage assessment methodology. In: Güneysu, T., Handschuh, H. (eds.) CHES 2015. LNCS, vol. 9293, pp. 495–513. Springer, Heidelberg (2015). doi:10.1007/978-3-662-48324-4_25

18. Standaert, F.: How (not) to use Welch's T-test in side-channel security evaluations. IACR Cryptol. ePrint Arch. 2017, 138 (2017)

Mind the Gap: Towards Secure 1st-Order Masking in Software

Kostas Papagiannopoulos[1] and Nikita Veshchikov[2(✉)]

[1] Radboud Universiteit, Nijmegen, Netherlands
[2] Quality and Security of Information Systems, Département d'informatique,
Université Libre de Bruxelles, Brussels, Belgium
nveshchi@ulb.ac.be

Abstract. Cryptographic implementations are vulnerable to side-channel analysis. Implementors often opt for masking countermeasures to protect against these types of attacks. Masking countermeasures can ensure theoretical protection against value-based leakages. However, the practical effectiveness of masking is often halted by physical effects such as glitches and distance-based leakages, which violate the *independent leakage assumption* (ILA) and result in security order reductions. This paper aims to address this gap between masking theory and practice in the following threefold manner. First, we perform an in-depth investigation of the device-specific effects that invalidate ILA in the AVR microcontroller ATMega163. Second, we provide an automated tool, capable of detecting ILA violations in AVR assembly code. Last, we craft the first (to our knowledge) *"hardened"* 1st-order ISW-based, masked Sbox implementation, which is capable of resisting 1st-order, univariate side-channel attacks. Enforcing the ILA in the masked RECTANGLE Sbox requires 1319 clock cycles, i.e. a 15-fold increase compared to a naive 1st-order ISW-based implementation.

Keywords: Masking · AVR · Verification tool · Simulator · Independent leakage assumption · Distance-based leakage · RECTANGLE · SCA

1 Introduction

Nowadays, the explosive growth of the "Internet of Things" (IoT) is reshaping modern society, pervading its infrastructure and communications. The rapid price drop in IoT components has transformed everyday products, enhancing them with network connectivity and information exchange capabilities. Amidst this new status quo, devices, ranging from cheap sensors to expensive vehicles, are required to maintain a heightened level of theoretical and physical security.

The work described in this paper has been supported by the Netherlands Organization for Scientific Research NWO under project ProFIL (628.001.007).

This work is the result of a short term scientific mission that has been supported by ICT COST Action IC1204: "Trustworthy Manufacturing and Utilization of Secure Devices".

© Springer International Publishing AG 2017
S. Guilley (Ed.): COSADE 2017, LNCS 10348, pp. 282–297, 2017.
DOI: 10.1007/978-3-319-64647-3_17

For instance, side-channel attacks (SCA) allow adversaries to recover sensitive data, by observing and analyzing the physical characteristics and emanations of a cryptographic implementation [19]. Such physical attacks motivated research towards countermeasures that perform noise amplification, thus hindering the adversary's recovery capabilities. A common choice for provably secure, noise-amplifying software countermeasure is masking [9,18]. Masking employs secret-sharing techniques that establish theoretical security against the value-based leakage model. Rephrasing, masking secures implementations against adversaries that can only extract information about the value being processed at a given time. This underlying assumption is often referred to as the *independent leakage assumption* (ILA) [24]. Unfortunately, the exact values under manipulation are not always visible at a given layer of abstraction, e.g. at assembly code and such a limited adversarial model is not applicable in many practical, software-based scenarios. For instance, devices often exhibit *distance-based leakages*, which can reduce the security of the masking countermeasure [1,14]. Likewise, coupling effects [24] and glitches [20] can pose similar security hazards.

This work attempts to bridge the gap between theory and practice in the masking countermeasure with the following threefold contribution. First, we investigate several effects that violate ILA in an ATMega163 microcontroller and subsequently, we establish solutions that mitigate these issues. Second, we use this knowledge in order to build an assembly-oriented tool that is capable of detecting ILA violations in AVR-based masked implementations. Third, assisted by the developed tool, we craft the first (to our knowledge) 1st-order masked implementation in ATMega163 that is capable of resisting 1st-order, univariate attacks. In other words, we enforce the ILA in order to severely limit the informativeness of 1st-order leakages, forcing the adversary to resort to 2nd-order attacks. As a proof of concept, we develop a "hardened" 1st-order, ISW-based [18], bitsliced Sbox for the RECTANGLE cipher [35]. The "hardened" implementation requires 1319 clock cycles, a 15-fold increase compared to a "naive" 1st-order, ISW-based, bitsliced Sbox of the same cipher.

The rest of this paper is organized as follows. In Sect. 2, we provide preliminaries w.r.t. masking, the experimental setup and the evaluation techniques we employ. In Sect. 3 we offer a detailed description of all the ILA-breaching effects that we have identified in ATMega163. Section 4 discusses the development of the assembly verification tool. Section 5 details the construction of a "hardened" RECTANGLE, 1st-order masked Sbox for ATMega163. We conclude and discuss future work in Sect. 6.

2   Background

2.1   Boolean Masking and Order Reduction

Chari et al., Goubin et al. and Messerges [9,15,21] were among the first to suggest splitting intermediate values with a secret sharing scheme, in order to force attackers to analyze higher-order statistical moments. Analytically, a dth-order Boolean masking scheme splits a sensitive value x into $d+1$ shares $(x_0, x_1, \ldots, x_d)$, as shown below.

$$x = x_0 \oplus x_1 \oplus \cdots \oplus x_d \tag{1}$$

The shares $(x_0, x_1, \ldots, x_d)$ are also referred to as the $(d + 1)$-family of shares corresponding to x [26]. Given that the ILA holds and assuming sufficient noise, it has been shown that the number of traces required for a successful attack grows exponentially w.r.t. the order d [9,23]. Several implementation options have been suggested for the masking countermeasure, ranging from lookup-table techniques [11,32] to GF-based circuits [8,16,18,26].

In parallel with the development of masked implementations, side-channel research focused on the practical evaluation of the countermeasure. Balasch et al. [1] put forward the concepts of value-based and distance-based leakages, as well as the notion of order reduction. We briefly restate their definitions below.

Value/Distance-Based Leakage Function. A leakage function $L(.)$ consists of a deterministic part $L_d(.)$ and random additive noise N. The leakage function is *value-based* if $L_d(.)$ can only take arguments from the set of intermediate values produced by the masking scheme. The leakage is *distance-based* if $L_d(.)$ can take arguments from the set that contains all possible pairwise combinations of intermediate values. The combination can imply operations such as XOR, concatenation, etc.

Order-Reduction Theorem. A dth-order secure masking scheme under value-based leakages is $\lfloor \frac{d}{2} \rfloor$th-order secure under distance-based leakages.

The applicability of the order-reduction theorem has been verified experimentally by Balasch et al. [1] for orders $d = 1, 2$ in AVR-based and 8051-based devices. De Groot et al. [14] have verified experimentally the theorem's applicability for orders $d = 1, 2$ in the ARM Cortex-M4.

2.2   Experimental Setup and Evaluation

The implementation and SCA evaluation is performed on a smartcard equipped with an 8-bit, AVR-based ATMega163 microcontroller[1]. The device features a 4.4 MHz clock, 1024 bytes of SRAM and 17 Kbytes of Flash memory. The acquisition of power traces is carried out using the Riscure PowerTracer[2] and the Picoscope 5203 oscilloscope. The sampling rate is set at 31.5 MSamples/sec and the only post-processing applied is signal alignment.

The evaluation of the *actual* security order of a masking scheme is, in general, an open problem. We often face the *limited attack scope*, i.e. a given attack may not be able to exploit the available leakage due to e.g. an unsuitable choice of intermediate values or an incorrect power model. To address this problem, generic side-channel distinguishers and extensive profiling techniques have been developed [3,28,33]. In this work, we opt for the leakage detection methodology [10] which prioritizes leakage detection over leakage exploitation, speeding

---

[1] http://www.atmel.com/images/doc1142.pdf.

[2] https://www.riscure.com/security-tools/hardware/power-tracer.

up certain evaluation aspects. In detail, we employ the random vs. fixed, non-specific, 1st-order t-test. We perform a random vs. fixed acquisition and obtain two distinct tracesets S_{fixed} and S_{random}, under the same encryption key. The input plaintext for S_{fixed} is set to a fixed value, while for S_{random}, the input is uniformly random. The implementation receives the fixed or random plaintext in a non-deterministic and randomly-interleaved way (as recommended by Schneider et al. [27]). Following the data acquisition, the 1st-order t-test will assess whether the two sets S_{fixed}, S_{random} stem from the same population, using the following statistical test.

$$H_{null}: \quad \mu_{fixed} = \mu_{random}$$
$$H_{alt}: \quad \mu_{fixed} \neq \mu_{random} \tag{2}$$

$$w = \frac{\mu_{fixed} - \mu_{random}}{\sqrt{\frac{\sigma_{fixed}^2}{n} + \frac{\sigma_{random}^2}{m}}} \qquad v = \frac{\left(\frac{\sigma_{fixed}^2}{n} + \frac{\sigma_{random}^2}{m}\right)^2}{\frac{\sigma_{fixed}^4}{n^2(n-1)} + \frac{\sigma_{random}^4}{m^2(m-1)}} \tag{3}$$

Parameters μ_x and σ_x^2 are the estimated mean and variance of set x; n and m denote sizes of sets S_{fixed} and S_{random} respectively. The null hypothesis H_{null} is rejected at a given level of significance α (often set to 0.99999), if $|w| > t_{\alpha/2,v}$, where $t_{\alpha/2,v}$ is the value of the Student t distribution with v degrees of freedom. In the evaluation context, rejecting H_{null} implies leakage detection, i.e. potential evidence of an ineffective masking scheme.

In this paper, we will use the t-test as a detection tool w.r.t. ILA-breaching effects and their solutions (see Sect. 3). Still, we will also employ 1st-order CPA methods [7] in order to demonstrate the exploitability of such effects. In order to reduce the computational cost of the evaluation, we use the memoryless formulas suggested by Schneider et al. [27] and the incremental approach for CPA by Botinelli et al. [6].

3   ILA-Breaching Effects

In this section, we present three effects identified in the ATMega163 microcontroller that breach ILA and pose a hazard to any masking scheme's security. Analytically, the effects below demonstrate that independent computations *do not* necessarily lead to independent leakages and thus, the order-reduction theorem can become applicable.

Every effect (Sects. 3.1, 3.2 and 3.3) is described as a standalone, assembly-based scenario that manipulates two 4-bit shares x_0, x_1 originating from the sensitive, key-dependent, 4-bit value x, such that $x = x_0 \oplus x_1$. The shares x_0, x_1 are always manipulated in a theoretically sound manner, adhering to the masking scheme's requirements, i.e. we never combine the shares directly (e.g. via an exclusive-or instruction `eor x0, x1`).

For all the described scenarios, that are theoretically sound, we show experimentally that ILA is not fulfilled by employing 1st-order, univariate techniques.

Namely, we perform correlation-based analysis [7], computing the correlation coefficient ρ between the Hamming weight of the sensitive, key-dependent value x and the experimentally acquired traceset. To maintain a wide attack scope, we also use the leakage detection methodology [10,27] and compute the 1st-order, random vs. fixed t-test. We conclude every scenario by suggesting possible solutions that enforce ILA. Restating Balasch et al. [1], as we are always limited by the traces at hand, we cannot rule out the existence of 1st-order leakages, yet we establish that their informativeness is limited compared to 2nd-order leakages in the target device. Note that extra care is taken in order to assess all effects independently, i.e. we use the suggested solutions so as to isolate the effect under discussion from the rest.

The analyzed effects can manifest in several data storage units (e.g. registers, SRAM/Flash memory cells, I/O buffers, etc.) and may relate to different instructions of the AVR ISA[3], leading to a very large number of potential scenarios. In order to maintain a feasible scope, we limit our discussion to storage units and instructions that are often encountered in the context of cryptographic implementations, i.e. SRAM memory accesses and logical instructions.

3.1 Overwrite Effect

The overwrite effect is observable when a share gets overwritten by a different share from the same family. For instance, if share x_0 in a data storage unit (register, memory cell, etc.) gets overwritten by share x_1, then the power consumption correlates with the number of bits switched i.e. $x_0 \oplus x_1$. This effect was observed by Daemen et al. [30] and later revisited by Coron et al. [12].

Below, we address the most common situations in which overwriting arises during a cryptographic implementation. We perform two experiments: a register-based overwrite via the instruction mov x0, x1, and a memory-based overwrite via the instruction st SRAM_x0, x1. The experiments are described in Listings 1.1 and 1.2. Their analysis follow in Fig. 1.

We confirm that overwriting is indeed an ILA-breaching effect, manifesting both in registers and SRAM memory. Note that the exploitability of the effect varies according to the data storage unit: in ATMega163, register-based overwriting can be exploited with roughly 500 traces (Fig. 1a), while memory-based requires at least 40k traces (Fig. 1c). Preventing register and memory-based overwrites is straightforward: the corresponding register (or memory cell) needs to be cleared in advance.

```
1 ; share x0 in r17
2 ; share x1 in r23
3 mov r17 , r23
4 ;
5 ;
```

Listing 1.1. Register overwrite experiment.

```
1 ; share x0 in SRAM 0x0080
2 ; share x1 in r17
3 ldi r27 ,0x00
4 ldi r26 ,0x80
5 st X , r17
```

Listing 1.2. Memory overwrite experiment.

---

[3] http://www.atmel.com/images/Atmel-0856-AVR-Instruction-Set-Manual.pdf.

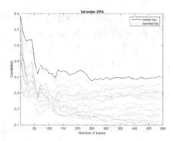

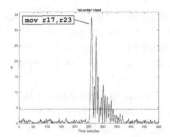

(a) Register overwrite, 1st-order CPA, HW model, 500 traces.

(b) Register overwrite, 1st-order t-test, 5k random vs. 5k fixed.

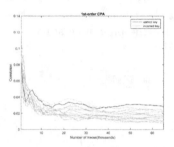

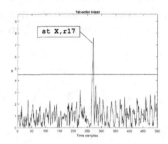

(c) Memory overwrite, 1st-order CPA, HW model, 65k traces.

(d) Memory overwrite, 1st order t-test, 50k random vs. 50k fixed.

Fig. 1. Register/memory-based overwrite effects

3.2 Memory Remnant Effect

The memory remnant effect is a leakage originating from consecutive SRAM accesses to shares of the same family. Assume that shares x_0, x_1 are stored in SRAM cells and get accessed sequentially. Naturally, the first access leaks share x_0 (value-based leakage), yet it also creates a "remnant" of x_0. The second access will leak the transition of the share x_1 and the remnant x_0, reducing the security.

```
1  ; share x0 in 0x0080
2  ; share x1 in 0x0090
3  ldi  r27,  0x00
4  ldi  r26,  0x80
5  ld   r17,  X
6  ldi  r27,  0x00
7  ldi  r26,  0x90
8  ld   r20,  X
9  ;
10 ;
```

Listing 1.3. Memory remnant experiment.

```
1  ; share x0 in 0x0080
2  ; share x1 in 0x0090
3  ldi  r27,  0x00
4  ldi  r26,  0x80
5  ld   r17,  X
6  ldi  r17,  0x00
7  ldi  r26,  0x85
8  ld   r17,  X
9  ldi  r26,  0x90
10 ld   r20,  X
```

Listing 1.4. Clearing remnant experiment.

We address the remnant scenario with two experiments. Listing 1.3 demonstrates how two consecutive SRAM accesses `ld rA, SRAM_x0`, followed by `ld rB, SRAM_x1` produce the remnant effect. Second, in Listing 1.4, we show how clearing the register and accessing an unrelated SRAM address (0x0085) can remove the remnant.

As shown in Figs. 2a and b, consecutive SRAM accesses can potentially lead to ILA violations. Exploiting (in a univariate manner) the memory remnant effect in ATMega163 needs less than 500 traces with our setup. Preventing the effect requires the clearing of the register and the insertion of a dummy SRAM access. Alternatively, the implementor could ensure that same-family shares are not accessed sequentially. Note also that the `st` instruction produces a similar effect. We speculate that the memory remnant effect is caused by the structure of the memory access mechanism and potentially, the pipelining stages.

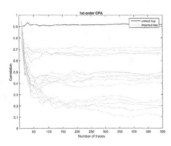

(a) Memory remnant effect, 1st-order CPA, HW model, 500 traces.

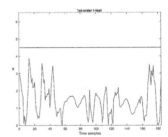

(b) Memory remnant effect, 1st-order t-test, 5k random vs. 5k fixed.

(c) Clearing remnant effect, 1st-order CPA, HW model, 100k traces.

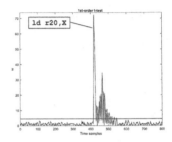

(d) Clearing remnant effect, 1st-order t-test, 100k random vs. 100k fixed.

Fig. 2. Memory-based remnant effect

3.3   Neighbour Leakage Effect

The neighbour leakage effect implies that accessing or processing the contents of a data storage unit will cause leakage in another unit as well. For example, assume that share x_0 is stored in register `rB` and share x_1 is being processed in register rA. Assume also that the registers rA, rB are subject to the neighbour leakage effect. Processing rB will produce a value-based leakage of x_0. At the same time, the neighbouring leakage effect will cause rA to leak the value of x_1, resulting in transition between shares and the recovery of sensitive value x. The following two experiments (Listing 1.5) verify the neighbour leakage effect between registers r2, r3, i.e. a share stored in r2 leaks when manipulating r3 and vice-versa.

```
1  ; clear all registers
2  ; sensitive 'x' is in the selected register (r2 OR r3)
3  mov r0, r0
4  nop ; 5 times
5  mov r1, r1
6  nop ; 5 times
7  mov r2, r2
8  nop ; 5 times
9  mov r3, r3
10 nop ; 5 times
11 ...
12 mov r31, r31
```

Listing 1.5. Neighbour leakage experiment for r2 and r3.

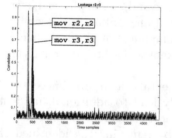 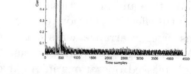

(a) Correlation $\rho(HW(x_0), traceset)$, r2-r3, 5k traces.

(b) Correlation $\rho(HW(x_0), traceset)$, r3-r2, 5k traces.

Fig. 3. Neighbour-based leakage effect

As shown above, we use the same code from Listing 1.5, but in the first time we put the sensitive variable x into register r2 (*only* line7 should result in leakage). In the second time, we put the sensitive value into the register r3 (*only* line 9 should leak). However, Fig. 3 shows that both register accesses leak. As a result, we have identified a pair of data storage units (r2,r3) that exhibit the neighbour leakage effect. Note that in this case the effect is symmetrical, i.e., r2 triggers r3 and vice-versa (Figs. 3a and b). We also observed that the effect is persistent, i.e. the mov instructions will trigger the same behavior, even if performed later (not necessairly in order as in Listing 1.5). We run the same experiment in order to identify all possible neighbour leakages in the register file (all pairs in set $\{r0, \ldots, r31\}$). The results are available in the Appendix, matrix R. The issue mostly affects consecutive registers, although exceptions exist, e.g. register r0. We did not identify a similar effect in SRAM memory, yet our experiments were limited to a small region of cells. Neighbour-like effects have been observed in consecutive instructions, yet it remains open whether they are cause by proximity or they stem from other effects. We speculate that they relate to the structure of the register file and likely involve the storage and multiplexing mechanism of the registers. Given the pairwise manifestation of the effect, we speculate a pair-based organization of the register file. Still, note that it is hard to link architectural options at the hardware layer directly to side-channel effects. As a solution to the neighbour effect, the developer can opt to avoid storing shares in hazardous registers and keep a safety distance between consecutive instructions. Alternatively, he can store all shares in SRAM, except for the ones currently in use.

Summing up, we stress the following focal points regarding the ILA-breaching effects and their solutions:

- All identified effects are device-dependent, i.e. there is no hard guarantee that they are observable and reproducible in different AVR-based microcontrollers, let alone different architectures such as ARM, TI, PIC etc. Both intra-AVR and inter-architectural observability of the effects remains open.
- The effects are often counter-intuitive when viewed in the assembly layer of abstraction. They originate from the hardware and/or the physical layer, thus can only be detected via experimental evaluation. Linking the assembly ILA-breaching effects to a particular hardware component or physical phenomenon is non-trivial [24,29], especially without knowledge of the underlying chip architecture and properties.
- Since the effect's detection requires experimental evaluation, different instructions or code arrangements can potentially lead to additional, unidentified ILA-breaching effects. Still, we maintain that it is possible to construct "hardened" masked operations in ATMega163 by removing the identified effects (see Sect. 5). It remains open whether the suggested solutions are computationally optimal or more efficient clearing techniques can be identified.

The takeaway message of this section is that assembly-level soundness cannot enforce ILA and hence 1st-order security, due to the nature of the breaching effects. However, it is possible to acquire sufficient knowledge about effects and solutions in a particular device. These non-intuitive checks discussed above can be subsequently integrated into a code-checking tool which can identify such effects in assembly code.

4   Leakage Detection Tool

Several tools that can help designers of cryptographic systems were already suggested and discussed in literature.

SILK[4] presented in 2014 [31] can be used to generate simulated traces based on C++ code. Thus, it allows to generate tracesets during the early stages of development, in order to test an implementation against any attack. However, SILK works only with high level, C++ source code and can not take into account reordering or replacement (even removal) of instructions that is often used by compilers during optimization. Also, this tool does not detect flaws in implementations, it only allows to easily generate simulated traces.

A tool based on formal verification was presented at EUROCRYPT in 2015 [2], capable of detecting design flaws in masking schemes. This tool can analyze programs written using the EasyCrypt framework and its language and it requires the designer to transform the original implementation (e.g. assembly or C code) to EasyCrypt. Unfortunately, errors could potentially be introduced during this process and there is no guarantee that the program written using EasyCrypt will be equivalent to the program in the original programming language. To the best of our knowledge, free automated tools that can transform C or assembly (or other languages that are often used for development of cryptographic software in embedded systems) programs to EasyCrypt do not exist. Moreover, this tool is not opensource and thus can not be used by the developer community.

A simulation-based tool that can be used to analyze masking implementations was presented at FSE in 2016 [25]. It can be used with software and hardware implementations and it requires only high-level implementation source code, such as C language. Due to this fact it can also be blind to re-arrangements of operations (which can lead to side-channel leakage) created by the compiler. Until today, the source code of this tool also remains unavailable[5].

4.1   ASCOLD

In order to assess the security of implementations at the assembly level, we developed a tool called ASCOLD, standing for Assembly Code Leakage Detection tool. The tool is written in **python** and the source code is available on our website[6]. ASCOLD uses assembly code as its input in order to run a simulation while checking for potential issues that can cause side-channel information leakage. The tool is compatible only with assembly code (which can be used as is during the development or extracted from the compiled binary file). Thus it is possible to be sure that the executed code will be exactly the same as the code which is analyzed. Otherwise, it becomes impossible to provide any guarantees on the quality of the analysis, i.e. be sure that no additional issues are introduced during compilation.

The simulation run by the tool does not use an instance of an execution i.e., we do not use specific values in order to run the program. ASCOLD starts a program in an initial state and propagates all changes such as combinations

---

[4] http://www.ulb.ac.be/di/dpalab/download/SILK_v0.1.zip.

[5] We have contacted the author, there is intention to eventually publish the code.

[6] https://github.com/nikita-veshchikov/ascold.

of values, their modifications and replacements of one value by another. More precisely, it keeps track of which shares or combinations of shares are stored in each register (or memory cell). During any arithmetic or logical operation, shares stored in different operands are verified, specifically we check whether we combine different shares of the same family without randomizing beforehand. Note that not all combinations are hazardous, yet we opt for such a conservative approach in order to speed-up the verification process.

In the same way, we verify the implementation for the device-specific distance-based leakages for every arithmetic/logical operation, SRAM store or load instruction that is executed. Analytically, we verify whether the previously stored value and the new value cause the overwrite effect Sect. 3.1. Similarly, our tool checks the load/store instructions for remnant effects discussed in Sect. 3.2. In addition, it features the matrix R of *neighbours*, which represents registers that can leak while another register is used (neighbour leakage effect, Sect. 3.2). In order to bootstrap the whole simulation, the developer needs to provide a configuration file. The configuration file is a simple text file that contains information about the initial state of the system i.e., it describes which registers or addresses in memory contain different secret shares of sensitive values. As the result of the simulation, ASCOLD prints out a line number and the rule that was violated by the program.

ASCOLD works with the AVR family of microcontrollers, it implements the most common memory instructions such as load and store as well as a set of commonly used (in cryptography) instructions such as arithmetic operations (add, mul, ...) and logical operations (and, eor, or, ...). The same core principles can be applied in order to build a similar tool for a different instruction set or to add new AVR instructions supported by newer microcontrollers.

Limitations. The current version of our tool incorporates our findings which are based on the ATMega163, other models of microcontrollers might have slightly different (even additional) issues that cause unintentional information leakage. Among other things, leakage described in Sect. 3.2 is more likely to be different (affecting different sets of registers) in other models of AVR microcontrollers. ASCOLD does not take into account the effects of pipelining which might be an issue in case of a microcontroller which can potentially handle two different shares of the same sensitive value (at different stages of the pipeline) during the same clock cycle. We did not implement all AVR instructions, most importantly the current version of ASCOLD does not support loops. However, we implemented the most commonly used instructions and new instructions/rules can be added due to the tool's extensibility. The lack of jump instructions (loops) can be disregarded via loop-unrolled implementations.

5   Hardened 1st-Order Masked Sbox for RECTANGLE

We have discussed the ILA-breaching effects in Sect. 3 and integrated these observations in the ASCOLD tool, described in Sect. 4. The current Section builds up on these advances by putting forward a "hardened", 1st-order masked, ISW-based

RECTANGLE Sbox. The desired aim is to produce an assembly-based, lightweight Sbox implementation that is secure against 1st-order, univariate attacks, hence forcing the attacker to resort to 2nd-order and/or multivariate techniques.

Our implementation opts for a bitsliced [5, 13] representation, due to both the bitsliced structure of RECTANGLE and to the $GF(2)$-oriented nature of the ISW countermeasure. We employ a bitslicing factor of 2, i.e. we exploit the 8-bit AVR architecture in order to process two 4-bit Sboxes in parallel (nibble-slicing). The Sbox is decomposed into $GF(2)$ operations which can be accelerated by via SIMD-like, 8-bit assembly instructions. The decomposition suggested by Zhang et al. [35] is optimal w.r.t. $GF(2)$ multiplicative complexity, since Grosso et al. [17] established that the minimum number of non-linear operations required by 4×4 Sboxes is 4.

In order to "harden" the Sbox, we use the solutions suggested in Sect. 3 and follow two approaches: efficient and conservative. In the *efficient* approach, after processing any share, we clear the registers on a need-to basis and insert dummy ld instructions to avoid overwrite and remnant effects. We avoid neighbouring leakage effects by always storing the shares in SRAM, i.e. the register file contains only the shares used by the current instruction. In the *conservative* approach, we perform all the afore-mentioned clearing techniques. In addition, we insert dummy st instructions and perform thorough register/memory clearing. Both efficient and conservative approaches are applied to every single instruction of the implementation, i.e. the cost is linear w.r.t. the number of instructions that manipulate masked shares. The resulting computational overhead is significant: the efficient "hardened" Sbox implementation runs in 993 clock cycles, i.e. almost 12 times slower compared to the "naive" 1st-order, ISW-based RECTANGLE Sbox, which runs in 87 clock cycles. The conservative "hardened" Sbox implementation requires 1319 clock cycles, i.e. it is 15 times slower. Table 1 contains a comparison between "naive" 1st-order, "naive" 2nd-order and efficient/conservative "hardened" 1st-order bitsliced implementations of the RECTANGLE Sbox in AVR assembly.

Using the random vs. fixed t-test, we evaluate the efficient and conservative "hardened" 1st-order Sboxes, as well as the "naive" 1st-order Sbox. Using a 25k random vs. 25k fixed t-test does not yield any statistically significant leakage in the efficient "hardened" version (Fig. 4a). However, we note that a 50k random vs. 50k fixed t-test is able to detect leakage, i.e. trying to reduce the cost of

Table 1. Masked Sbox comparison in ATMega163

Order d	Hardened	Latency cycles	Throughput bits/cycle $\times 10^{-3}$	RNG bytes
Unprotected	No	32	250	0
1st order	No	87	91	4
	Yes (eff.)	993	8	4
	Yes (cons.)	1319	6	4
2nd order	No	775	10	12

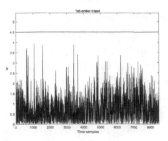

(a) Efficient hardened Sbox, 1st-order t-test, 25k random vs. 25k fixed.

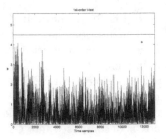

(b) Conservative hardened Sbox, 1st-order t-test, 100k random vs. 100k fixed.

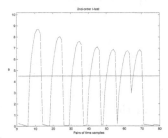

(c) Consevative hardened Sbox, 2nd-order t-test, 25k random vs. 25k fixed.

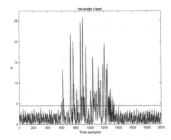

(d) Naive Sbox, 1st-order t-test, 1k random vs. 1k fixed.

Fig. 4. Hardened and naive Sbox evaluations

enforcing ILA can have a detrimental effect on security. For the conservative "hardened" Sbox, a 100k random vs. 100k fixed t-test does not detect any leakage (Fig. 4b). Note that a *2nd-order* 25k random vs. 25k fixed t-test on a chosen sample window is able to detect leakage. Therefore, we conclude that for the given device, the informativeness of 1st-order attacks is substantially limited and a 2nd-order attack is the preferable adversarial strategy (Fig. 4c). Naturally, the "naive" 1st-order version rejects the null hypothesis (Fig. 4d) due to the ILA-breaching effects and the 1st-order leakage can be easily exploited.

So far, the only way to guarantee the actual security order of a real-world implementation was to increase the scheme's theoretical order d, in order to ensure that the implementation attains an actual order of $\lfloor \frac{d}{2} \rfloor$ [1,14]. Clearing the ILA-breaching effects requires a significant overhead and is device-dependent, yet it is the only technique known to us that can enforce 1st-order, univariate security. In addition, hardening does not increase the scheme order d, thus *the random number generation (RNG) cost is not increased*. The previous suggestions require a higher scheme order, hence a significant overhead, since both the implementation cost and the RNG cost are quadratic w.r.t. the order. We compare the "hardened" 1st-order and "naive" 2nd-order implementation costs (in clock cycles) and we observe that hardening the 1st-order Sbox is slower than increasing the scheme's order from 1 to 2 (both in the efficient and in the conservative case). Still, the solution requires no extra RNG and we maintain that removing these effects can also be beneficial to higher-order implementations, i.e. it is complimentary to masking. The extent to which higher-order implementations can benefit from removing such effects remains an open problem.

6   Conclusions

This work investigated the hazards in software masking, suggested a verification tool and established a secure, 1st-order masked Sbox implementation against 1st-order, univariate attacks. Still, several important questions for future work arise. We demonstrated that removing the ILA-breaching effects is feasible, yet identifying the best clearing mechanism and minimizing the overhead is a topic for further exploration. Similarly, the current work is limited to AVR ATMega163 and needs to be extended to different devices and platforms. It could be done by using ASCOLD tool as a base for this kind of work. Moreover, higher-order evaluation techniques are still nascent and in this work we did not focus on 1st-order, yet multivariate attacks such as those that exploit horizontality [4]. In addition, note that the ILA effects are observable throughout an implementation. Not only the cipher-related operations but any manipulation of shares during I/O, RNG routines etc. can create hazards. Thus, there is need for effort towards a fully hardened implementation. Last but not least, we stress that the effects identified depend on the architecture and the physical layer, thus preventing them in the assembly layer is, in principle, less efficient and prone to errors. Future work can strive towards custom-made microcontrollers that enforce ILA in hardware. Ideally, such a microcontroller should be able to guarantee ILA without additional countermeasures such as threshold implementations [22].

A   Appendix

Below, we include the 32×32 matrix R that is generated experimentally, while investigating all possible neighbouring leakage effects in the ATMega163 register file (by performing 32 experiments similar to Listing 1.5). Value '1' denotes the presence of leakage and '0' the absence. The tool ASCOLD uses R in order to detect neighbour-based ILA violations.

References

1. Balasch, J., Gierlichs, B., Grosso, V., Reparaz, O., Standaert, F.-X.: On the cost of lazy engineering for masked software implementations. In: Joye, M., Moradi, A. (eds.) CARDIS 2014. LNCS, vol. 8968, pp. 64–81. Springer, Cham (2015). doi:10.1007/978-3-319-16763-3_5
2. Barthe, G., Belaïd, S., Dupressoir, F., Fouque, P.-A., Grégoire, B., Strub, P.-Y.: Verified proofs of higher-order masking. In: Oswald, E., Fischlin, M. (eds.) EURO-CRYPT 2015. LNCS, vol. 9056, pp. 457–485. Springer, Heidelberg (2015). doi:10.1007/978-3-662-46800-5_18

3. Batina, L., Gierlichs, B., Prouff, E., Rivain, M., Standaert, F.-X., Veyrat-Charvillon, N.: Mutual information analysis: a comprehensive study. J. Cryptol. **24**(2), 269–291 (2011)
4. Battistello, A., Coron, J.-S., Prouff, E., Zeitoun, R.: Horizontal side-channel attacks and countermeasures on the ISW masking scheme. In: Gierlichs, B., Poschmann, A.Y. (eds.) CHES 2016. LNCS, vol. 9813, pp. 23–39. Springer, Heidelberg (2016). doi:10.1007/978-3-662-53140-2_2
5. Biham, E.: A fast new DES implementation in software. In: Biham, E. (ed.) FSE 1997. LNCS, vol. 1267, pp. 260–272. Springer, Heidelberg (1997). doi:10.1007/BFb0052352
6. Bottinelli, P., Bos, J.W.: Computational aspects of correlation power analysis. IACR Cryptology ePrint Archive 260 (2015)
7. Brier, E., Clavier, C., Olivier, F.: Correlation power analysis with a leakage model. In: Joye, M., Quisquater, J.-J. (eds.) CHES 2004. LNCS, vol. 3156, pp. 16–29. Springer, Heidelberg (2004). doi:10.1007/978-3-540-28632-5_2
8. Canright, D., Batina, L.: A very compact "perfectly masked" s-box for AES (corrected). IACR Cryptology ePrint Archive 11 (2009)
9. Chari, S., Jutla, C.S., Rao, J.R., Rohatgi, P.: Towards sound approaches to counteract power-analysis attacks. In: Wiener [34], pp. 398–412
10. Cooper, J., DeMulder, E., Goodwill, G., Jaffe, J., Kenworthy, G., Rohatgi, P.: Test Vector Leakage Assessment (TVLA) methodology in practice (2013). http://icmc-2013.org/wp/wp-content/uploads/2013/09/goodwillkenworthtestvector.pdf
11. Coron, J.-S.: Higher order masking of look-up tables. In: Nguyen, P.Q., Oswald, E. (eds.) EUROCRYPT 2014. LNCS, vol. 8441, pp. 441–458. Springer, Heidelberg (2014). doi:10.1007/978-3-642-55220-5_25
12. Coron, J.-S., Giraud, C., Prouff, E., Renner, S., Rivain, M., Vadnala, P.K.: Conversion of security proofs from one leakage model to another: a new issue. In: Schindler, W., Huss, S.A. (eds.) COSADE 2012. LNCS, vol. 7275, pp. 69–81. Springer, Heidelberg (2012). doi:10.1007/978-3-642-29912-4_6
13. Daemen, J., Govaerts, R., Vandewalle, J.: A new approach to block cipher design. In: Anderson, R. (ed.) FSE 1993. LNCS, vol. 809, pp. 18–32. Springer, Heidelberg (1994). doi:10.1007/3-540-58108-1_2
14. Groot, W., Papagiannopoulos, K., Piedra, A., Schneider, E., Batina, L.: Bitsliced masking and ARM: friends or foes? In: Bogdanov, A. (ed.) LightSec 2016. LNCS, vol. 10098, pp. 91–109. Springer, Cham (2017). doi:10.1007/978-3-319-55714-4_7
15. Goubin, L., Patarin, J.: DES and differential power analysis the "Duplication" method. In: Koç, Ç.K., Paar, C. (eds.) CHES 1999. LNCS, vol. 1717, pp. 158–172. Springer, Heidelberg (1999). doi:10.1007/3-540-48059-5_15
16. Goudarzi, D., Rivain, M.: How fast can higher-order masking be in software? Cryptology ePrint Archive, Report 2016/264 (2016). http://eprint.iacr.org/2016/264
17. Grosso, V., Leurent, G., Standaert, F.-X., Varıcı, K.: LS-designs: bitslice encryption for efficient masked software implementations. In: Cid, C., Rechberger, C. (eds.) FSE 2014. LNCS, vol. 8540, pp. 18–37. Springer, Heidelberg (2015). doi:10.1007/978-3-662-46706-0_2
18. Ishai, Y., Sahai, A., Wagner, D.: Private circuits: securing hardware against probing attacks. In: Boneh, D. (ed.) CRYPTO 2003. LNCS, vol. 2729, pp. 463–481. Springer, Heidelberg (2003). doi:10.1007/978-3-540-45146-4_27
19. Kocher, P.C., Jaffe, J., Jun, B.: Differential power analysis. In: Wiener [34], pp. 388–397

20. Mangard, S., Schramm, K.: Pinpointing the side-channel leakage of masked AES hardware implementations. In: Goubin, L., Matsui, M. (eds.) CHES 2006. LNCS, vol. 4249, pp. 76–90. Springer, Heidelberg (2006). doi:10.1007/11894063_7
21. Messerges, T.S.: Securing the AES finalists against power analysis attacks. In: Goos, G., Hartmanis, J., Leeuwen, J., Schneier, B. (eds.) FSE 2000. LNCS, vol. 1978, pp. 150–164. Springer, Heidelberg (2001). doi:10.1007/3-540-44706-7_11
22. Nikova, S., Rechberger, C., Rijmen, V.: Threshold implementations against side-channel attacks and glitches. In: Ning, P., Qing, S., Li, N. (eds.) ICICS 2006. LNCS, vol. 4307, pp. 529–545. Springer, Heidelberg (2006). doi:10.1007/11935308_38
23. Prouff, E., Rivain, M.: Masking against side-channel attacks: a formal security proof. In: Johansson, T., Nguyen, P.Q. (eds.) EUROCRYPT 2013. LNCS, vol. 7881, pp. 142–159. Springer, Heidelberg (2013). doi:10.1007/978-3-642-38348-9_9
24. Renauld, M., Standaert, F.-X., Veyrat-Charvillon, N., Kamel, D., Flandre, D.: A formal study of power variability issues and side-channel attacks for nanoscale devices. In: Paterson, K.G. (ed.) EUROCRYPT 2011. LNCS, vol. 6632, pp. 109–128. Springer, Heidelberg (2011). doi:10.1007/978-3-642-20465-4_8
25. Reparaz, O.: Detecting flawed masking schemes with leakage detection tests. In: Peyrin, T. (ed.) FSE 2016. LNCS, vol. 9783, pp. 204–222. Springer, Heidelberg (2016). doi:10.1007/978-3-662-52993-5_11
26. Rivain, M., Prouff, E.: Provably secure higher-order masking of AES. In: Mangard, S., Standaert, F.-X. (eds.) CHES 2010. LNCS, vol. 6225, pp. 413–427. Springer, Heidelberg (2010). doi:10.1007/978-3-642-15031-9_28
27. Schneider, T., Moradi, A.: Leakage assessment methodology. In: Güneysu, T., Handschuh, H. (eds.) CHES 2015. LNCS, vol. 9293, pp. 495–513. Springer, Heidelberg (2015). doi:10.1007/978-3-662-48324-4_25
28. Standaert, F.-X., Malkin, T.G., Yung, M.: A unified framework for the analysis of side-channel key recovery attacks. In: Joux, A. (ed.) EUROCRYPT 2009. LNCS, vol. 5479, pp. 443–461. Springer, Heidelberg (2009). doi:10.1007/978-3-642-01001-9_26
29. Stöttinger, M.: Mutating runtime architectures as a countermeasure against power analysis attacks. Ph.D. thesis, Darmstadt University of Technology, Germany (2012)
30. Keccak Team: Note on side-channel attacks and their countermeasures. http://keccak.noekeon.org/NoteSideChannelAttacks.pdf
31. Veshchikov, N.: SILK: high level of abstraction leakage simulator for side channel analysis. In: Preda, M.D., McDonald, J.T. (eds.) Proceedings of the 4th Program Protection and Reverse Engineering Workshop, PPREW@ACSAC 2014, New Orleans, LA, USA, 9 December 2014, p. 3:1–3:11. ACM (2014)
32. Wang, J., Vadnala, P.K., Großschädl, J., Xu, Q.: Higher-order masking in practice: a vector implementation of masked AES for ARM NEON. In: Nyberg, K. (ed.) CT-RSA 2015. LNCS, vol. 9048, pp. 181–198. Springer, Cham (2015). doi:10.1007/978-3-319-16715-2_10
33. Whitnall, C., Oswald, E., Mather, L.: An exploration of the Kolmogorov-Smirnov test as a competitor to mutual information analysis. In: Prouff, E. (ed.) CARDIS 2011. LNCS, vol. 7079, pp. 234–251. Springer, Heidelberg (2011). doi:10.1007/978-3-642-27257-8_15
34. Wiener, M. (ed.): CRYPTO 1999. LNCS, vol. 1666. Springer, Heidelberg (1999)
35. Zhang, W., Bao, Z., Lin, D., Rijmen, V., Yang, B., Verbauwhede, I.: RECTANGLE: a bit-slice lightweight block cipher suitable for multiple platforms. Sci. China Inf. Sci. 58(12), 1–15 (2015)

Author Index

Printed in the United States
By Bookmasters